Invisible Connections

Instruments, Institutions, and Science

SPIE Institutes for Advanced Optical Technologies series

Transformations in Optical Signal Processing, William T. Rhodes, James R. Fienup, Bahaa E. A. Saleh, Editors, 1984, SPIE Volume 373 (Out of print)

Optical and Hybrid Computing, Harold H. Szu, Editor, 1987, SPIE Volume 634

Photonics: High Bandwidth Analog Applications, James Chang, Editor, 1987, SPIE Volume 648

Large-Area Chromogenics: Materials and Devices for Transmittance Control, Carl M. Lampert, Claes G. Granqvist, Editors, 1990, Volume IS 4

Dosimetry of Laser Radiation in Medicine and Biology, Gerhard J. Müller, David H. Sliney, Editors, 1989, Volume IS 5

Future Directions and Applications in Photodynamic Therapy, Charles J. Gomer, Editor, 1990, Volume IS 6

Automatic Object Recognition, Hatem N. Nasr, Editor, 1991, Volume IS 7

Holography, Pál Greguss, Tung H. Jeong, Editors, 1991, Volume IS 8

Invisible Connections: Instruments, Institutions, and Science, Robert Bud, Susan E. Cozzens, Editors, 1992, Volume IS 9

Cover art: Scanning tunneling microscope by WA Technology, Cambridge, U.K. 1986. The first tunneling microscope to be produced commercially. Photograph by Robin Chaphekar. Courtesy Trustees of the Science Museum.

SPIE INSTITUTES

Volume IS 9

Invisible Connections

Instruments, Institutions, and Science

Robert Bud
Science Museum London

Susan E. Cozzens
Rensselaer Polytechnic Institute

Editors

Roy F. Potter,
General Editor, SPIE Institutes for Advanced Optical Technologies

SPIE OPTICAL ENGINEERING PRESS

A Publication of SPIE—The International Society for Optical Engineering
Bellingham, Washington USA

Library of Congress Cataloging-in-Publication Data

Invisible connections : instruments, institutions, and science : 12–14
April 1991, London, England / Robert Bud, Susan Cozzens, editors ;
Roy F. Potter, general editor.
p. cm. — (SPIE institutes for advanced optical technologies
; v. IS 9)
Papers from the SPIE Institute on Instruments, Institutions, and
Science
"A publication of SPIE—the International Society for Optical
Engineering."
ISBN 0-8194-0767-4 (hardcover). — ISBN 0-8194-0768-2 (softcover)
1. Scientific appliances and instruments—Congresses. 2. Science
and state—United States—Congresses. 3. Research—Management—
Congresses. I. Bud, Robert. II. Cozzens, Susan E. III. Potter,
Roy F. IV. Series.
Q185.I68 1992
502.8—dc20 92-15563
CIP

Published by
SPIE—The International Society for Optical Engineering
P.O. Box 10
Bellingham, Washington 98227-0010

Printed in the United States of America

Introduction to the Series

The concept of the Institute for Advanced Optical Technologies developed out of SPIE's desire to foster increased interaction and collaboration among researchers working in emerging optical technologies. The Institutes provide a forum for experts in these areas to analyze and document the state of the art and to point toward future trends and applications. Institute topics are selected for their timeliness as well as for their significance to future progress in the application of optics. Institute organizers invite selected experts to participate as paper contributors and discussion participants. It is intended that the interaction generated by the small-group structure in a retreat-like setting will foster productive discussions that are beyond the scope and possibility of a regular conference format.

Each Institute has two primary objectives: first, that the interactions and dialogue stimulate technical advancement, and second, that the publicaton of the Institute volume results in an authoritative collection of significant papers covering key topics in the field. While each editor and committee has unique criteria for determining the acceptability of contributions, it is intended that the Institute process itself will establish the worth and appropriateness of the individual contributions.

The Institute on Invisible Connections: Instruments, Institutions, and Science was held at the Science Museum London in April 1991. Starting with the premise that in the past the role of instruments in the development of science has been largely invisible, Robert Bud and Susan Cozzens invited participants from the USA, France, Canada, and Britain to meet for three days to examine the changing and increasingly visible role instruments are now playing.

A diverse group of authors involved in areas ranging from policy analysis to the history and sociology of science as well as practicing scientists in the forefront of NMR techniques met to examine the common ground of concern: the relationship between scientific instruments and the political and economic context of science. This collection of papers is intended to develop a more coherent view of the historic and current influence of instruments and illustrates why their role in scientific advance should not remain invisible any longer.

Roy F. Potter
General Editor
SPIE Institutes for Advanced Optical Technologies

Table of Contents

Part III National Concerns

List of Contributors

Lusin Bagla
Groupe d'Etudes sur les Sciences
et la Technologie
Maison des Sciences de l'Homme
54 Boulevard Raspail
F-75006 Paris, France

Stuart Blume
University of Amsterdam
Dept. of Science Dynamics
Nieuwe Achtergracht 166
NL-1018 NV Amsterdam, Netherlands

Robert Bud
Science Museum London
South Kensington
London SW7 2DD, UK

Susan E. Cozzens
Rensselaer Polytechnic Institute
Dept. of Science and Technology Studies
Troy, NY 12180-3590, USA

David O. Edge
Science Studies Unit
Department of Sociology
Edinburgh University
34 Buccleuch Place
Edinburgh EH8 9JT, UK

James Feeney
National Institute for Medical Research
Medical Research Council
The Ridgeway, Mill Hill
London NW7 1AA, UK

Luke Georghiou
University of Manchester
PREST/CASR
Oxford Road
Manchester M13 9PL, UK

Iskender Gökalp
Laboratoire de Combustion
et Systèmes Repactifs
Centre National de la Recherche
Scientifique
IC Ave. de la Recherche Scient
F-45071 Orléans Cédex 2, France

Peter Halfpenny
University of Manchester
PREST/CASR
Oxford Road
Manchester M13 9PL, UK

C. N. M. Jansz
Stichting voor de Technische
Wetenschappen
Technology Foundation
P.O. Box 3021
NL-3502 GA Utrecht, Netherlands

John Krige
European University Institute
Dept. of History and Civilization
San Domenico di Fiesole
Florence, Italy

Carlos Kruytbosch
Science Indicators Ltd.
National Science Foundation
Washington, DC 20550, USA

Everett Mendelsohn
Harvard Univ.
Dept. of History of Science
Science Center 235
Cambridge, MA 02138, USA

Peter G. Morris
University of Nottingham
Dept. of Physics
University Park
Nottingham NG7, UK

Chandra Mukerji
University of California/San Diego
MS 0503, Communications
and Sociology Departments
La Jolla, CA 92093-0503, USA

C. le Pair
Stichting voor de Technische
Wetenschappen
Technology Foundation
P.O. Box 3021
NL-3502 GA Utrecht, Netherlands

Yakov M. Rabkin
Université de Montréal
Dept. d'Histoire
CP 6128, Succ. A
Montréal, Quebec, Qc H3C 3J7, Canada

Yves-André Rocher
Ecole Nationale Superieure des Mines
Ctr. de Sociologie de l'Innovation
60 Boulevard Saint-Michel
F-75006 Paris Cedex, France

Simon Schaffer
University of Cambridge
Dept. of History
and Philosophy of Science
Free School Lane
Cambridge, UK

Jeffrey K. Stine
National Museum of American History
Smithsonian Institute
Washington, DC 20560, USA

John Yates
University of Manchester
PREST/CASR
Oxford Road
Manchester M13 9PL, UK

Introduction

The role of instruments in the development of science was once strangely invisible. Now, however, in two separate spheres of activity, instruments have become the focus of attention. In science policy, the increasing cost of research, driven especially by skyrocketing costs for its capital base, has prompted questions about spending on instruments: How much is enough? How should instruments be distributed among research organizations? Should central governments intervene, or should all decisions be left to institutions? At what point do shared instruments become cost effective?[1] Meanwhile, as part of the effort to correct an earlier over-emphasis on theory as a driving force in scientific change, historians, philosophers and sociologists of science have taken a new interest in the experimental process.[2] The history and sociology of science is already establishing a canon incorporating studies of modern research as well as seventeenth-century science. The potential for links with the policy debate has become clear, for the two lines of questions share a concern with the relationship between scientific instruments and the political and economic context of science. Yet the two traditions still have much to learn, and to contribute, from greater familiarity with each other's problems.

This volume presents papers first discussed at a 1991 workshop held in London's Science Museum, intended to explore the areas of common interest. The contributors include science policy analysts from the USA, France, Canada and Britain, as well as sociologists, historians, and practicing scientists in the forefront of NMR techniques. Individually the contributions join other recent work in drawing attention to instruments in the laboratory, particularly breaking new ground in the perspective they provide on industrial innovation. Together, the papers are intended as a step in bringing instruments back into coherent view, for policy makers and for those who analyze science.

The papers have been organized according to three broad themes. The first set explores the complex interaction between laboratory and industrial instruments, and the parallels between industrial organization and the organization of the scientific workplace as revealed through constellations of equipment. Next, a set of modern case studies introduces important concepts in the analysis of instruments in context: standardization, differentiation, and diffusion, among others. A final group of papers describes the emergence of large-scale and middle-range instruments as a problem in national science policies, along with recent attempts to address those problems. This introduction points to cross-cutting themes.

Each section is also prefaced by editorial comments intended to highlight local themes and issues. The interface between the variety of disciplines represented in this volume is too complex and dynamic to enable a definitive overview. However, three major issues are highlighted by the papers represented here. The reader will be constantly reminded of the tension between instruments as malleable experimental devices and as objects of consumption; their pivotal role between the cultures and economies of science, on the one hand, and, on the other, of much broader industrial markets; and of the number of issues fused within the "need" for an instrument.

1. The Scientific Workplace

Historians of scientific instruments have all too often reserved their analysis for the nineteenth century and earlier. By contrast, the authors of this book concentrate upon the twentieth century. They describe an independent realm of instrument innovation and diffusion, a realm which intersects fruitfully and often with other realms of technological innovation to exert a steady and significant force for change in scientific ideas. This is the practical importance of instruments, which form the material basis of laboratory life. They have, too, a cultural significance: their perceived role in science has been closely allied with an increasingly technological vision of the future.

While the authors use the term instrument comfortably, they make no naive assumption that they are dealing with simple, easily-bounded phenomena when they do so. Instrument systems are internally heterogeneous, as exemplified by Gökalp et al.'s description of laser Doppler anemometry, an experimental approach that combines several optical and electronic technologies. Furthermore, any instrument merges in practice with procedures and techniques, both practical and theoretical. Indeed, these authors claim, instruments and theories have more in common than the usual descriptions indicate. Theoretical techniques, like instruments, must be worked into new contexts through craft knowledge and local adaptation; take for example the diffusion of the Fourier transform.

The local adaptation of instruments emerges from these papers as a prominent feature. Edge's description of mosaic arrays in infrared astronomy both illustrates and explains this phenomenon. On the one hand, the mosaic arrays are differentiated by local circumstances—sociologists would call them contingencies. But they are also differentiated through the pattern of competition in the field; each team needs to find its "niche" and tunes its instrument to that task. As a result, the configuration of skills and instruments in a laboratory becomes what Mukerji calls a laboratory "signature". These factors push scientific instruments toward a state of "smallest possible market"—each laboratory's needs being ever-so-slightly different from the next's.

This differentiation is intrinsic to their role. Instruments provide for exploration and play in the experimental sciences. They allow scientists to explore tentatively at the edge of what is known, without the constraints of deductive theorizing. Mukerji's paper posits a parallel between ocean scientists and children, both of whom playfully push the limits of what their peer groups know. Both use tools to do this, although the oceanographers often find theirs in the tool chests of their military sponsors rather than the toy chests of the nursery.

Conspicuous consumption is a counterforce to the process of differentiation. Laboratory equipment is more than functional; it is also an important element in the power and prestige of its owners. This factor can set in motion a wave of diffusion of a standard instrument, whether or not it will be efficiently used. Blume points to such a pattern in hospital technologies, but laboratories are unlikely to be immune from the syndrome.

Several of these authors (Blume, Rabkin) claim that instrumentation is becoming more standardized—that the spread of black box equipment overshadows the pattern of differentiation. Blume goes further in claiming that the trend toward standardization takes its force from a long-term shift in work organization in the sciences. Mendelsohn's arguments, however, lead to the opposite conclusion. He calls attention to the increasing scale of instrumentation in science, a factor which can only lead to more unique instruments and fewer standard ones. The true balance between the differentiating forces of competition and the homogenizing forces of conspicuous consumption is of course a matter for empirical investigation: both patterns are theoretically plausible, and have actually been observed.

2. Markets and Innovation

The authors in this volume share the assumption that instruments, as technologies, are not disconnected from the larger worlds of consumer and industrial technology, even though as a market, science is dwarfed in size by those worlds.

First, techniques diffuse across the boundaries of industrial and scientific technologies, and spark new combinations in both areas. Research can change the technical basis of the technologies that address a certain practical problem, as in the second wave of NMR innovation (see Feeney and Morris). Likewise, new technologies can radically alter the access of a community of scientists to its phenomenon of study. Throughout, the book provides examples of technological innovation stimulating scientific change, with technological needs often calling forth scientific efforts. Feeney's description of the research that went into the second wave of NMR development again illustrates just such a process. Edge describes how the mosaic array allowed infrared astronomers new levels of data acquisition; and the laser Doppler velocimeter, which had no antecedent equivalent, opened a whole new window on the process of combustion.

As with many types of technological innovation, the cases reported here show that people are an important element in spanning the institutional boundaries between the laboratory and the industrial firm. Scientists clearly do get involved in the development of instruments, in particular because of their ability to merge scientific and technical aims in the process of scientific work. Instrument makers, likewise, do interact with the laboratory as they develop and refine new products.

Because of the close historic connection between science and the military, and the constant level of technological change in modern warfare, the military has been a major supplier of new technologies for science. Mukerji's descriptions of oceanographers suggest how unlikely it is, in a military-dominated science, that change would flow from science into technology. Her example, and indeed the military market, is not unique. Thus several authors describe how techniques developed for other purposes first dribble into science, and then scientists scramble to individualize them for competitive purposes.

This perspective raises important questions for policy. Is innovation through instruments a major contribution of science to technology, or not? Is it increasing or decreasing? Which set of forces is stronger—those for standardization, or those for tinkering and differentiation?

3. Needs

The central science policy interest in instruments has been how to respond to scientists' calls for expenditure. This volume examines the pressure as a phenomenon to be measured, evaluated and examined critically. Some authors take off from the vision that informs contemporary science policy. A new instrument is equivalent to a major but passive scientific resource. The advance of science demands them. If policymakers fail to provide adequate resources to supply this need, they risk both losing out in the competition between scientific nations and violating the technological imperative of progress. The papers by Stine and Kruytbosch chronicle the emergence of this viewpoint, which dominates policy discussions in all industrialized nations today. Georghiou and Rocher illustrate the resulting efforts to gather data on instrument stock and flow.

Policy makers have taken scientific needs as given; however, they are a complex category. Their relations to instrumentation are explored by other papers which question the assumption that scientific advance, defined without reference to its content, is an adequate definition of the goal of policy. The authors examine the process of choice among goals, and therefore explore the question, "What kind of scientific progress is being provided?" The cost of instrument-driven scientific developments can itself carry overtones of importance; and it can sometimes seem that the reductionism implicit in experimental control produces more valuable knowledge than the study of emergent properties of complex systems.

The history of accelerators (Krige tells part of it in this volume) illustrates the relationships particularly clearly. As his story shows, the emergence of an accelerator takes place in a context of scientific and national competition. The community of high energy physics is up against nature in its demand for higher energy ranges for their accelerators; but it only takes one machine to meet this "need". There have been numerous proposals, however, to build a World Accelerator, but they have never been implemented. Why? One workshop participant observed that "if there were a World Accelerator, there would be no high energy physics". He implies that the connection between technology-based science and national pride is not just a peripheral factor; it is constitutive of the science itself. Would that we could see that element of our own culture as clearly as Simon Schaffer helps us see the values of Victorian England in the development of metrology.

In conclusion, the papers in this volume richly illustrate how scientific instruments can be looked at together, by scientists, specialist policy makers and academic science studies communities. If their interests remain distinct, this volume suggests they have much to share.

Robert Bud
Science Museum London

Susan E. Cozzens
Rensselaer Polytechnic Institute

June 1992

[1] The literature relating to policy makers' concerns has been largely in the form of reports. The contributions of Georghiou, Kruytbosch, and Stine to this volume bring together much of this literature.

[2] As recently as 1985, a distinguished historian of chemical instrumentation, Robert Anderson, could write "Historians of science have paid relatively little attention to scientific instruments and apparatus over the past twenty years. Chemical apparatus has fared particularly badly, and it is difficult to think of a single major contribution over this period." Robert Anderson, "Instruments and Apparatus," in Recent Developments in the History of Chemistry, ed. Colin A. Russell, London: Royal Society of Chemistry, 1985. Since then, social historians have been interested in the ways facts and achievements have come to be established and the nature of experiment. Works which have proved particularly stimulating include Steven Shapin and Simon Schaffer, Leviathan and the Air Pump: Hobbes, Boyle and the Experimental Life, Princeton: Princeton University Press, 1985; and Peter Galison, How Experiments End, Chicago: University of Chicago Press, 1987. The ways in which the process of measurement in a technological context can be opened up are demonstrated by Donald Mackenzie, Inventing Accuracy: A Historical Sociology of Nuclear Missile Guidance, Cambridge, Massachusetts: MIT University Press, 1990; and the development of technological measurement techniques has been recounted by Stuart Bennett, for instance in his A History of Control Engineering, 1800-1930, Stevenage: Peter Peregrinus, 1979. Many recent contributions have been made through conference proceedings such as David Gooding, Trevor Pinch and Simon Schaffer, eds., The Uses of Experiment: Studies in the Natural Sciences, Cambridge: Cambridge University Press, 1989; and Frank A. J. L. James ed., The Development of the Laboratory: Essays on the Place of Experiment in Industrial Civilization, Basingstoke, Hampshire: Macmillan, 1989. The generally more detailed approach taken by papers read to the Scientific Instruments Commission of the International Union of History and Philosophy of Science is represented by Christine Blondel, Françoise Parot, Anthony Turner and Mari Williams, eds., Studies in the History of Scientific Instruments, London: Roger Turner Books Ltd., 1989. The interests of scientists in their work is expressed in such work as Carol L. Moberg, The Beckmann Symposium on Biomedical Instrumentation, New York: Rockefeller University, 1986.

Part I
The Industrial Connection

Introduction to Part I

Research instruments often facilitate the interpenetration of science and industry. This section explores the dynamics of that interaction from several angles. First, Mendelsohn and Schaffer, using different cases, show how research practice can reflect or incorporate industrial problems and forms. As a broad statement, this observation may not be unique to the analysis of research instruments. But the connections become particularly visible when empirical attention focuses on instruments, as these papers demonstrate. The first rule of method in the study of industry-science interaction may well be to "follow the instruments."

Mendelsohn adopts this strategy in the first paper in the section, following first the microscope in the seventeenth century, next physiological instruments in the nineteenth, then the instruments of big physics in the twentieth. In each case, the forms of manufacturing are reflected in the way instruments are incorporated into scientific practice. In the seventeenth century, craft knowledge still predominates, and craft skills characterize experimental practice as well. How did the natural philosophers of the seventeenth century learn to use the new instruments? How did they "pick them up," Mendelsohn asks; where did they turn for advice? The new literary technology played some role, but even the writers of manuals acknowledged that making instruments work was difficult if not impossible to do from books. No wonder then that "the first 25 years of microscopy were marked by frustration." What did the natural philosophers do with the instruments? They intervened in the material: they cut it open, sliced it, and (in later centuries) stained it. Manual manipulation made "inspired seeing" possible.

By the nineteenth century, the machine was replacing the hand. The age of industrial machines and industrial production was accompanied, Mendelsohn tells us, by the notion that the body was a machine, and was best understood through machines. When the machines were too expensive, less expensive ones were invented, so that the style of investigation they embodied could be spread through the classroom. Still, when the machines were shared, they were shared by a few. By the twentieth century, sharing had outgrown the classroom table of the physiologist, with its central electrical tower for six to eight students. The symbol of the power of the twentieth century science is the high energy physics laboratory, now so large that it is more likely to be called a "facility" or "installation." A fine-grained division of labor becomes necessary when hundreds work on an experiment, and scientific workers become as alienated from the products of their labor as industrial workers were claimed to have become a century earlier. The atmosphere of the corporation invades research through hierarchical control of research teams, an activity much closer to industrial management than to the activities of the solitary bench scientist.

In Schaffer's case, the establishment of the Cavendish laboratory at Cambridge as a center for standards, industry is more explicitly incorporated into research. Metrology, the establishment of standard units for natural quantities, was essential to the working of telegraphs, one of the long-distance technologies that held the British empire together in Victorian times. From this politico-economic need arose the call for an "absolute standard," a measure of resistance which was independent of material. James Clerk Maxwell began to transform the laboratory at Cambridge into a factory-like unit that could produce and maintain such a standard, and his successor, Rayleigh, completed the job. This intrusion of the workshop style into a university caused comment, but that comment was answered with the moral significance of the project. In Victorian England, the discourse of metrology was, "profoundly moralistic. A culture convinced of the need for accurate data and stern surveillance as the right means of social management would also see precision measures as an ethical good." Thus standardization contributed to liberal education. In the end, the goal was achieved. British standards, spreading communications technology, and the work which went into creating this situation disappeared into the taken-for-granted. As Schaffer summarizes, "Metrology involves work which sets up values and then makes their origin invisible."

Both Mendelsohn and Schaffer describe the influence of research instruments in shaping the thought styles of students, who then carry the instrumental style into other spheres. A generation of medical scientists and practitioners were trained in the nineteenth century physiology laboratories. This generation did not have to be told to understand the body as a machine: they could not complete their laboratory work if they assumed otherwise. Likewise, at Cambridge, "the student was never to escape the sense that he was an initiate in the worldwide network of electro-technology, and the local culture of absolute measurement." The third paper in the section, by Rabkin, also highlights the influence of the instrumental world view through the movement of professional scientists, trained in specific instrumental skills, into government and industry. Frequently, he writes, scientists "choose the same instruments they have used in their formative years," thus not only transferring the "moral culture" of science, but also creating demand for specific instruments.

Rabkin's main theme, however, is a more explicit effect of industrial practice on research instrumentation than the subtle ones described by Mendelsohn and Schaffer. Government analysts may like to think of the diffusion process for scientific instruments as starting in the laboratory, then spreading to industry. But Rabkin points out that this is much less frequently the case than its opposite: that industrial practice changes and creates a new technical base from which specialized research instruments evolve. What underlies Rabkin's argument is that both research and industrial practice share a common material culture, expressed in part through measurement techniques. Thus both industrial dynamism and research vitality depend on an innovative instrument industry. This is apparently true at the national level. Rabkin connects the decline of the French chemical industry to the transfer of leadership from French to German instrument-making firms, and also the weakness of the field of chemistry in the Soviet Union to its isolation from the moving front of world technology.

Jansz and le Pair provide further support for the essential connection between industrial modes of operation and instruments, by observing that instrument innovators, even when they are scientists, act like technologists in relation to their instrument inventions: physical artifacts are their accomplishments rather than published papers; their greatest rewards are invitations to conferences where they can gesture and sketch with others who practice their skills; and their paper trails end up in the basements of their institutes as mold-gathering reports rather than in journals and libraries. Citation-counting does not capture their impact; instead one must "follow the instruments" themselves.

The message from these students of history to government research managers is clear. The connection between instrumentation and industrial practice is so strong, in so many directions, that resource allocation questions concerning research instruments should be considered part of technology, not science, policy.

-SEC

Chapter 1

The social locus of scientific instruments

Everett Mendelsohn
Harvard University

The organizing question I chose for this paper goes to the heart of my conception of the relations of instruments to science: what would you expect to find in the work organization or system for the production of scientific knowledge of late 20th-century advanced science? And how would instruments be utilized in this process? How would they be integrated? Similarly, of course, you can ask the same for earlier periods of the history of modern societies in which science was being produced. Recent studies, including several important ones in this volume, have addressed the question of the production of science in the period of the Scientific Revolution, the advanced industrial revolution of the 19th century, and the large-scale science after post-World War II decades.

1. THE MICROSCOPE: INSTRUMENT AND PRACTICE

Let me turn first, however, to some early examples as a way of identifying elements to which I will return later. The age of classical microscopy recalls a special era in science. The microscope entered the scene with praise, full praise, of all sorts:

> "Of all the Inventions none there is surpasses the noble Florentine's Dioptrick glasses.
> For what better, fitter gift could be in this world's aged Luciosity,
> To help our blindness so as to devise a pair of new and artificial eyes,
> By whose augmenting power we now see more than all the World has ever done before."

Henry Power, the physician, microscopist, experimenter, clearly took the instrument and put it in the context of providing instrumental augmentation to other activities, to those other facilities and faculties that humans had.[1] Robert Hooke, another user of the instrument and refiner of it, recognized a discontinuity with the past. The instruments in this sense--for Hooke and for Power--were creating a revolution in science. "We have far exceeded all the ancients," Hooke wrote.[2] To him it seemed clear that the future of science was to be found in new instruments and new technologies. And indeed, the context of the 17th century was one in which there was a deep tension between the turning to nature with aids, with helps, with instruments, and the fear or distrust of theory. This tension was to mark the early period for many reasons. But the seventeenth-century context was shaped by the belief systems, systems of authority in state and church, and the primitive systems of manufacture which set modes and constraints on which things would happen and how they could occur. Instruments cost money; the microscope, we are told by many of its historians, was a relatively expensive instrument. Certainly telescopes and vacuum pumps were equally costly. Steve Shapin referred to these as "big science" for their day.[3] (This is an observation worth further critical examination. I am not wholly convinced.) But there is little doubt that the costliness of these aids, which were becoming ubiquitous, was a characteristic of the entrance of machines and technologies into science.

The costliness also created a dependence. Experimenters were dependent on instruments to do the things which they set out to do and the necessity for paying for instruments forced scientists to become dependent upon some form of support, whether it be patronage or private wealth. In turn, the instrument became essential to keeping up with the activity it had helped initiate. A process of scientific investigation was being developed which integrated the investigator into systems which constrained and shaped what could be done and how. One way of dealing with the problem was the establishment of societies and academies which acquired instruments and provided the institutional setting, the rules for their use, and to some extent provided for shared access to costly instruments. The academies certainly provided a public form and forum of practice, and they provided modes of certification of new found knowledge.

On the other hand, there is a continuity through the century of the lone practitioner. Antoni Van Leeuwenhoek is a classic example of someone operating in and out of the science system as it emerged. A draper and minor civil servant, he worked within what I would call a guild production system (the lens grinders), but also made forays into the "high science" of his day. Who makes observations and how do they achieve a status as public knowledge? Leeuwenhoek made fascinating observations which were unrepeatable by the British colleagues to whom he sent his reports for publication in the Transactions of the Royal Society.[4] What do you do when an observation made with an instrument is unrepeatable, and yet it has great potential importance? In the seventeenth century some alternate means of certification were sought; you write a letter to a friend in Amsterdam and say "Is this a reputable individual whose word we can take?" You want a character witness to provide reassurance that the instrumental observations were made by a plausible user of the microscope. Steve Shapin points out that Toricelli's experiments also could not be repeated in Paris.[5] Instrument-aided investigation could not provide the certainty of knowledge which later became associated with instruments, and consequently they had to be embedded in systems or organizations which could raise the level of credibility.

The institutional context in which experimentation and use of instruments was embedded did provide contours and boundaries for forms of empiricism. It provided mental contexts, and, as Lorraine Daston is pointing out in a series of new studies on objectivity and facticity, the ways of thinking were being molded to accompany the ways of acting which were being carried out.[6] Thinking and acting were being conditioned upon each other.

How does someone know what to do when they get a microscope or when they see an instrument of this sort? How do they know where to pick it up and where to turn? An attempt is made to gain some sort of guidance. The first illustration in Robert Hooke's Micrographia is well known to most historians of science. The plate is a diagramatic illustration of a microscope and implied rules for its use. Notice what he has done; he is giving instructions; he is saying, "Here is my manual in which I am going to tell you many things. I'm going to tell you how to think about nature as well as how to do things with nature. But I am going to give you fairly detailed instructions, because otherwise I know how difficult it is to get these things working, and without some of these instructions you won't be able to know what I have done; to appreciate the skill [and the skill was great] or to be able to replicate and to know whether what is being said had meaning or not." The technique, then, of instrument use was as important with the simple instrument as it was with the NMR or the variety of other instruments which cascaded through the sciences since. The question, then, of the skill, of knowing what are the important variables, for example, the factor of light is absolutely critical, and how you get the right amounts of light in the right place without burning the specimen, especially if you are dealing with a fine cut of a slice of a root cell or some other similar specimen, is essential. There is also almost a form of play, teasing the instrument, that Hooke engages in. As Michael Dennis indicates in a fascinating recent article, Hooke presented the Micrographia to the Royal Society; the Royal Society both liked it and was suspicious of it simultaneously because of the nature of work and what it is doing.[7] The book itself is in part a recruiting document for the Royal Society, as Dennis points out. It was a way to show people some of the interesting things that can be done in the sciences and perhaps recruit them to the Society; the volume was aimed at the London virtuosi who the Royal Society hoped to capture.[8] Dennis argues that deploying of the microscope was part of the reformation of philosophy; the instrument was being used as a form of changing the nature of mental practice as mental practice turned to deal nature itself. It was a very conscious deployment of a technique, even though it often involved accidental technical discovery and accidental technical work.

Hooke himself turned out to be quite a special figure, although in some ways similar to Leeuwenhoek. He was a mechanic, an instrument maker and designer on the one hand, but also a natural philosopher on the other. He worked within an advanced craft tradition of production. But there were limits to that tradition at the time. One could not find a shop with a shelf full of implements which you could just assemble and create a working instrument; it was necessary to design the parts themselves. Hooke's microscope was complex in design; Leeuwenhoek's was remarkably simple. The instrument collectors with whom one talks most often are be able to tell you about the instrument makers, but are not nearly as good at telling you about the instrument users or the conditions of use. Most catalogues of instruments focus on the device itself, and illustrations seldom if ever show people using instruments. The instruments have been pulled out of the context of their use, pulled out of the systems which gave them meaning. The instrument was juxtaposed with other artifacts rather than being juxtaposed with practice. This is a fascinating comment on what we have done most often with our instruments.

One of the interesting things about the early instruments, the microscopes for example, is that no two of them were identical. The lenses and the mechanisms in each varied slightly. We know, for example, that Leeuwenhoek had hundreds of instruments, all small, about the size of a silver dollar or slightly larger. His were crudely made, with all his attention focused on the spherical lens and little care given to the mechanisms.[9] Specimens were often permanently mounted, with the microscope seemingly designed specifically for a given specimen or type of specimen. This, of course, posed interesting and difficult problems when later uses of instruments are considered. Hooke's microscopes, from this point of view were more manageable. But, he himself tells us, there were many difficulties for proper utilization.

Magnification was an aid to the senses; the small was expected. What is interesting, though, is that there were no fixed units in microscopy at this time; at best one was dealing with comparative units or examples. There was no fixed measure of the small, which became difficult since comparison must always be done in linguistic terms rather than in someother fixed notation. Leeuwenhoek estimated bacteria size by the number which could fit under a grain of sand.

In practice Leeuwenhoek was engaged in "inspired seeing". Bacteria that he described others were unable to see for another hundred years: he drew red blood cells including the nucleus, an extremely difficult observation not soon repeated. His inspired seeing clearly stretched right to the edge, if not past the edge, of the ability of his instruments. He used both an imagination and imaging of uncanny ability to fill in details, seeming to infer what was not absolutely visible. By contrast Hooke engaged in what Dennis has called "disciplined seeing," observing the same object in different lights and situations, thus "guaranteeing the translation of Hooke's private experience into public knowledge." Hooke tells us, "I never began to make any drafts, before many examinations in several lights, and in several positions to the light, I had discovered the true form. For it is exceedingly difficult in some objects to distinguish between a prominency and a depression, between a shadow and a black stain, or a reflection and a whiteness in color."[10] He could go on at length about the difficulties that there were in seeing what he wanted to see. In addition, however, Hooke was engaged in a shared practice, working in the context of a group who wanted to see what he was seeing and representing in his drawings. They seldom looked through his microscope;

instead he would produce the drawings and share them. He represented; he reported; he discussed. What he was using was, of course, an "imaging instrument"--something which produced images which seemed a normal extension of what you could do with the eye or magnifying lens, even a fairly powerful one; for Hooke there was a continuity. Another variety of other instruments, are what Peter Galison and others have called logic instruments; they provide a series of signals, of one sort or another, and the practitioner using reason constructs them into either an image or an equation; the image and the equation in this case are interchangeable because they are both logically constructed rather than derived through a series of directly observed images.[11] The history of science as a field, as the papers in this volume indicate, is engaged in revising its foci. If theory had been the hero of the history of science, practice had until recently been the anti-hero, often ignored or degraded. Shapin has told us about the invisible technicians; in addition, there are even almost invisible scientific practitioners.[12] All but the heroes were readily lost, often I believe in their own time. Concentration was given to giants standing on the shoulders of earlier giants. Some historians of science, of course, denigrated experiment to being nothing but a marginal help to theory building. Alexandre Koyré talked about Galileo's thought experiments; and it is the thought experiment that Koyré liked best. Even William Harvey's "thought" experiments were raised to prominence. The real experiments were much messier. Koyré may be right that some of the things that Galileo wanted to show could not be demonstrated in real experiment, but there is little doubt that for Galileo it was real experiment that was the model rather than theory retrodicted into experimental form.

What does one do with instruments? It was hard to know what one did; for some they were toys. Hooke tells us quite clearly that in his practice there was an active element in using an instrument rather than a passive one; his was an image of engagement with an instrument. One learned to see with a microscope by doing.[13] You intervene in the material; you cut it open; you slice it; you stain it (although not this latter in Hooke's time); but you intervene to make the material most available to the technique you have for observing. This represents something of a change in the mode of looking at nature, and Hooke in his way personifies something of this intervention--intervention with instruments and intervention at an epistemologic level. The role of theory in these instruments was slight. Turner makes the claim that "in the case of the microscope the development to 1870, that is the pre-Abbé period, saw the limit of resolution closely approached by empirical means."[14] Somewhere between 1 micromometer to O.2 micrometers. For the first 200 years of the use of the instrument it was empirically guided.

How do you know how to use an instrument? Theory was not needed to use a microscope, albeit at times observations may be driven or constrained by theory of what was being observed. Much of the discussion of globules in the mid-19th century had to do with poor microscopes used up to, if not past, the limits of resolution; but the mental constructs had been developed in both German and French physiology and morphology that the surfaces of living things were round, whereas the surfaces of inanimate objects--crystals, etc.--had square angles, a fascinating notion which pervades a good deal of the literature of mid-19th century--biology.

There were limits to instruments, and the limits in part were technical, sometimes severe: light, chromatic aberrations, spherical aberrations, preparation of materials: slicing, staining and fixing. It turns out that you have to prepare almost anything which you are going to use an instrument to observe or to interact with--whether it be an NMR or a simple microscope. The preparation required forms of inventiveness. So far as organisms were concerned, most often the contact with the organism involved killing it, involved taking it to some state which was not its normal state.

The first 25 years of microscopy was marked by frustration: frustration with the optics, frustration with the preparation. Hooke himself complained in the mid-1690s that microscopical observation had largely been abandoned by scientists. Only Leeuwenhoek, he said, was still making useful observations. Instead, what had happened is that the instrument had become a toy in polite culture. If the century opened with praise, the classical period of microscopy closed with ridicule. "So naturalists observe a flea has smaller fleas that on him to prey, and these have smaller still to bite them, and so on ad infinitum." Thus, Jonathan Swift's satirical response to the microscopical image--Hooke's drawing of the flea. What is fascinating about the illustration is that the organism is repeated over and over and over again by almost all the early microscopical observers. There was clearly a form of piracy. But also, I would suggest, was easy availability of this insect as people could pick them off their bodies or their domesticated animals almost at will. Familiarity may have peaked interest.

The microscope as an instrument may at times have been very functional; at other times it was non-functionally elaborated. Many of the expensive elements of the fancy brass microscopes of the eighteenth and nineteenth centuries were probably not needed. But they were part of an artistic or esthetic embellishment to an instrument that served other purposes. If one were to line up a series of microscopes of the seventeenth through the nineteenth centuries, the obvious feature is an upright tube, partially functional but not essential. By the time the electron microscope was developed in the mid-decades of the twentieth century, its orientation was historically established: the upright tube. It could, of course, have been upside-down, sideways, or turned around at any angle, but the orientation is one clearly borrowed from the continuity of design of the light microscope. Perhaps the user felt better, perhaps the builder unconsciously constructed it in just that form.

The seventeenth-century context within which instruments were developed? Class ridden, patronized, invisible technicians, partially visible small scientists, court jealousies and competitions, mixed financing, uncertain forms of training and recruitment, advanced guild and craft style in the production of goods, knowledge and machines, national forms of organization looming very large at that point, and national forms of problem choice--there were very blurred lines existing between the practitioners and the public. And when it came to instruments, very different ones were found in different sectors. If indeed there were a period when the sector of science in which one was working made a significant difference, it would be in the seventeenth century when the immediate predecessor of the scientific instrument was almost always an instrument used in one of the productive fields then in action. This no doubt explains why during this century chemical instruments were the most pervasive.

2. PHYSIOLOGY IN THE AGE OF INDUSTRIAL MACHINES.

Let me turn briefly to the issues of what I call physiology befitting the age of industrial machines and industrial production: Harvard's physiological instruments.[15] In 1883, in the celebration of the centenary of the Harvard Medical School, Oliver Wendell Holmes presented an oration as he opened the Medical Faculty's new building at the corner of Boylston and Exeter Streets in the newly developed section of Boston. "You will see extensive apartments," he said, "destined for the practical study of chemistry and physiology. But these branches are no longer studied as of old by merely listening to lectures.... Physiology as now studied involves the use of much delicate and complex machinery.... Never was the human body as a machine so well understood; never did it give such an account of itself as it now does in the legible handwriting of the cardiograph, the sphygmograph, the myograph, and other self-registering contrivances with all of which the student of today is expected to be practically familiar."[16] The forms of industrial productivity were being taken into the laboratory, and particularly into the training of the new generation of medical scientists and practitioners. The organism and the instrument were now being conflated, and the organism becoming a machine. The instrument was identified as an analog for the organism, and the organism was being interpreted as an instrument. The machines produced mechanical images, for example, the recordings on the smoked drum kymograph. To ask the question of whether they were holistic machines is almost an oxymoron. That element of discussion did not exist. From the period of Harveian physiology, one of direct investigation and manipulation, some limited quantification used as a heuristic the nineteenth century witnessed the move to physiology as a chemical-physical science with explanatory models being chemical and/or physical. Quantification and replication became critical, and the imagery set forth, while perhaps not always as strongly reductionist as the program of the Berlin group of 1847, did establish the folding of biology and the organism into the physical. Although the vivisectionist programs of Claude Bernard were not the same as the reductionist programs of Carl Ludwig, scholars such as Oswei Temkin and others have suggested the image of an organic machine. Even Claude Bernard inferred machine-like regulatory systems which controlled and governed organism as a whole. The strategy was there, a complex living phenomenon had to be analyzed into the sequences of cause and effect which were available through the devices then being built. The instruments and apparatus were being adapted from the physical sciences: the kymograph, the revolving drum recorder became a prototype for physiological instruments of the day. Transient physiological activities were recorded and quantified; graphic registration was used for nerve conduction and muscular contraction. The science was going beyond the dissecting knife to reproducing, in machine-readable form, physiological phenomena. These, it was believed, could be replicated at will using the new instruments brought to physiology. The modes of thinking and of practice were brought from the continental laboratories of experimentalists and teachers such as Carl Ludwig and Claude Bernard. The German influence was the one most strongly felt. As Charles W. Eliot, the President of Harvard, remarked concerning Henry Pickering Bowditch's return from Ludwig's laboratory, he helped "inoculate" the University with "German science."[17] He brought back instruments, techniques, and a whole new pedagogy based on laboratory practice. As Merriley Borell indicates in her recent study, Bowditch helped redefine what it was you were to do in physiological training and education, the mode by which you were to proceed and act. "In physiology's general laboratory next to the lecture hall, 6-8 students could be accommodated at tables for 'practical courses for special work.' ... A large central 'tower' interrupter case held apparatus for regulating the 'galvanic current'; ... a varnishing apparatus based on the design of Hugo Kronecker of Berlin was employed to preserve the tracings. ... Two researchers could work at a Ludwig-style kymograph".[18] Additional instruments were described in detail: "a new interrupter with platinum contacts based on a model described by Christian Bohr had been constructed by the German instrument maker Baltzar and imported from Leipzig." This description of the place in which the student was put was a surround of machines which could replicate the functions of the organisms which were to be the study of the physiologist and the medical student: the respiration apparatus that Bowditch himself had developed, the kymograph, and the cages filled with dogs, cats, and rabbits which were kept in the laboratory as experimental animals (much to the dismay of the local Boston neighbors who objected to this handling of what they often saw as pets), the mercury table, a meat juice press (to draw out the fluids which you were going to analyze using the techniques then available in biological chemistry), a constant temperature chamber, a reflecting galvanometer, a balance.

Here were instruments, as teaching devices and as mind-shaping devices; for socializing, for orienting, for creating the profession and creating its practices and its exclusionary elements. The initiates learned to use these instruments as they entered the profession. Those who were a non-users were outside the profession and remained outside. A distance was created between general knowledge and special knowledge and practice. William Townsend Porter, Bowditch's successor as Professor of Physiology at Harvard, faced a large problem. Teaching through the laboratory was to be the hallmark, and he further redesigned and built a new laboratory. His first class, however, had 208 men (and they were all men--women's admission to Harvard Medical School came several decades later). As he put it, at the time the laboratory needed 104 experimental set-ups.[19] The imported Balzar kymograph, essential to all the work, cost $200 each, in those days a very sizeable sum. Porter did what one might expect of the inventive designer. When the scale in numbers became the critical restraint, he invented a new kymograph. Smaller, it could be lifted in one hand and, by contrast, cost $16. Similarly he invented a inductorium to gain the electric impulses needed to cause the nerve action and the muscle contraction. What he was engaged in doing was very specifically extending the range of what was available. He was enunciating a philosophy of instruments. "New apparatus was invented; old apparatus was redesigned; the strictest regard was paid to economy of weight into the smallest possible 'number of operations....' Reliable instruments were turned out at one-tenth the previous cost. Wholesale laboratory teaching became possible."[20] He was also introducing an explicit national mentality or outlook in regard regard to instruments. "American quantity production requires economy of material, the fewest operations, and the making of large numbers of identical pieces at one time. The saving in price (per item) is magical," Porter reported. There was also a surplus of the instruments produced which could be sold cheaply to other institutions, thereby accelerating the transformation of laboratory teaching in physiology. Max von Frey, a Ludwig disciple who visited Harvard, praised these instruments; "I am astonished," he wrote to Porter, "that such a useful instrument can be sent out at so small a price."[21] Once he was engaged in producing 104 instruments, why not 204? Thus the Harvard Physiological Apparatus Company was established (not unlike the Cambridge Instrument Company being established in the other Cambridge at about the same time). The company at first went outside and used commercial manufacturers for most of their productions; for example, Waltham Clock Company, one of America's oldest clock makers, took on the major task, but it didn't prove satisfactory. They did not produce on time or at low enough cost. The production was pulled back into Harvard, and the Mechanics of Harvard Physiological Laboratory took over manufacturing. They set up shop alongside the laboratory where they were building and maintaining instruments for students and researchers. For Robert Hooke the instruments were part of the reform of philosophy; for Porter the instruments were part of a reform of physiology, and with it the reform of medicine and medical education, with the instrument becoming the hallmark of a new form of medicine. The scientist was experimenter, teacher, instrument designer, and entrepreneur.

The success of the instrument company turned out to be embarrassing for the University. It made money, a lot of it![22] Ultimately the propriety of using university property and space for manufacturing was challenged and the company was moved off campus, where it became independent. This was an early case of the entrepreneurial spirit of science outrunning what was admitted as the proper bounds for the organization of the institution which had given it birth. A seeming case of conflict of interest which has been repeated over and over again as the sciences became useful and the mode of practice became mechanized and industrialized. The apparatus was used as part of a system of reform. The standardized experiment became possible because there were standardized instruments available at low enough cost that they could be widely distributed and widely used. Bowditch's or Porter's texts described their experiments, and scientists could conduct the same work in laboratories a thousand miles away, say, at the newly established University of Chicago. The entrepreneurial element of instrument building was clear then, it became even clearer later. Porter, like Hooke, produced a text explicitly to tell readers how to use the instrument in practicing the science on which they were working.

3. SCIENCE AND INSTRUMENTS IN THE ERA OF HIGH TECHNOLOGY

John Krige's description of CERN and the international effort which built the big machines, and in comparison the national competitiveness over the smaller, but still significant detectors, is a fascinating and instructive story.[23] He recounted the 200-member work groups which were brought together to cooperate on experimental design, practice, and application. Are they ever on the experimental floor together? I doubt it. Spatially it would be almost impossible. Instead, there is a highly developed division of labor and assigned positions which are accepted by the team members. The assigning may be done in semihierarchical fashion, or it may be done in consensual fashion. The order of names on published papers is according to an agreed upon set of canons and rules which have grown up through the period of transition from the solitary to the corporate system of the production of knowledge. The long-term efforts of these groups, because of the expense of the instruments they are building, become an indicator of a constraint and maneuverability may be limited. going to be limited.

Erwin Chargaff, the man who first spelled out the chemical composition of DNA, lamented a nostalgic lament: " A pall of monotony has descended on what used to be the liveliest and most attractive of all scientific professions. The noble study of botany, for example, has been all but banished from many universities. Before, the biological sciences had their characteristic faces and their distinct spheres of interest into which they drew different types of scientists. Now I go through a laboratory, be it of virology or developmental physiology, and there they all sit before the same high speed centrifuges or scintillation counters, producing the same superposable graphs. There has been very little room left for the all important play of scientific imagination. <u>Homo</u> <u>ludens</u> has been overcome by the seriousness of corporate finances."[24] Chargaff was angry; he was angry at his profession; he was angry at what he saw as the increasing arrogance of those organizing it and directing it; he was angry personally for never having gotten a Nobel Prize when he thought he deserved one. (His own view was that Nobel Prizes were now being given for something other than the science itself which he nostalgically reconstructed in his mental image.) Instead, there was a different form of governance of modes of laboratory design and practice. Corporate finance was identified as the culprit. I suspect what he really means is not only the financial element alone, but also the corporate model of organization, the corporate-industrial form of reorganizing the workplace, developing the relationship between the practitioner and the instrument. And what he notes so clearly in his description is the assembly line nature of the activity now being carried out in the biological laboratory.

E. O. Lawrence joked on the occasion of his receiving the Nobel Prize in 1940: Perhaps the difficulties in the way of crossing the next frontier in the atom are no longer in our laboratories; we have handed the problem over to the president of the university.[25] Expense. The father of all machines, the father of all cyclotrons which he wanted to build was going to be very, very expensive. Lawrence was aware, of the large-scale and expensive nature of the activity in which he was engaged. Alongside him, almost parallel in time, George Ellery Hale was engaged in financing and building large telescopes, first at Mt. Wilson and then at Mt. Palomar. These men provide the image of the entrepreneurial scientist entering an age which dramatically stretched the boundaries of the instruments which were available for use, and with them of the data one could collect and the areas to which one could turn to collect data. There is, however, a second interesting element in the award of the Nobel Prize to Lawrence. The prize was given for an instrument, as Heilbron and Seidel put it in their recent study, and also for the creation of the environment necessary to exploit it. The work scene which could use a large instrument had to be created. The committee, Heilbron found when he turned to the interior records of the Nobel Prize, liked the scaling-up in the production of radioactive isotopes, the gamma rays, etc., that they could get from them. The scaling-up was being rewarded. As they note, however, Lawrence's was the first award for the development of an instrument of physics. They leave aside Wilson's cloud chamber and note, after all, Geiger, who had been proposed on numerous occasions, never received a prize. Cockcroft didn't get his Nobel award until fifteen years after Lawrence, even though if they were just looking for a simple machine and discoveries from it, Cockcroft could have gotten his prize earlier. Heilbron and Seidel conclude that the award went to Lawrence in part because of the increased popularity of nuclear physics and because of the scale of the machines and laboratories that the new instrument required.[26] Oliphant interpreted the Prize Committee's meaning directly to Lawrence in case he hadn't understood it himself: "It is extremely encouraging to find that the Nobel Prize Committee, in common with many other authorities, is now recognizing the tremendous importance of technique in scientific investigations.... The technical side of the subject is now recognized as equally important with the advances that follow from the use of these techniques and more important, I hope, then the theories that endeavor to explain them.... It is certain that you have no difficulty now in raising funds for your' father of all cyclotrons.'"[27] Notice the switch which has occurred; the implication is that the nostalgic image of the doing of science, which Chargaff and others celebrated, the lone worker with the string and sealing wax instruments, was now being replaced, even in the mentality of the Nobel Prize committees by large-scale machines and instruments themselves, as well as for what they might produce. The research systems they brought together were gaining explicit recognition.

The link to military utilization was made explicit. Many people had commented, that the cyclotron was a parade ground for military metaphors. As the Berkeley campus newspaper put it when Lawrence won his prize, "The atom smasher wins highest scientific honors." Ernest O. Lawrence, they said, "has discovered, a blasting technique far more potent than anything Alfred Nobel ever dreamt of." An elder statesman of science, of physics, wrote in private to Lawrence: "As you are laying the foundations for the cataclysmic explosion of uranium (if anyone accomplishes the chain reaction) I am sure old Nobel would have approved."[28] Right there the linkage seemed clear enough to those looking on. Physicists were prepared to interpret and respond to the discovery by Otto Hahn in December, 1939.

Peter Galison, in a fine set of papers, has explored the emergence of large machines--the accelerators and the detectors necessary for the questions that they now found possible to answer. He makes clear that the upscaling came in part from the military experience in World War II, and the expectation of being able to gain large funding for projects designed by physicists. He points to the extensive collaboration with engineers that became necessary in order to build machines which were beyond the capability of the experimentalists themselves to design and to build. As he put it, "The new generation of detectors also altered the kind and quantity of data that had to be analyzed." The traditional role of the physicist was changing with a removal of the physicist from the apparatus itself, the specialization of tasks, the increased role of computation, and the establishment of hierarchial collaborations which have now become hallmarks of high energy physics experiments.[29] What counted as an experiment in 1953 was not the same, he writes, as what counted as an experiment by 1966. In some cases the changes were being consciously sought, and the past experience of some of the physicists, particularly in their wartime efforts, was being extrapolated to new post-war situations.

Louis Alvarez was willing to adopt engineering criteria for the new machines; a machine which was "good enough" was a concept, he said, which physicists would have to embrace. An instrument that was merely "good enough," was an idea which would have been anathema to a physicist at an earlier period.[30] Further, Alvarez recognized the transition to a corporate form practice and he was willing to accept it. He reminded his physicist colleagues that it was well known among industrialists that little companies often go bankrupt trying to pass from an organization in which individuals handled many jobs to a situation in which high efficiency comes from production line operation with very expensive production tooling.[31] Alvarez, believed that the changes in corporate structure necessary to tool up and scale up in industrial production, had to be adopted within science itself.

It was just this scaling up, the move to corporate assembly line activity in physics, that finally drove Donald Glaser out of physics. He rejected the big machines and all they implied for the scientific workplace: "There was a psychological side to this. I knew that large accelerators were going to be built and they were going to make gobs of strange particles. But I didn't want to join an army of people working at big machines.... I decided that if I were clever enough, I could invent something that could extract the information from cosmic rays and you could work in a nice peaceful environment rather than in a factory environment with big machines."[32] This was a crucial step in his attempt to design and develop the bubble chamber. It was a magnificent design, won him a Nobel Prize, and provided an exit from physics as it became clear that the scale of activity that was to be involved in the new detectors was one that Glaser himself did not like.

But if Glaser didn't like the move to a corporate productivity model, a factory model of scientific activity, Walter Gilbert, the physicist turned molecular biologist, saw this new system of work and work organization leading to a major paradigm shift in biology. The human genome project was the largest ever designed for biology itself. The key project is a direct mechanical activity, the sequencing of DNA-base pairs. The scale is large; some 3,000,000,000 base pairs to be sequenced, located in some 50,000 to 100,000 genes, equal in terms of information to 500 bound volumes, with a thousand pages each, and each page with 1000 six-letter words on it.[33] The project has been designed in the United States as a fifteen-year effort, costing in the initial estimate $5 billion. Small by the scale of the big physics instruments, enormous by scale of biological projects. The mode of operation was clearly entrepreneurial.

Walter Gilbert moved from a Nobel Prize in molecular biology and Professorship at Harvard to the presidency of Biogen, a Swiss-financed biotechnical firm with offices both in Switzerland and Cambridge, Massachusetts, and back again to the university (albeit remaining on Biogen's board). Gilbert, restless, with a background in progressive politics, became the quintessential entrepreneur in the biological sciences. He urged that the sequencing of DNA be done by the private sector, and in fact he contemplated setting up a corporation to do to do the job. He wanted to avoid the slow moving government bureaucracy which would take much too long in the task. This threatened move was all that was needed to have the other part of the biological establishment, led by Jim Watson, energize the government to take on the project. Somewhat surprisingly for the outsider, the Department of Energy said that it would become a sponsor of the effort. Why the Department of Energy? Because of its interest in the genetic effects of radiation, the DOE had a long history of doing genetics. Perhaps, with the end of the Cold War and the search for new non-bomb tasks, the DOE was ooking for new things to do. The decline of the cold war was coming to the aid of biology. The DOE effort prodded the the NIH (National Institute of Health), and the competition got a major effort started. James Watson emerged as the czar and the enforcer of new modes of action.

"Sequencing is boring," wrote Walter Gilbert.[34] And he is exactly right. But the changes that had occurred during the past 15 years now made sequencing possible; previously, the task would have taken "forever," or at least more than the life of a single generation of biologists. New technologies allowed new questions to be posed and to be answered. The specific new technology that seemed to be available to speed the project along was an adapted form of gel electrophoresis. The ability to tag the base pairs meant that they could be rapidly read. Of the 3,000,000,000 pairs, a single machine might now read 15,000 a week. This led Gilbert to the notion of a new paradigm for biology. "In the current paradigm," he says, "the attack on the problems of biology is viewed as being solely experimental. The 'correct' approach is to identify a gene by some direct experimental procedure determined by some property of its product or otherwise related to its phenotype, to clone it, to sequence it, to make its product and to continue to work experimentally so as to seek an understanding of its function."[35] This indeed is not a bad description of the classic biology of genetics, even in its molecular biological form. "The new paradigm now emerging" Gilbert writes, "is that all the 'genes' will be known (in the sense of being resident in data bases available electronically) and that the starting point of a biological investigation will be theoretical." The "use of this flood of knowledge which will pour across the computer networks of the world means biologists must not only become computer literate, but also change their approach to the problem of understanding life." Gilbert is setting out a strong, new program which indeed will become, is becoming, the natural mode of action of the advanced forms of biology, and will set the pattern which other sectors will follow.

What we see as we look back, are major changes in scale. Not that all things up-scale at the same time and rate--they don't. But a change in scale of instrument and a change in scale of the modes of knowledge production--both larger instruments and/or large numbers of smaller instruments working simultaneously similar to the factory floor. There has been a triumph of industrial forms, including a triumph of the corporate management forms. If Louis Alvarez, Walter Gilbert and James Watson are correct, the success includes industrial forms of management, funding of organization and structure of the work place, and indeed governance of the enterprise of science. This does not mean an exact mimicry in a simple sense of corporate industrial structures and procedures, but rather an adaptation of these forms to the system of science. This is not just science becoming high tech factory production, that would be a misreading and a misidentification, but it means adopting the modes and procedures to a different place and insisting that the enterprise adopt these same procedures, modes, and norms of action. They are highly compatible to industrial modes of action; they are able to interface with industry in production, and in financing. Venture capital flows into the new biology as venture capital flows into other advanced industrial activities. The interface must have a match if this kind of interaction is to occur. In this sense Chargaff correctly identified corporate financing as setting the terms of the new biology. If Walter Gilbert is right, and he's in the middle of it, biology will have been reconstructed so that it can become part of a system which involves today's advanced industries. There will, of course, be increased competition involving parallel attempts to solve problems--competing groups, competing laboratories, even as there are competing industrial firms. There will also be increased cooperation within large work groups which are dependent on large and expensive machines and laboratories, in many ways similar to the industrial system. There will, in addition, be national competitiveness, even as it exists at the industrial level. There will be international agreements and international regulations as well.

National states are willing to finance very large-scale and expensive projects in science and in high technology--accelerators, space probes, human genome initiatives, in part because the governments have become familiar with the mode of operation of the sciences beginning in the wartime efforts and in the long cold war involvement of science and managers of science in engagement with the government and the military. The interaction is pushed along because of the extent to which government, military and industry have developed a high reliance on science and high technology for commercial and security goals. There has been a thorough interpenetration of modes of organization and thinking so that the interface is not a difficult one to cross at this period, and indeed any number of significant scientists, have marched back and forth on a regular or permanent basis.

Are these large-scale activities which we have examined simply scaled-up "normal science," or are the large-scale systems (because they are systems; they are not instruments and a few people working with them, they involve instrument designers, builders, engineers, funding agencies, users, and regulators) or are they something fundamentally different? Will these new complex systems be slowly miniaturized and down-scaled and become the model for other sectors of the scientific system? My own sense is this latter model may well be what's happening as we watch the transformation in several key areas of science today.

NOTES

1. See Henry Power, Experimental Philosophy in Three Books; New Experiments Microscopical Mercurial, Magnetical. With Some Deductions and Probable Hypotheses, raised from them... London, 1664.

2. Quoted from Robert Hooke, Micrographia, "Preface," 1665, in Michael Aaron Dennis, "Graphic Understanding: Instruments and Interpretation in Robert Hooke's Micrographia," Science in Context, vol. 3, 1989, p. 310.

3. Noted in Dennis, "Graphic Understanding," where Shapin is specifically citing the air pump, p. 310; Dennis briefly discusses costs on p. 324.

4. See the detailed new study of Leeuwenhoek's instruments, techniques, and practice, Brian Ford, The Leeuwenhoek Legacy, Bristol and London: Biopress and Farrand Press, 1991.

5. Steven Shapin, as reported in a lecture by Lorraine Daston, Harvard University, February, 1991.

6. Lorraine Daston, "Baconian Facts, Academic Civility, and the Prehistory of Objectivity," Annals of Scholarship (in press); Lorraine Daston and Peter Galison, "The Image of Objectivity," Representations (in press).

7. Dennis, "Graphic Understanding."

8. Ibid, p. 311.

9. See Ford, Leeuwenhoek Legacy, pp. 141-181 for numerous photographs and illustrations of the instruments and a detailed account of their development through Leeuwenhoek's years of use.

10. Dennis, "Graphic Understanding," see pp. 319, 323-5.

11. See Peter Galison, "Bubbles, Sparks, and the Postwar Laboratory," pp. 214-215.

12. Steven Shapin, "The Invisible Technician," American Scientist, vol. 77, 1989, pp. 554-563.

13. See Ian Hacking, Representing and Intervening, Cambridge: Cambridge Univ. Press, 1983, p. 189.

14. Gerard L'Estrange Turner, Essays on the History of the Microscope, Oxford: Senecio Pub., 1980, p. 19.

15. Merriley Borell's important study has been of great use to me: "Instruments and an Independent Physiology: The Harvard Physiological Laboratory, 1871-1900," in Gerald L. Geison (ed.), Physiology in the American Context, 1850-1940, Bethesda, MD: American Physiological Society, 1987, pp. 293-321.

16. Quoted in W. Bruce Fye, The Development of American Physiology, Scientific Medicine in the Nineteenth Century, Baltimore: Johns Hopkins Univ. Press, 1987, pp. 124-125.

17. Quoted in Borell, "Instruments and an Independent Physiology," p. 296.

18. Ibid, p. 297-298.

19. Ibid, p. 301.

20. Ibid, p. 302.

21. Quoted in Ibid, p. 301-2.

22. Ibid, p. 303.

23. See John Krige's paper in this volume; for a fuller account of the development of CERN see the collaborative volume by Armin Hermann, John Krige, Ulrike Mersits, and Dominique Pestre, A History of CERN, Amsterdam: North-Holland, vols. I and II,

24. Erwin Chargoff, Voices in the Labyrinth, Nature, Man and Science, New York: Seabury Press, 1977, p.61.

25. See the slight variant of the remark in the new definitive study by John L. Heilbron and Robert W. Seidel, Lawrence and His Laboratory: A History of the Lawrence Berkeley Laboratory, Vol. 1, Berkeley: Univ. California Press, 1989, p. 492.

26. Ibid, p. 490.

27. Ibid, p. 492.

28. Ibid, pp. 492-493.

29. Peter Galison, "Bubble Chambers and the Experimental Workplace," in O. Hannaway and P. Achinstein (eds), Experiment and Observation in Modern Physical Science, Cambridge: MIT Press, p. 311; see in addition, "Physics Between War and Peace," in E. Mendelsohn, M.R. Smith and P. Weingart (eds.), Science, Technology and the Military, Sociology of Sciences Yearbook, Vol. XII, pp. 47-86; "Bubbles, Sparks and the Post-War Laboratory," in Laurie M. Brown, Max Dresden, Lillian Hoddeson, (eds.), Pions to Quarks: Particle Physics in the 1950's, Cambridge: Cambridge Univ. Press, 1989, pp. 213-251.

30. Alvarez put the idea in almost provocative form: "The idea of 'good enough,' comes from engineering, although it may upset instrument designers, I feel it is a concept we must embrace." Quoted in Galison, "Bubble Chambers," p. 346.

31. Ibid, p. 347.

32. Quoted in Galison, "Bubbles, Sparks," pp. 215-216.

33. Leroy Hood, "Biotechnology and Medicine of the Future," JAMA, vol. 259, Mar. 25, 1988, p. 1841.

34. Walter Gilbert, "Towards a Paradigm Shift in Biology," Nature, vol. 349, 10 January, 1991, p. 99.

35. Ibid.

Chapter 2

Late Victorian metrology and its instrumentation: a manufactory of Ohms

Simon Schaffer
University of Cambridge (United Kingdom)

"The original B.A. standards of resistance are at the Cavendish Laboratory...When the place is in working order, I hope to be able to verify any standards already constructed and sent to me, but I do not expect or think it desirable that a manufactory of "ohms" should be established in the building" (James Clerk Maxwell, 14 January 1875)

Much scientific work helps make experiments into governable instruments. For example, during the 1660s trials by Wren and Hooke using the height of mercury in glass tubes to study the relation between the air's spring, tidal height and weather resulted in barometers and weather glasses. In the 1770s, Priestley's experimental technique for estimating the phlogiston content of airs by mixing samples with nitrous air was transformed by Italian chemists into a workable eudiometer. During the 1820s, Oersted's dramatic experiments on the magnetic effect of a current-carrying wire became embodied in the routine laboratory use of galvanometers.[1] Experiments with many possible messages are changed into tools with apparently straightforward, unproblematic functions.

Recent studies of the laboratory workplace have indicated that institutions' local cultures are crucial for the emergence of facts, and instruments, from fragile experiments. In seventeenth-century laboratories, gentlemanly codes of honour and trust governed the capacity of natural philosophers to render experience and testimony as reliable reports of natural order. Technologies of machine-minding and reporting were developed to make experimental philosophy fit Baroque social mores. Similarly, in Victorian observatories workers were regulated within factory discipline to guarantee their scrutiny of the stars. New machinery, like galvanic clocks, chart recorders and telegraphic relays were introduced between the eyepiece and the observer to make an ensemble capable of valid astronomical experience.[2] These stories indicate the extreme localisation of scientific practice: in these examples, experiments and observations are understood in terms of the features of specific workplaces. But if facts depend so much on these local features, how do they work elsewhere ? Practices must be distributed beyond the laboratory locale and the context of knowledge multiplied. Thus networks are constructed to distribute instruments and values which make the world fit for science. Metrology, the establishment of standard units for natural quantities, is the principal enterprise which allows the domination of this world. Metrology matters because of this relationship between the lab interior, in which values are established and reproduced, and the networks which link labs, in which values are translated and transformed.[3]

Metrology and the integrity of telegraphs

Standardisation is therefore an obvious concern for historians of science. The physical values which the laboratory fixes are sustained by the social values which the laboratory inculcates.[4] Metrology has not often been granted much historical significance. But in milieux such as those of Victorian Britain the propagation of standards and values was the means through which physicists reckoned they could link their work with technical and economic projects elsewhere in their society. Instrumental ensembles let these workers embody the values which mattered to their culture in their laboratory routines. Intellectualist condescension distracts our attention from these everyday practices, from their technical staff, and from the work which makes results count outside laboratory walls. In an unprecedented manner, late Victorian scientists joined networks in which standard machines, values and practices were distributed worldwide: imperialism, mass production and metrology dominated their universe. They saw an immediate connexion between imperial standards and the integrity of their laboratory work. Their great accomplishments, such as electrotechnology, proved their command along the cables upon whose working the Empire's integrity depended. In 1887 a British MP told the Royal Colonial Institute that there was no reason to fear that the Empire would "break up and dissolve like its predecessors". He insisted on the role of "the postal and telegraphic services", a "cohesive force which was utterly lacking in former cases". "The two or three slender wires that connect the scattered parts of her realm" were of more use than the military, the administration, or "the unswerving justice of Queen Victoria's rule". The new British physics laboratories of 1860-1880 were part of this imperial communications project, either through their managers' work for the British association campaign for electromagnetic standards, or through their training of expert telegraphists, instructors and electrotechnologists. [See Figure 1] [5]

In Cambridge, where the Cavendish Laboratory was built between 1871 and 1874, proponents carefully argued for the combination of electrotechnology with liberal pedagogy. Kalley Miller, a fellow of Peterhouse, was well placed to present this case, having been William Thomson's locum at Glasgow and then mathematics professor at the Royal Naval College. The Glaswegian ideal was a difficult but important topic in Cambridge, for it indicated how the social life of the university laboratory could be multiplied throughout the Empire. In autumn 1870 Miller told the University that "the success of the Atlantic cable is in great measure the results of years of patient work in the Glasgow Laboratory". He referred to students' work "testing and perfecting the numerous exquisite instruments of Sir William Thomson's invention. Students are trained in Glasgow for home and foreign telegraphic service". He had to show that the proper enclosure of the laboratory, which would fit it for the genteel Cambridge milieu, would nevertheless allow vital links between the commanding heights of Victorian electromagnetism and thermodynamics, and the practice of metrology. Miller placed equal weight on standardisation and on instrument testing: "the determination of the wavelength of light, of the mechanical equivalent of heat, and of the relation between the electrostatic and electromagnetic units are among other vitally important practical points which have been successfully experimented upon. I may add that the excellent electrometers turned out by the Glasgow makers owe much of their value to the fact that each one has been carefully tested and regulated in the University laboratory before it is sent out for service".[6] Within a decade, the Cavendish itself became a centre of Victorian electrotechnical metrology, certifying electrometers and resistance boxes for the cable industry and the nascent network of physics laboratories. In March 1874, Clerk Maxwell, the new Cavendish professor, ordered "one or more of the original standard Ohms to verify commercial Ohms withal", and eventually such firms as the instrument company Elliott Brothers asked the Cavendish to verify its resistance boxes.Teaching regimes changed in company with research practice, and this was by no means a straightforward manoeuvre. By January 1875, in the letter quoted at the head of this paper, Maxwell stated that when the laboratory "is in working order, I hope to be able to verify any standard already constructed and sent to me". But he now added a crucial caveat: "I do not expect or think it desirable that a manufactory of "ohms" should be established in the building".[7] So Maxwell outlined a metrological programme for the new Cavendish Laboratory, a programme for the verification of others' resistances and devices and for the production of new, revised standard instruments. But he also pointedly distinguished this project from the work of a factory.

This was a crucial institutional distinction. Workshop culture sat ill with the liberal University. Maxwell stressed the point in his inaugural lecture of October 1871: because "modern experiments...consist principally of measurements", Maxwell explained, many had come to believe that "in a few years all the great physical constants will have been approximately estimated, and that the only occupation which will then be left to men of science will be to carry on these measurements to another place of decimals". This statement has often been cited as though it were Maxwell's own view. But Maxwell reckoned that this view was doubly wrong: wrong because physics had much more than precision measurement as its task, and wrong because this picture would render the University an inappropriate home for the new laboratory. "Our Laboratory may perhaps become celebrated as a place of conscientious labour and consummate skill, but it will be out of place in the University, and ought rather to be classed with the other great workshops of our country". By 1882, under Lord Rayleigh, the Cavendish was earning a good income from fees charged for verification and testing electromagnetic standards. At the same moment, Rayleigh also insisted on a fruitful "division of labour" between experimentalists and mathematicians.[8] The division of labour mattered to the possibility of making a standards laboratory; and the creation of standards dominated the linkage between the Cavendish and its various constituencies.

The purpose of the paper which follows is to explore the cultural setting in which these patterns of labour and skill were established, and to use these cultural factors to analyse the relationship between the Cavendish Laboratory and its outside world. There were many late Victorian metrological institutions: the Board of Trade, Kew Observatory, and the Royal Observatory, for example. In Cambridge, a precision measurement standards laboratory was set up within an existing academic institution. We gain unusual access to the resources needed to establish a site of calculation and vigilance. We also begin to understand the social conflicts which metrology needed to resolve. No Victorian savant could view metrology with condescension. Its public discourse was profoundly moralistic. A culture convinced of the need for accurate data and stern surveillance as the right means of social management would also see precision measures as an ethical good. In late Victorian culture, it has been observed, "diligence, good time keeping and good behaviour were rewarded. Time measured out by the machine, not that dictated by the Sun or the seasons, had come to impose a different and totally new discipline on the British people". [9] Standardisation showed the integrity of the system, its continuity in moral action and physical space and time. Edwin Chadwick concluded his notorious <u>Report on the Sanitary Condition of the Labouring Classes</u> (1842) with a paean to "the advantages of uniformity in legislation and in the executive machinery, and of doing the same things in the same way (choosing the best) and calling the same officers, proceedings and things by the same names". A monument to this ethos was Charles Babbage's project for an all-embracing <u>Table of the Constants of Nature and Art</u>. Even the phrase "constant of nature" was a neologism which Babbage skilfully used. He urged the production of a collaborative encyclopaedia which would contain "all those facts which can be expressed by numbers in various sciences and arts". This project would "call into action a permanent cause of advancement towards truth, continually leading to the more accurate determination of established facts and measurement of new ones".[10]

Babbage helped develop the important notion of "absolute measurement". The career of submarine cables demonstrated this importance. William Thomson showed that telegraph performance depended on a precise knowledge of the resistance per unit length of the cable wire. So in 1855 he told the Cambridge professor George Stokes that "it will be an economical problem...to determine the dimensions of wire and covering which, with stated prices of copper, gutta percha and iron, will give a stated rapidity of action with the smallest initial expense".[11] This knowledge had to be expressed in units which manufacturers and telegraphists could reliably use. The failure of the first Atlantic cable showed how hard it was to achieve this reliability. In summer 1871, Thomson reminded the British Association that the combined material interests of "those who perilled and lost their money in the original Atlantic telegraph", and the British Association's standards committee set up in 1861 to "supply their electricians with methods for absolute measurement" had ensured the profitability of the cable firms, the culture of "accurate measurement in every scientific laboratory in the world", and had started "a train of investigation which now sends up branches into the loftiest region and subtlest ether of natural philosophy".[12] The integrity of the ether, of electromagnetism, of the cables, and of the Empire all hinged on these standards. Many physicists said so in numerous tomes on standard systems. J.D.Everett's treatise, commissioned in 1875 by the British Association, approved at the vital international electrical congress in Paris in 1881, and edited by Maxwell and Carey Foster, argued that "the existence of quantitative correlations between the various forms of energy, imposes upon men of science the duty of bringing all kinds of physical quantity to one common scale of comparison".[13] The British made absolute units depend on the integrity of the scale of work. So the link with imperial integrity was secured: the spread of these systems of units and instruments made labs like Glasgow and the Cavendish centres of calculation for world electrotechnology, discrediting rival entrepreneurs. The technology embodied in the cable network hinged on such assays.

The British Association standards programme

The British Association standards committee was an important forum for this development. It was established in 1861 following rival reports by the Siemens brothers and by Latimer Clark and Charles Bright on the need for a system of units in the cable industry. Siemens were engaged in cable repair and maintenance using "stringent electrical tests which the authors are charged by the British government to apply". They were using the Thomson-Siemens formula for specific induction and a new in-house Siemens resistance unit, based on the resistance of an arbitrary length of mercury. Workers such as Clark and Bright, alert to "the wants of practical telegraphists", also demanded an adequate system.[14] But in the event the committee was made up of a professorial group who commandeered the campaign for reliable units. Many of the new physics labs were managed by committee members, including Maxwell. In the preface to his Treatise, completed in February 1873, Maxwell noted that this electrotechnology gave "a commercial value to accurate electrical measurements", and also allowed tests "on a scale which greatly transcends that of any ordinary laboratory".[15] In this setting, the question of workshop culture and of the morality of pedagogy became vital. Before taking up the Cavendish chair, Maxwell had spent the previous decade at King's College London and, after 1865, collaborated with other London workers from his retreat at Glenlair. As a preamble to this campaign, Maxwell told the British Association that "between the student's mere knowledge of the history of discovery and the workman's practical familiarity with particular operations which can only be communicated to others by direct imitation, we are in want of a set of rules, or rather principles, by which the laws remembered in their abstract form can be applied to estimate the forces required to effect any given practical result". He urged that training and standards could only survive in company: "whenever many persons act together, it is necessary that they should have a common understanding of the measures to be employed".16 Throughout the 1860s, Maxwell worked hard to make a reliable resistance standard. His collaborators included the telegraphist Fleeming Jenkin and the instrument maker Charles Hockin, a Cambridge fellow. Links were forged with other members of the British Association committee, and with the manufacturers, notably Elliott Brothers, where Becker made Maxwell's coils into reproducible instruments. The salient issue was whether standards made in the London labs could be disseminated through cable networks, and, thus, whether the standards of conduct in the labs could match those in cable laying ships or telegraph stations.

A workable absolute system had to satisfy these constraints. An early effort to make a system which would co-ordinate international research was that of Wilhelm Weber, developed at Göttingen as part of the geomagnetic crusade of the 1830s and 1840s, published in 1851. Weber's units were absolute, and independent of any particular material, but they were therefore abstract and hard to reproduce. The British agreed that "absolute permanency" was a prerequisite of a resistance unit. Yet they shared a different view from Weber on the relation between work, force and electromagnetic action. Thomson reckoned that any system of units must display the work performed when unit current flowed across unit resistance at unit potential difference. So they sought to reform Weber's system and to rule out the conventional Siemens system. Werner Siemens fought back. He reckoned that "those cases in which the expression of absolute measure is of advantage occur very seldom and only in purely scientific exercises". Defending his use of standard mercury columns, the great German electrotechnologist argued that "every other definition would not only burden unnecessarily the calculations which occur in modern life, but also confuse our conception of the measure". He urged that the modern life of telegraphy be freed from British energy physics, whereas the British Association committee urged that the cable networks be liberated from German industrial control. Their solution was to set up an absolutely permanent standard, weld it into international telegraphy, and then change telegraphic practice to display it as a success for British electromagnetic skill.17

Maxwell's team was part of this effort. He adopted a Glaswegian technology: a spinning coil revolving round its vertical axis in the geomagnetic field, driven by a Huygens gear and a new governor designed by Jenkin, would produce a constant deflection on a galvanometer needle at its centre [See Figure 2]. The angle of deflection would depend on the coil's diameter, the rate of spin, and the coil's resistance. The resistance could therefore be established by measuring the other three parameters. This set up placed major demands on precision engineering and experimental skill. The magnet was suspended on a three yard silk fibre. It was claimed that the governor was so good that Thames steamers outside the College had more effect on the galvanometer needle than did variations in the coil's speed. Jenkin's design relied on an ingenious brake band mechanism. Maxwell took the chance to discuss this feedback system in an important theoretical paper for the Royal Society.18 The superintendent of Kew Observatory, Balfour Stewart, a master of precision measurement, designed a sound-and-clock method for the timer. The aim of mobilising these telegraphic and quantitative resources was to make a coil whose resistance was as close as possible to the unit of resistance set up in the electromagnetic system, 10^9 cm/sec. Maxwell had to defend this project against Siemens. According to William Thomson, the German's success in producing a length of mercury which could be measured to within 0.05% simply "affects the question of the best material for the construction and reproduction of the standard", but "leaves the question of its magnitude quite untouched".[19] This spinning coil soon became the emblem and the instrument through which standards could be propagated. The tough question was whether this could be achieved without compromising those standards.

There was a subtle compromise inherent in the British Association programme of which the King's College subcommittee were all too aware. Their group dismissed Siemens because his Berlin-based system was not absolute. It did not embody work and energy relations which they considered essential to the value of the standards programme. But the point of absolutism was that only if its standards were reproducible would the network evince the superiority of this physics. The choice between reproducibility and absolutism was a choice between the standards of laboratory physics and of the cable networks. So in 1864 Maxwell, Jenkin and Hockin repeated their spins at King's, changing most of the important elements including wire winding and rotation speed.They now claimed an accuracy of 0.08%. The Committee recognised that coils had already begun to be produced, by Elliott Brothers and by Siemens, for distribution to to the Indian telegraph and elsewhere. Such coils had not been certified, and "meanwhile practical standards of resistance are urgently required...Defective systems are daily taking root". Maxwell's team acknowledged that further research might yield more accurate coils. But instead they opted for the demands of the cable network: "it has never been proposed that the British Association standard should be considered as representing exactly an absolute measurement". By March 1865, Maxwell told the Edinburgh physicist Peter Guthrie Tait, "the true origin of electrical resistance as expressed in BA units" had become "Fleeming Jenkin, Esq., price 2 pounds 10 shillings in a box".[20] What was boxed here, however, was not the absolute unit. Siemens, for one, pointed this out very soon: "by what means the Committee holds itself justified, in the face of such differences between even the means of their single observations, in concluding upon an error of only 0.1%, I am totally unable to imagine".[21] It could be successfully alleged that the artefactual character of the Associations' boxed units, which erred noticeably among themselves and from any notional absolute, testified to the failure of laboratory standards among Maxwell's London group. They had to work out ways of making their experiments count as instruments.

Jenkin dealt with these puzzles in a series of lectures in spring 1866 at the Society of Arts. He spoke both as Maxwell's London collaborator and as a veteran of seven years' hard work with Thomson's Glasgow group and with the Birkenhead firm of R.S.Newall on patents for improvements in telegraphic communication. This was a moment of great risk for the telegraph industry, at the very point where confidence in the forthcoming laying of the second Atlantic cable was in question. Jenkin reckoned he could turn the tables on Siemens, by showing how the British Association Unit could be made to work at sea and on board ship. Inaccuracies in London would pale into insignificance if cable telegraphists could be won to the British Association system. These units were good means of cable fault finding, since if the tester knew the resistance per unit length of the line he could locate any cable break by balancing the potential across a standard resistance against that of the unknown length of cable up to the fault. But this meant knowing that cable materials were made of reliably standardised wire, and thus knowing that laboratory values had been transferred to cable works. [See Figure 3] Jenkin lectured on existing fault finding methods, especially those of Siemens. They needed 20 minute tests of insulation, resistance and signal transmission, using Siemens' resistance boxes, calibrated to his mercury tubes.[22] Since these standards were supposed to enhance ship-shore communication, it was deleterious to allow such long breaks during testing. Jenkin announced a new regime, developed with Thomson, using the new British Association boxes and Glasgow-made galvanometers and electrometers. These were just the instruments Kalley Miller lauded at Cambridge in 1870. Ship-shore links would be continuous, the integrity of the line could be checked and, significantly, the continuity between lab skills and telegraphic skills could be assured through new patterns of training in this electrotechnology. "Undoubtedly this test is not of so simple a nature that it could be executed by an ordinary clerk", Jenkin conceded, "but it is interesting to know that a test does exist by which even a fault which has hitherto baffled electricians can have its position fixed with mathematical certainty...More operators can at present be found who are familiar with galvanometers than with electrometers; so that clerical error would not be so likely to occur as with [another] plan". The same message was stressed in the report the British Association committee, to which Jenkin was secretary, issued that year. Costs of the research were high; but the alleged success of the new, absolute, "coherent" system had moved skill from pure science to real practice. "If resistance only were to have been practically measured, the advantage of adopting the coherent system would have been wholly confined to men of science; but now that measurements of capacity, currents and electromotive force are becoming common, the practical advantages of the coherent system are very considerable". To back this up, the report listed all those agencies and electricians who were now using the new method and the new standard, including India, Australia, a number of major companies, and the leaders of British electromagnetic research.[23]

Jenkin's gamble at the Society of Arts and through the British Association paid off, as did the Atlantic Telegraph Company's deals with Jenkin and his colleagues, Thomson and Cromwell Varley. The Company agreed that if they were offered any instruments which seemed better than the galvanometers, keys, electrometers and test devices designed by the team, then Jenkin's group would be given a year to come up with their own improved instruments. By linking the status of the Association's units with cable integrity, Jenkin and Thomson were playing a powerful hand. On 8 September 1866, just after the British Association meeting, the great Transatlantic cable had been secured from Valentia to Newfoundland. Under the leadership of John Pender, British industrialists now rose to international domination of the system.[24] The cable's success was perceptibly linked to electromagnetic wisdom, and the security of laboratory-ship linkage was seen as vital to this success. Just as the projectors of the first, failed cable of 1858 had violated theoretical integrity by ignoring Thomson's principles of electromagnetism, so the warrant of the integrity of the 1866 cable was the integrity of the vast system of instruments and values which sustained and were vouchsafed by it - resistance boxes developed by Maxwell and Jenkin and made by Becker; the conduct of the workshops and labs at London, Birkenhead and Glasgow; the truth of the electromagnetic connexion between resistance, current and work. Jenkin responded that his instruments' credit rested on "submarine cables, in the examination and manufacture of which, the practical engineer soon found the benefit of a knowledge of electrical laws".[25]

The events of the 1860s resulted in a temporary conciliation between the values of the laboratory and of commerce. Towards the end of the decade, freed from the rigours of his London chair, Maxwell reworked the standards project as part of his electromagnetic physics. Maxwell's experimental resources were compromised because Jenkin and Hockin were recruited to the Atlantic cable team. When Hockin was released from cable duty, however, he could help: Maxwell reported that Hockin "performed everything, except the actual observation of equilibrium, which I undertook myself". And through his contacts with Elliott Brothers, Maxwell had access to the British Association condensers and resistances. The firm's engineer, Becker, designed his new electrical balances, and Maxwell could occasionally command J.P.Gassiot's London laboratory with its great batteries.[26] The trials performed there were designed to ascertain the relative values of the electrostatic and electromagnetic units, a task laid down in the mid-1860s for the British Association and adumbrated in his celebrated 1865 "Dynamical Theory" as part of the linkage between light and electromagnetic radiation. It was a fine example of the way in which "purely scientific" problems could be made to seem commercially crucial. The set up involved balancing the force developed between two oppositely charged discs against the repulsion between two current carrying coils. Maxwell changed the gap between the two plates until balance was reached, while Hockin varied the resistance in the circuit until a galvanometer linked to the coils was undeflected. Sometimes the two men changed places. They reckoned that high precision tooling would allow a good estimate of the ratio of the two units and hence the value of Maxwell's electromagnetic programme: "as it seemed probable that the time occupied in the construction and improvement of these instruments would be considerable, I determined to employ a more direct method".[27]

There were reasons for haste. Maxwell understood that "the importance of the determination of this ratio in all cases in which electrostatic and electromagnetic actions are combined is obvious. Such cases occur in the ordinary working of all submarine telegraph-cables, in induction coils, and in many other artificial arrangements". This was conventional work for the British Association Committee. But he also insisted that "a knowledge of this ratio is I think of still greater scientific importance when we consider that the velocity of the propagation of electromagnetic disturbance through a dielectric medium depends on this ratio and hence, according to my calculations, is expressed by the very same number".[28] The status of the claim that light was related to electromagnetic radiation was made to hinge on the standards programme. In 1866 Jenkin had made telegraph engineering validate Glaswegian electrotechnology. Soon afterwards, Maxwell used this engineering to link electromagnetism and light, not just signal propagation and work. So the significance of the cable system's integrity varied. In 1868, Maxwell found very large differences between his calculated velocity, V, and that of light. He told Hockin to find ways of increasing accuracy. Hockin told him "the box containing B.A. units did not at all agree...I shall do them again...If the error comes from a bad contact anywhere the result will not come out the same. This brings 'v' from 22×10^7 to 24.5×10^7 [metres/second]".[29] Eventually Maxwell managed to force the value up to 28.8×10^7 metres/second, still too far from Foucault's value of a light speed of 29.8×10^7 metres/second. He did this by following Hockin's suggestion, suppressing almost one-third of his runs, "on account of the micrometer being touched during the observation of equilibrium", and unremittingly blaming the value of the British Association resistance unit. He could try this latter strategy because his best value for V was like those now being made elsewhere in London and Glasgow. Recalling his project of the 1860s, Maxwell stressed "the hunting for sources of discrepancies, the hitches in the working of new apparatus, the mathematical difficulties, and not least the physical difficulty of keeping up the spirit of accuracy to the end of a long and disappointing day's work". Needing larger batteries, possibly Grove cells, a better workforce and much better resistance boxes, his work stalled. It could not be revived until, and may well have prompted, his move to Cambridge in 1871.30

The Cavendish as a metrological institution

Maxwell's work with the British Association committee provided him with one set of resources when he reached Cambridge. But it cannot completely explain the appearance of a standards laboratory inside the University. The experience of the 1860s dramatised the links between cable telegraphy and electrotechnical metrology. But this did not make an electrotechnical standards programme a self-evidently appropriate enterprise for a liberal educational institution. The establishment of the Cavendish was not a moment when the culture of the workshop effortlessly found a place within academic life. Instead, two complementary features of the deployment of electrotechnical instrumentation in Cambridge must be stressed. On the one hand, Maxwell and his allies worked hard to make metrology look like a moral enterprise, and so a programme fit for the university; on the other hand, the laboratory managers sought to set up work relations within the Cavendish which would make its values reliable and precise measures of useful parameters. In neither of these strategies can we deduce laboratory life solely from the simplified intentions of its professoriate. Life outside the institution, and outside its leadership, mattered at least as much to the metrological programme developed there.

During the early 1870s, Maxwell tried to match exploitation of the instruments provided by the British Association and the cable engineers with the concerns of other scientific practitioners. Amongst these were the new astrophysicists. Maxwell claimed that observers like William Huggins, based at Tulse Hill, were also providing evidence of the universal standardisation of the performance of the ether, because they showed that light and matter behaved the same way in deep space and on Earth. In a celebrated paper attacking fashionable German models of action at a distance, Maxwell argued for the homogeneity of the ether, "this wonderful medium, so full, that no human power can remove it from the smallest portion of space, or produce the slightest flaw in its infinite continuity". The integrity of the ether warranted the integrity of star spectra: "the medium receives the impulses of these vibrations and after carrying them in its immense bosom for three years, delivers them in due course, regular order and full tale into the spectroscope of Mr Huggins at Tulse Hill".31 And the regularity of spectra in its turn connected astrophysics, laboratory physics and theology. Maxwell peddled this moral argument about metrology throughout the 1870s in public lectures, addresses at the British Association and in encyclopaedias. Citing John Herschel, he claimed that the fact that molecules behaved identically everywhere proved they were "manufactured articles". Molecules' vibrations were like soldiers' drill or Whitworth's screws. This was a fine metaphor, linking the values of the production line with those of God and of each molecule. The enterprise which documented their standardisation must contribute to human knowledge of the Creator. "From the ineffaceable characters impressed on [molecules] we may learn that those aspirations after accuracy in measurement, truth in statement and justice in action...are ours because they are essential constituents of the image of Him who in the beginning not only the heaven and the earth but the materials of which heaven and earth consist".32 The implication was that a standards programme would be fit for varsity science, and that thus electromagnetic physics would connect cables, circuits, spectroscopes, the tutorial room and the pulpit.

It was necessary to place the Cavendish Laboratory's in this network with some circumspection. No exact precedent for this enterprise existed. The example of Glasgow, used by Maxwell, Kalley Miller and its manager William Thomson in the early 1870s, was a highly ambiguous resource. A good case of the contrast with Glaswegian values is provided by Maxwell's searching interrogation of Jenkin and Thomson in summer 1872, a year after his arrival in Cambridge and during the completion of the Treatise on Electricity and Magnetism. Maxwell needed information for the book about the range of new instruments, including Latimer Clark's voltaic standard of electromotive force: "the electromagnetic [system] is of the greater use to those practical electricians who are occupied with electromagnetic telegraphs". Maxwell also needed details of Clark's electrodynamometer, which he reproduced in the Treatise.[33] He took the chance to point out what he judged a bizarre feature of telegraphists' jargon: where Clark used electromagnetic terms, such as ohm or farad, to refer to instruments, Jenkin used them to refer to quantities of fluid. So Maxwell asked Thomson whether he agreed that a 'Farad' should be defined as "the capacity of a very large condenser", rather than as "a dose of electricity". Thomson responded brutally that units must be embodied as real instruments with commercial values. "When electrotyping, electric light, etc. become commercial, we may perhaps buy a microfarad or a megafarad of electricity...If there is a name given to it, it had better be given to a real, purchaseable, tangible object, rather than to a quantity of electricity"[34]. Before the end of the decade, Thomson would be involved with the distribution of "potted energy", via his association with Joseph Swan and Camille Faure. His blunt and characteristic reply to Maxwell in 1872 revealed one set of values which these instruments were designed to embody: those of the late Victorian marketplace.

These were not, however, values which could be easily absorbed in Cambridge. The experience of the young wrangler John Strutt, later Lord Rayleigh, showed how hard it was to adopt Glaswegian models. It was recalled that after he had completed the Mathematics Tripos in the later 1860s "there was no experimenting [in Cambridge] except Stokes [the Lucasian professor] and Liveing [the chemistry chair]. Three precious years of experimenting work were lost for want of knowing how to set to work. Glasgow ? He had never even heard that you could work there". The establishment of the new experimental physics chair largely resulted from the campaign of Stokes, Rayleigh and their allies for reform of this Tripos problem. Rayleigh indicated he would take the new job if Maxwell would not.[35] So when considering the post in spring 1871, Maxwell asked both Rayleigh and Thomson for advice. He laid out plans for the lab and told Thomson that "we should get from the BA some of their apparatus for the standards committee. In particular the spinning coil and the great electromagnetometer. If not we must have a large coil made and measured with the utmost precision to be used as a standard". The standards programme was central to the new vision - but it was very hard to set up. Hence when, in March 1871, Maxwell tried to keep Rayleigh committed to Cambridge experimental physics, he also told the new baron that "it will need a good deal of effort to make experimental physics bite into our University system which is so continuous and complete without it. To wrench the mind from symbols and even from experiments on paper to concrete apparatus is very trying at first, though it is quite possible to get fascinated with a course of observation as soon as we have forgotten all about the scientific part of it". The worry was that the physicists might "corrupt the minds of youth" until they no longer made it to the status of wranglers:

"we may bring the whole university and all the parents about our ears".36 The work on electromagnetic standards meant running the lab like a workshop. It needed new skills and new instrument-makers. The localisation of these skills and workers makes up the power of laboratory culture.

So managerial problems dominated the institutionalisation of Cavendish physics. Initial provision of instruments was hopeless. Maxwell joked to Thomson that the University boat crew might be used as source of power. These jokes had political implications, because they seemed to subvert the moral order of the University. Thus Maxwell told his friend Monro that in the new standards experiments "the desideratum is to set a Don and a Freshman to observe and register (say) the vibrations of a magnet together, or the Don to turn a winch, and the Freshman to observe and govern him". Governance would always be a trouble. More seriously, Maxwell had to budget £2500 for restocking, and kept lobbying the founder-benefactor, the Duke of Devonshire, for help.37 This was why the transfer of the British Association instruments agreed during summer 1871 mattered. When the lab was finished in 1874, Maxwell ordered Jenkin to "get laid on the scent" of more standards devices: "the revolving coil and governor, scale and telescope, and the provisional resistance coils, nearly equal to the spinning coil, also one or more of the original standard Ohms to verify commercial Ohms withal". They reached Cambridge by autumn 1874, but on their own these transfers could not summon a standards laboratory into existence. Hockin's memories of their work of the previous decade, and a copy of his notebook, were equally insufficient. 38 Maxwell's own laboratory assistant, Pullin, was spectacularly incapable of working on the project - good at ordering railway employees around but no technician. When Rayleigh assumed the chair in 1879, Pullin told him that in his entire career "I soldered two wires together and made some oxygen for Mr Glazebrook". Rayleigh dispensed with him. Maxwell's other coworker, William Garnett, was also a problem. He sustained Maxwell's lectures, and student numbers, a source of income, languished. So at the end of Maxwell's regime, Rayleigh judged that the Cavendish was "very imperfectly provided with apparatus", that Garnett should be forced to start laboratory demonstrations, and that major reform would be needed to start the standards programme.39

The institutionalisation of science is dominated by rites of passage. Cambridge experimental physics was no exception, but training was famously chaotic and hotly disputed. Arthur Schuster, introduced by Garnett as a researcher in 1876 allegedly to work on Ohm's law but soon to switch to gas spectroscopy, recalled that "the instruments suffered more than the students by their being so freely allowed to come unchaperoned into mutual contact". Richard Glazebrook, Rayleigh's adjutant, saw advantages and disadvantages in such indiscipline. He recognised that "with a large class of students, all of whom have to be taught certain leading laws and facts in a limited time, organisation...is absolutely necessary", but he reckoned that under a rigid scheme "the assistance given to students is far too great. The apparatus is so nicely adjusted, the connexions to be made so clearly indicated, that all that remains to be done is to press the button and the answer will come out".40 The new regime led by Rayleigh and Glazebrook backed the manufacture of physicists, just as this process was fiercely resisted by the ci-devant mathematics coaches. Their spokesman was Isaac Todhunter, chief celebrant of the Mathematics Tripos, literary executor of the Tripos' architect William Whewell, and assailant of Maxwell's ally Peter Tait on the issue of examination and experiment. Todhunter simply denied that experimenters could be trained. He urged that in an endless repetition of classroom instrumental work "we can never recover any of the interest which belonged to the first trial". He concluded that "it may be said of the experimenter, as of the poet, that he is born and not manufactured". Todhunter and his allies made it all the more difficult to produce trained physicists and a standards lab within the liberal University.41

The laboratory managers also had to solve the problems of instrumentation. The gap between laboratory and workshop was crucial. In July 1873 Maxwell reported that "it is impossible to procure many of the instruments as they are not kept in stock, and have to be made to order". Maxwell was engaged in replicating some of the experiments of Henry Cavendish, his patron's eighteenth-century kinsman. He looked in "old instrument shops" for eighteenth-century devices with which to calibrate these trails. But standards instruments needed personal supervision during construction: "their whole value depends on their fulfilling conditions which can as yet only be determined by trial, so that it may be some time before everything is in working order". Technology transfer and design needs settings which cultivate tacit skills. These cannot be developed in long-distance, formal, communications between maker and user. So, as Maxwell argued in 1877, the cultural and practical distance between operative and researcher had to be effaced. "The experimenter has only occasional opportunities of seeing the instrument maker, and is perhaps not fully acquainted with the resources of the workshop, so that his instructions are imperfectly understood by the workman. On the other hand the workman had no opportunities of seeing the apparatus at work, so that any improvements in construction which his practical skill might suggest are either lost or misdirected".42 Changing this situation was slow and tough. When Garnett was hired he soon "introduced into the Laboratory workshop a few engineer's and joiner's tools". At least three resources existed for developing more effective skills: closer links with London and Glasgow engineers; recruitment from the new University engineering department, under James Stuart's controversial leadership; and hiring in-house laboratory technicians. In the mid-1870s the technician Robert Fulcher from Stuart's works in Balgonie was switched to the Cavendish. Stuart's own shop began to make instruments for the lab, and this helped the new Cambridge Scientific Instruments Company expand from its status as supplier for Michael Foster's physiology laboratory into a general scientific instruments concern. The Whipple Museum in Cambridge still holds a tuning fork which Fulcher built in the late 1870s, perhaps for the standards project as a means of calibrating coil speed, as well as a number of galvanometers designed in Stuart's workshop. The shop began to provide the Cavendish with equipment and jobs for students.43

The skills of men like Fulcher were crucial for the electromagnetic standards project. Maxwell began designing new methods for redetermining the resistance unit and for assessing the value of V. They needed high quality precision skills. He reckoned that a better value of V could be obtained with a commutator linked to an air condenser: The current strength in the circuit could be measured in electromagnetic units using a balancing constant current, while the electrostatic value of the quantity of charge could be found from the commutator's speed and the dimensions of the condenser. This method still needed a good value for the British Association resistance standard. So the laboratory was put to work. The aim was to match his support staff to the problems of eliminating the troubles his team had had at King's in the 1860s. He devised a null method with little dependence on battery output variation and no need to calibrate the ballistic galvanometer. Surviving ballistic galvanometers from this period include one designed by Maxwell and then used in the standards project, and one obtained by the Cavendish from the Cambridge chemistry professor in early 1875. Once again, commutators played a central role. [See Figure

4] Maxwell told the high wrangler George Chrystal that "I am glad the commutator with the crank worked well. Such a commutator is very much wanted for many experiments. Of course, there must be some kind of counter for absolute measurements of time". To ensure very fast current decay when the commutator was running, he wanted to fit a narrow radial gap filled with an insulator inside the gunmetal former which held the mutual inductance coils. Maxwell's very last lecture at Cambridge was devoted to an investigation of the shape of such a decaying current in a mutual inductance system. Ambrose Fleming's notes on this lecture were posthumously included in the next edition of the Treatise.[44]

Good coils and formers were needed. In summer 1877 they could still not be made in Cambridge. Maxwell ordered them from the London engineers Thomas Horn. His troubles with distant, and thus unverifiable, makers began here. Fulcher would solve them by making the coils on site. Supervision and hierarchy mattered too. In January 1878 Hockin was asked again about other labs' work, including that of Carey Foster in London, where temperature effects on resistance were used. Fleming, then a new student at St John's College, was given the task of getting the London method working in Cambridge.[45] Chrystal was given the surveillance job. During 1878 he told Maxwell of the troubles he faced when winding coils onto the gunmetal formers. He filled a dozen pages of the lab notebook with details of the winding, suggested means for regulating the commutator's resistance, and reported that "the wire we got was not what it should be". The main problem was technical support: "owing to the dearth of assistance I could get no further than a preliminary trial which promised success". Maxwell ordered Fulcher onto the job, and even tried to persuade Pullin to help: "I think he is converted". By July Chrystal confirmed that "I propose having the coils wound by Fulcher in my immediate presence, so that I can give account of every layer if necessary". Because of this new in-house resource, linking Stuart's technicians and Maxwell's wranglers, the professor was able to tell the University that at last "much of the apparatus to be used in Cambridge which has hitherto been ordered from London may in future be constructed in Cambridge and tested while in the maker's hands by those who are to use it".[46] So the troubles of long-distance calibration which had plagued the isolated laboratory could be solved with a new pattern of workshop skill developed inside the Cavendish.

This solution involved the integration of instrumentation and training: the induction coils, magnetometers and commutators became emblematic of laboratory initiation. Maxwell's Treatise helped explain this status. The text described and rationalise instrumentation. Maxwell's brilliant analysis of ballistic galvanometry was one example. Elsewhere, chapters on instruments outlined a hierarchy of tools, ranging from "null or zero methods" such as the electroscope, through registration devices and scale readings, to "instruments so constructed that they contain within themselves the means of independently determining the true values of quantities". Maxwell here explained the doctrinal and practical significance of absolute measures. His influential programme notes on the South Kensington scientific instruments exhibition of 1876 served a similar purpose. Instrument design embodied the principles of physics they were designed to illustrate and then to investigate. "There must be a prime mover or driving power, and a train of mechanism to connect the prime mover with the body to be moved". Dynamics was basic in theory and energy transmission basic in design. Electrical instruments were to be classed into sources, channels, restraints, reservoirs, overflows, regulators, indicators and, finally, scales, such as the Ohm. By making instruments look like machines, Maxwell helped link workshop with laboratory practices.[47]

This machine vocabulary aided the introduction of the British Association standards into the laboratory. In the 1870s, Garnett's custom was to introduce tyros to the Cavendish by getting them to work on the Kew magnetometer, as it "afforded practice not only in reading scales and making adjustments, but also in time observations, counting the beats of a watch while observing the vibrating magnet".[48] The status of the British Association machinery was also visible in the laboratory layout.The magnetic room occupied the east end of the ground floor, a site of honour, with solidly based stands and a relatively large space for ancillary equipment. These devices were central to the rites of entrance to the laboratory. Richard Glazebrook began work at the Cavendish in 1876. He was ordered to help Chrystal by using a quadrant electrometer to supervise the constant e.m.f. of a set of Daniell cells. Glazebrook soon gave up, revealing the difference in values between the standards programme and the varsity <u>cursus honorum</u>: "the measurement which I was set to do did not seem likely to afford material for a Fellowship Dissertation".[49]

This was one way in which it was hard to form a cadre of skills. There were others. In the very first year of operations, in spring 1874, Maxwell made his class use the magnetometer to measure the geomagnetic field strength and direction. Then he checked with Kew Observatory to compare his group's results and to receive instruction on the best way of setting up the great instrument.The same problems developed in his exchanges with Jenkin about the best ways of running the British Association equipment. He asked Jenkin for the recipe for the governor he had designed in the mid-1860s, since "I must have a speedy friction governor essentially on your principle but with some few alterations to secure good balancing". Maxwell did not know where to get one. He also needed lessons on the design of the old equipment, such as the guard ring condensers which were to play a crucial role in his redetermination of the Ohm. Importantly, Maxwell began to develop a theoretical technology using the <u>Treatise</u> to interpret the behaviour of his ballistic galvanometer. "We are learning to distinguish between shoves, kicks and jerks in the galvanometer needle". Initially this learning process was ruthlessly empirical. "Rule - set the needle swinging with an amplitude equal to the jerk, then when it is at at the extremity of its swing, make or break contact so that the jerk brings it to zero". By the following autumn, however, Maxwell's team was rapidly gaining in competence by linking his work in the <u>Treatise</u> on current change with their work on the galvanometer in the mutual inductance circuit: "we are now going ahead with the comparison of coils, and several men are able to rig up the various combinations out of their own heads, now that they know by inspection the difference between a

shove α current

kick $\alpha \int$ current. dt

jerk $\alpha \iint$ current. dt dt

I have just got a clear case of double jerk from zero to + and back to zero and stop there".[50] This work is not to be interpreted as the simple-minded application of truths set out in Maxwell's principles. Theoretical and practical resources were produced together. In cases like these, the "men" were learning to see the laboratory bench electromagnetic skills as instances of more general principles of changing current and field strength. But, as his collaborators often emphasised, there was a very wide gap between visions of a trained workforce and the actual production of routine practical skills. For example, in 1874 W.M.Hicks, inspired by Maxwell's teaching, tried to use a ballistic galvanometer balanced between two coils of different sizes set at different distances to measure the speed of electromagnetic waves: "of course, nothing came of it, but the practice was worth a great deal to me".[51] The problem the Cavendish faced was defining and distributing this fragile notion of practice.

The Resistance Unit Institutionalised and Internationalised

This was also the problem which the new regime faced on Maxwell's death in 1879. His successor, Rayleigh, had interests which were not exactly those of a reforming factory manager. He accepted the chair at least partly because the agricultural depression had reduced this income to a level where he could not afford the upkeep of his private laboratory at his country seat at Terling, where he spent much of the 1870s in important work on acoustics, resonators and experimental optics. He belonged to a tradition of private laboratory owners, of whom Stokes was the most celebrated and nearest example. Rayleigh's commitment to laboratory standards and making his laboratory a standards authority was overwhelming. He was well placed to make the interest needed to equip the Cavendish with what he reckoned it needed - between two and three thousand pounds' investment in instruments and staff. His political kinsmen, Arthur Balfour and Lord Salisbury, were enrolled and began participating in electromagnetic trials at Cambridge and at home in Terling. The new regime was characterised by careful discipline, centred in the classroom and in the standards project. Schuster reminded Rayleigh that "even a good man is, when he first comes into a Laboratory, a very helpless being". Under Maxwell, "too much time was wasted in consequence of apparatus set up for special experimental work being disturbed because a piece was wanted for an elementary lecture". Schuster recommended Chrystal, Richard Glazebrook and Napier Shaw to run the teaching programme. Student numbers rose from about 20 to over one hundred. New demonstrators were hired, notably Shaw and Glazebrook, and a research agenda imposed unattractive to most students but key for the managers' plans.[52]

Shaw had studied at Freiburg with Warburg and then with Helmholtz in Berlin. In summer 1879, Shaw returned to Cambridge to work on new hygrometric tests with Stokes. Rayleigh and his collaborators sought cheap and robust instruments for their classes, and searched desperately for new skills. There was a link between these skills and the working machines. Rayleigh told Chrystal that "I wish you were still in Cambridge as we have few workers with experience in experiment. We have been arranging a repetition of the BA experiment with revolving coil, but have not as yet got very far. There is a difficulty in insulating the suspended parts sufficiently from tremors, which has more effect in disturbing the needle than we would expect". Chrystal answered that he was pleased the British Association campaign would resume, though "I had some doubt at the time whether the trouble was likely to be repaid by the result". He confirmed that "Maxwell had considerable faith in the BA method - what you say about the effect of tremor is very curious". The demonstrators learned from Helmholtz's Berlin lab, from Pickering's physics teaching at M.I.T. and from the evening classes in telegraphy conducted by Grylls Adams at King's College London after Maxwell's departure. Their custom was to use the Cavendish magnetometer room to set up standard apparatus for pairs of students, with trials assigned on the basis of a rota displayed on the wall.[53] Standardised devices became indispensable - the Cambridge Scientific Instrument Company boomed. Manuscript volumes of good procedure, constantly updated, were corrected by the assistant staff and copied by students. These notes resulted in the basic new textbook, Practical Physics, issued by Glazebrook and Shaw in spring 1885. It reached a fourth edition in 1893. The book carried advertisements from the Cambridge Scientific Instrument Company and details of the standard equipment the Cavendish used. [See Figure 5] Glazebrook and Shaw specified that "the description, in order to be precise, must refer to particular forms of instruments, and may therefore be to a certain extent inapplicable to other instruments of the same kind...we have thought it best to adhere to the precise descriptions referring to instruments in use in our own Laboratory".[54] Laboratory assistants and company workers were now the force on which this rapid process of standardisation and institutionalisation depended, while the stipulations about specific instruments were designed to guarantee the spread of this style.

Textbooks, workforces and material standards were all needed to propagate the Cavendish's standards.This was especially important, for example, in the Cambridge engagement with the expansive world of German physics laboratories. After the problems of the 1860s, Maxwell had made some further efforts to contest the dominance of Siemens' resistance unit by distributing Cambridge resistance units to such important sites as H.Weber's laboratory at Zurich. Hockin and Maxwell welcomed the Zurich replication of the British Association unit which this export drive enabled. Elsewhere, however, notably at the physics laboratory at Göttingen, Cambridge values did not yet count. Both Cambridge and Göttingen were visited by the American physicist Henry Rowland in 1875-6. He gave Maxwell some encouragement. He attacked the efforts to produce good electromagnetic standards mounted by the team of Friedrich Kohlrausch at Göttingen. There Rowland saw the original apparatus used by Gauss and Wilhelm Weber to determine the first absolute electric and magnetic units in the early nineteenth century, and also Kohlrausch's new set-up for the determination of the absolute value of Siemens' unit. Rowland told Maxwell in March 1876 that despite impressive German achievements "it seems to me that the accurate measurement of resistance either absolutely or relatively is an English science almost unknown in Germany".[55]

This was good news. When Glazebrook and Shaw produced their new textbook in 1885, they used Kohlrausch's exemplary Leitfaden der Praktischen Physik (1870; fourth edition 1880) as their model. Kohlrausch wrote this work for Göttingen students in the late 1860s. He was encouraged to publish it by his boss Wilhelm Weber, the veteran of the work with Gauss on absolute units. Kohlrausch had some useful clues for the Cambridge men. He presented basic training as culminating in electric and magnetic standards determination. His book also had defects. He did not place dynamics at the heart of the lab's project, nor could he explicate the relation between the British Association system and the practical telegraphic units adopted in Britain and Germany. The Cambridge textbook did. It restated the faith that "the development of physical science on the lines indicated by the principle of conservation of energy has made more conspicuous the importance of experimental dynamics as the basis of experimental physics". The book relied on Maxwell's Treatise and on his 1876 text on instruments. The doctrines of the Treatise appeared in summary form in successive editions of Practical physics. The Treatise argued that "the determination of electrical resistance may be considered as the cardinal operation in electricity, in the same way that the determination of weight is the cardinal operation in chemistry'. Glazebrook and Shaw agreed: "standards of resistance have the advantage of material standards in general. The resistance is a definite property of a piece of metal, just as its mass is. The coil can be moved about from place to place without altering its resistance and so, from mere convenience, electrical resistance has come to be looked upon as in some way the fundamental quantity in connection with current electricity".[56] The term "convenience" was supposed to refer to Siemens' arbitrary unit and his German admirers. The shift to fundamentals was a characteristic of the Cavendish life style. The student was never to escape the sense that he was an initiate in the world wide network of electrotechnology, and the local culture of absolute measurement.

Laboratory technicians mediated between the two worlds. George Gordon, a Liverpool shipwright hired into the lab by Glazebrook in 1879, linked the standards laboratory and the teaching programme, making instruments and designing trials for both.[57] When Rayleigh asked his colleague Dewar for advice on assistant staff, the new Cambridge professor was horrified to be told that such men must be treated as personal hirelings and discouraged from their own research. Rayleigh's account of hierarchy was different. Gordon was surely his scientific valet, and was taken by Rayleigh back to Terling in 1884. He died two decades later in a lunatic asylum after attempting suicide.[58] But within the lab this status was changed into that of technical manager. This had implications for precision standards, where Gordon's workmanship was combined with the skills and labour of Schuster, the manager of the Cambridge Scientific Instrument Company Horace Darwin, and Rayleigh's family, including his brilliant sister-in-law Eleanor Sidgwick.[59]

Firing Pullin and hiring Gordon was part of the process through which the whole British Association standards initiative was recast. Between 1879 and 1882, in Cambridge and Terling, the team re-examined and then developed their own techniques for "spinning". A comparatively large network of skills and resources was mobilised. Young wranglers were recruited - J.J.Thomson among them. Instrument makers from Horace Darwin's firm were used. Schuster and Rayleigh personally supervised coil winding at Elliott Brothers, who built the new apparatus for Cambridge. [See Figure 6] Standard coils of this period of ranges between 1 and 100 ohms survive in the Whipple Museum.[60] In the aftermath of contacts with H.Weber and Rowland, efforts were made to test (and discredit) measures of resistance of mercury columns in the style of Siemens. Kohlrausch was cited, consulted and his work scotched. Rayleigh also, unwillingly, had to criticise his Cambridge predecessor. He studied Maxwell's London and Cambridge set ups. Based on information from Hockin via Schuster, Rayleigh reckoned that Maxwell's team had miscalculated the inductance of the spinning coil, which was speed-dependent. He tried to make the coil more massive in order to avoid these structural problems but the larger frame generated circulating currents.[61] The new team found that with access to Maxwell's own apparatus they could improve his accuracy simply by insisting on stricter surveillance in the workshop and among the makers of the coils. For example, Rayleigh judged that the original British Association committees simply failed to clarify whether the "breadth" of the coil was its axial or radial dimension. Gordon and Darwin played crucial roles. Instead of the old Huygens wheel they used a toothed drive with a Koenig tuning fork to time rotations. Bailey's Thirlmere water engine was used as the drive. The team found that the public water supply did not provide a constant head, so they had to build their own water pumps inside the Cavendish. Auxiliary magnetic needles with new Cambridge Scientific Instruments mountings were used. The Cavendish ground floor began to look like that of a factory.[62]

The project took nine months. In summer 1881 Rayleigh reported back to the British Association on major improvements in accuracy. He encouraged Carey Foster at London to replicate their measures. It was soon agreed that a systematic resistance testing regime must be initiated, since the variation of the old British Association standards was intolerably large. "Such a system would be analogous to the system adopted by the Kew Committee for the testing of meteorological instruments at the Kew Observatory. It has not yet been settled by whom this duty should be undertaken". In other words, a standards laboratory was not enough.Major changes in the practices of telegraphy and other electrotechnological undertakings would also be needed if the Cambridge values were to reach beyond the Cavendish's immediate control.[63]

Rayleigh's team understood this. In the second stage of their campaign, they worked on more robust and portable spins. A new Elliott Brothers apparatus "designed more on engineering lines" was commissioned. [See Figure 7] Rayleigh also ordered Glazebrook to try Henry Rowland's rival and complementary method. This involved refinements of Maxwell's proposal for determining a secondary induced current when a primary current in an induction coil was reversed. The method's accuracy depended on the reliability of a ballistic galvanometer, notoriously hard to calibrate. By 1882-3, all the Cavendish's resources were being used to win its rights as the world electromagnetic standards laboratory . Every new resistance method was assessed, including a scheme of Lorenz for spinning a brass disc between two coaxial coils.[64] A test resistance carried the current that passed through the coils, and its value was determined by balancing the induced e.m.f. against that at the resistance terminals. A good example of the surveillance which these trials needed is the case of the variation in the position of the galvanometer needle: "the maximum discrepancy between any two deflections" varied between 1/3000 parts at low speeds and 1/6000 parts at high speeds. Rayleigh concluded that his best estimate of the value of the British Association unit as 0.94865×10^{9} cm/sec was accurate to within 1/1000 parts. It was claimed that no other lab could compete with this stability.[65] Rayleigh and Mrs Sidgwick reckoned they should recheck all the factors in the best method, especially the pitch of tuning fork on which their timings relied. They tried to calibrate their fork interruptor against a pendulum, getting a variation of about 1/1000 parts through days of scrutinising the average frequency. "The procedure adopted was quite good enough for our purpose; but if it were desired to push the power of the method to its limit, the work should be undertaken at an astronomical observatory, and extended over the whole time to rate the clock by the observations of the stars". This gives some hint of the resources the Cavendish was prepared to mobilise in defence of its values, including those of Greenwich and Kew. In all these cases careful management of complex techniques could be used to make a reliable version of the standard. The whole status of British electromagnetism was at stake.[66]

By the early 1880s, the Cavendish had emerged as a plausible candidate for the tribunal of resistance. The head of postal telegraphy, W.H.Preece, announced that he now accepted the Cambridge value of the ohm and was "about to issue a circular to all stations at home and abroad to that effect". Preece needed Rayleigh "to confirm these figures for my instructions will affect some hundreds of observers". Even William Thomson, master of metrology, sent his resistance boxes there to be verified "in true or Rayleigh ohms", as the Glaswegian put it. "From what I have heard...about the agreement of the standard ohms which have been sent as accurate from the Cavendish Laboratory, I may be confident that the one I have at Glasgow is near enough to the BA unit for my purpose".[67] The process of displacement of control to Cambridge was both managerial and disciplinary. The Cavendish of 1883 mirrored an ideal workshop regime; hence the legitimacy of Rayleigh's pointed comparison with the perfection of the chronometric system of Victorian observatories, which under George Airy had become the cynosure of precision measurement. These methods were valuable because they, and they alone, allowed the recruitment of resources and labour. They generated the right value for the ohm. They made the Cavendish look like a plausible judge of others' techniques. There were troubles here too. No workshop, and no telegraph office, could hope to ape such an alleged utopia. Furthermore, no rival laboratory, especially German ones, was likely effortlessly to demur to the Cavendish regime. For all this to happen, robust technologies were needed which could travel outwith Cambridge. These technologies were material: it was intended that Rayleigh's team make and verify new resistance boxes and standard cells. They were literary: textbooks like Practical physics were designed to reform laboratory order and change the order of the teaching experiment. Finally, they were ineluctably social: hence the significance of changes in the order of the telegraph station, and of the great international conferences where the British physicists were put through their paces in the 1880s.

These technologies made the ohm seem natural, a property of nature rather than a property of Lord Rayleigh. They also made absolute standards into explicitly moral virtues. Exhibitions, conferences, textbooks and instruments all helped this happen. Between 1881 and 1893 at least six international rallies of electricians were held. Each included debates about rival values and standards, and each provided invaluable opportunities for publicity and propaganda on behalf of rival technical systems, including those of Siemens and Edison. In 1881, for example, the show at Paris, including Thomson, Siemens, Helmholtz and Clausius, provisionally settled problems of nomenclature; offered a compromise between the Siemens and the Cambridge systems; and firmly agreed the value of a system of units which expressed fundamental dynamical relationships.[68] The meeting the following year, in Rayleigh's absence, agreed that the term "ohm" would refer to the absolute unit itself, not to any approximate embodiment. Long public discussions debated averaging over values which differed so much in size and in credibility. Thomson told Rayleigh that "I suppose your ohm must be declared the one and true ohm for our generation". In the event, "it was the particular one or two standards called Siemens units which we had that caused the discrepancy". These were the opening shots in a long war. Carey Foster, returned from Paris in 1882, waspishly told Rayleigh that "apparently the conclusion come to was that more experiments were desired and that anybody should make them who would". Henry Rowland, by contrast, seized on the opportunity, raised funds for a new standards committee, and set out to check the Cambridge values both with his old and with a new, high precision apparatus. By 1884, it had been agreed that the legal ohm be defined in terms of a length of mercury, roughly 106 cm. This was the closest the delegates could then come to Rayleigh's standard. It amounted to a compromise between the value, which belonged to Cambridge, and the unit, defined in Siemens' terms.[69]

The power of a standards laboratory was not lost on any delegate. Werner Siemens, who arranged for the despatch of his mercury standards to Cambridge via his London-based brother William, boldly stated that "there was found no place in Germany for carrying out the difficult work of producing the absolute resistance unit of Weber". Lobbying for the establishment of the emergent Physikalisch-Technische Reichsanstalt followed.[70] The alliance of laboratory instrumentation and lawlike standards was also registered in Britain in successive Tory and Liberal Weights and Measures Acts, protocols of which contained recipes for making the ohm and the standard cell. Instructions on the ways of making copper wire of standard properties, which instrument makers supplied materials, and which labs provided checks, were also offered. Nor were these lessons lost on Glazebrook, Rayleigh's unsuccessful nominee to the Cavendish chair in 1884. Defeated by J.J.Thomson, Rayleigh and Glazebrook then worked hard to set up a state-funded National Physical Laboratory. Alliances with the Tory administration, via Rayleigh's brother-in-law Balfour, proved successful. Significantly, it was Siemens' Physikalisch-Technische Reichsanstalt which the British lobbyists now used as their ideal.[71] The standards programme survived the change in personnel and programmes at the Cavendish of the 1890s: reliable instruments needed reliable institutions in which to function.

The same process of institutionalisation of training and naturalisation of standards was at work in textbook production. A key resolution taken at Paris concerned the publication of an accessible text in which the results of the meeting be displayed and explained. The British took charge, producing Everett's Illustration of the C.G.S. system of units in revised form. A stream of works followed, each reprinting the Paris resolutions, each striving to make the ohm "part of the student's nature", as Ayrton put it in his own version of this campaign. Ayrton was a forceful admirer of Rayleigh: "practical men are really very grateful for your determination". These textbooks combined Thomson's absolutism with techniques developed at the Cavendish. Andrew Gray's massively popular Absolute measurements in electricity and magnetism (1884) reversed the traditional order of exposition of the science. He began with descriptions of Thomson's and Horace Darwin's instruments; then he compared their values with the absolute system, and ended by spelling out the embodied energy relations. Ayrton adopted the same strategy: "the subject of current is treated first, because in almost all the industries in which electricity is practically made use of, it is the electric current that is employed". Students were told that metrology would "enable results to be expressed in units which are altogether independent of the instruments, the surroundings and the locality of the investigator".72

William Thomson turned this thought about the independence of units into a splendid science fiction fantasy in a lecture given to the Civil Engineers in May 1883. After consulting Rayleigh for details on the very latest work at the Cavendish to include in his lecture, Thomson sketched the career of "the scientific traveller roaming over the universe", equipped with nothing but the authority of contemporary metrology. "For myself, what seems the shortest and surest way to reach the philosophy of measurement...is to cut off all connection with the earth, and think what we must then do, to make measurements which shall be definitely comparable with those which we now actually make in our terrestrial workshops and laboratories".Thomson's spaceman gets his unit of length through the wavelength of light determined via a diffraction grating, and thence "goes through all that Lord Rayleigh and Mrs Sidgwick have done" to get the ohm and the unit of time. Eventually even Maxwell's method for oscillating discharge was proposed as an ideal means of recapturing absolute units of time, and hence of the whole of physics. 73

This Thomsonian message was crucial. Immense labour had been performed to achieve the vanishing trick through which the local practices needed to make standards had simply disappeared. Ayrton, Gray, Glazebrook and Shaw told their pupils and customers where to buy the robust instruments, where to learn the right techniques, and what morals would be appropriate for laboratory workers. Then they stated that the absolute system depended on no particular instrument, or technique, or institution. This helps account for metrology's power. Metrology involves work which sets up values and then makes their origin invisible. This paper has attempted to historicise this process, by linking the instruments of metrology with their favoured institutions. The integrity of these instruments and institutions is intimately bound up with their capacity to represent and affect the world beyond their immediate surroundings.

I gratefully acknowledge help with the preparation of this paper from Graeme Gooday, Bruce Hunt, Otto Sibum, Guy Strutt, Andy Warwick, and Norton Wise. The Librarians of Cambridge University Library and Imperial College Archives have graciously helped with the location and reproduction of manuscripts in their possession.

[1] J.A.Bennett, "A viol of water or a wedge of glass", in D.Gooding et al., eds., The Uses of Experiment (Cambridge, 1989), 105-14; Simon Schaffer, "Measuring virtue", in Andrew Cunningham and Roger French, eds., The Medical Enlightenment of the Eighteenth Century (Cambridge, 1990), 281-318; David Gooding, "Experiment and Concept Formation in Electromagnetic Science and Technology in England in the 1820s", History of Technology 2 (1985), 151-76.

[2] Steven Shapin, "The House of Experiment in Seventeenth Century England", Isis 79 (1988), 373-404; Simon Schaffer, "Astronomers Mark Time: Discipline and the Personal Equation", Science in Context 2 (1988), 101-131.

[3] Bruno Latour, Science in Action (Milton Keynes, 1987), p. 251; Thomas P. Hughes, Networks of Power (Baltimore, 1983).

[4] For the social and the quantitative notions of value, see T.S.Kuhn, "The function of measurement in modern physical science", Isis 52 (1961), 161-90; M.Norton Wise, "Mediating Machines", Science in Context, 2 (1988), 81-117.

5 Henniker Heaton quoted in Daniel Headrick, The Tentacles of Progress (Oxford, 1988), p.97; see also Crosbie Smith and M.Norton Wise, Energy and Empire (Cambridge, 1989), pp. 128-35, 684-98; Graeme Gooday, "Precision measurement and the genesis of physics teaching laboratories in Victorian Britain", British Journal for the History of Science, 23 (1988), 25-52.

[6] R.Kalley Miller, "The proposed chair of Natural Philosophy", Cambridge University Reporter, 23 November 1870, 118-119.

[7] Maxwell to Jenkin, 17 March 1874, Cambridge University Library MSS Add 7655 / II / 240; Elliot Brothers to Maxwell, 31 October 1877, Cambridge University Library MSS Add 7655 / II / 144; Maxwell to Fleming, 14 January 1875, in J.T.Macgregor-Morris, The Inventor of the Valve: a biography of Sir Ambrose Fleming (London, 1954), p.22.

[8] W.D.Niven, ed., Scientific Papers of James Clerk Maxwell, 2 vols. (Cambridge, 1890), vol.2, p.244; Lord Rayleigh, "Address to the Mathematical and Physical Science Section", British Association Report (1882), 437-4, p.439. See Romualdas Sviedrys, "The rise of physical science at Victorian Cambridge", Historical Studies in Physical Science 2 (1970), 127-45, p.130.

[9] James Walvin, Victorian values (London, 1988), p. 139. See E.P.Thompson, "Time, work-discipline and industrial capitalism", Past and present 38 (1967), 56-97.

10 Edwin Chadwick, Report on the sanitary condition of the labouring population of Great Britain, ed. M.W.Flinn (Edinburgh, 1965), p.425; Charles Babbage, "On tables of the constants of nature and art", Smithsonian Institution Annual Report (1856), 289-302. For "constants" see Ian Hacking, The Taming of Chance (Cambridge, 1990), p.57.

11 William Thomson, "On the theory of the electric telegraph", Proceedings fo the Royal Society, 7 (1855), 82-99. For telegraphy and theory, see Bruce Hunt, "Michael Faraday, cable telegraphy and the rise of field theory", History of Technology 13 (1991) 1-19.

12 William Thomson, "Presidential address", British Association Report (1871), in Kelvin, Popular lectures and addresses: volume 2, Geology and general Physics (London, 1894), p. 162.

13 J.D.Everett, Illustrations of the C.G.S. system of units (London, 1891), p.18; see Smith and Wise, Energy and Empire, pp. 684-6.

14 Werner Siemens and William Siemens, "Outline of the principles and practice involved in dealing with the electrical conditions of submarine electric telegraphs", British Association Report (1860), 32-34; Latimer Clark and Charles Bright, "On the formation of standards of electrical quantity and resistance", British Association Report (1861), 37-38. See Werner von Siemens, Inventor and entrepreneur: recollections, 2nd ed. (London, 1966), pp.166-7.

15 James Clerk Maxwell, Treatise on electricity and magnetism, 3rd ed. (Oxford, 1891), vol.1, p.vii.

16 James Clerk Maxwell and Fleeming Jenkin, "On the elementary relations between electrical measurements", British Association Report (1863), 130-63, pp.130-1. For King's College see C.Domb, "James Clerk Maxwell in London 1859-1865", Notes and records of the Royal Society 35 (1980), 67-103, pp.86-90.

17 Fleeming Jenkin, Report to the Royal Society on units of electrical resistance, and The Cantor Lectures bound with Reports of the Committee on Electrical Standards (London, 1873), pp.193-4; Siemens protests in his "Suggestions for the adoption of a common unit in measurement of electrical resistance", British Association Report (1862), 152-55, p.154; the need for permanence is asserted in British Association Report (1862), p.126.

18 For the teamwork see Ian Hopley, "Maxwell's work on electrical resistance: the determination of the absolute unit of resistance", Annals of science 13 (1957), 265-72, pp. 266-8; the governor is discussed in Maxwell, Scientific papers, vol.2, p.112.

19 British Association Report (1863), p.118-22; compare Silvanus P. Thompson, Life of William Thomson, 2 vols. (London, 1910), vol.1, pp. 418-19.

20 James Clerk Maxwell and Fleeming Jenkin, "Description of a further experimental measurement of electrical resistance", British Association Report (1864), 350-1. The Committee discusses standards on pp. 346-7 of this Report. For Jenkin's prices see Maxwell to Tait, 7 March 1865, Cambridge University Library MSS Add 7655 / Ib / 3.

21 Werner Siemens, "On the question of the unit of electrical resistance", Philosophical Magazine 31 (1866), 328.

22 Robert Louis Stevenson, "Memoir", in S.Colvin and J.A.Ewing, eds. Papers by Fleeming Jenkin, 2 vols. (London, 1887) vol. 1, pp. clv-vi; Smith and Wise, Energy and Empire, pp. 698-701; C.A.Hempstead, "An appraisal of Fleeming Jenkin, electrical engineer", History of Technology, 13 (1991), 119-44, pp. 131-2.

23 Jenkin, Cantor Lectures, p.247; Report of the Electrical Standard Committee (Nottingham, 1866) (in Cambridge University Library, MSS Add 7655 / V / i / 12).

24 Headrick, Tentacles of Progress, pp. 101, 104-9; Smith and Wise, Energy and Empire, p.704.

25 Jenkin, Cantor Lectures, p.192; Hunt, "Faraday, cable telegraphy", pp. 8-14.

26 Maxwell on Hockin in Maxwell, Scientific papers, vol.2, p.127; cable duty, in Hockin to Maxwell, 27 July 1868, Cambridge University Library MSS Add 7655 / II / 31.

27 Maxwell describes the method in Scientific papers, vol.2, pp. 130-2, 135 and mentions the time required on p.127; see I.B.Hopley, "Maxwell's determination of the number of electrostatic units in one electromagnetic unit of electricity", Annals of science 15 (1959), 91-108, pp. 91-3.

28 Maxwell, Scientific papers, vol.2, p.126. Compare his "Experiments on the value of V, the ratio of the electromagnetic to the electrostatic unit of electricity", British Association Report (1869), 436-8.

29 Hockin to Maxwell, 15 May 1868, in Hopley, "Maxwell's determination", p.97.

30 Maxwell, Scientific papers, vol.2, pp.135-6. He remarks on "hitches and difficulties" in Maxwell to Tyndall, 23 July 1868, Imperial College Archives: a letter of reference on behalf of Hockin.

31 Maxwell, Scientific papers, vol.2, p.322.

32 Maxwell, Scientific papers, vol.2, p.483 (for "manufactured articles"), and p.377 (for "the ineffaceable characters"). See Paul Theerman, "James Clerk Maxwell and religion", American Journal of Physics 54 (1986), 312-317. The Cavendish was a big customer of Whitworth's standard gauge screws: see Maxwell to Vice Chancellor, 24 July 1874 Cambridge University Library CUR 39.33 53 (3).

33 Maxwell, Treatise, vol.2, pp.268-9, 370.

34 Maxwell to Clark, 16 July 1872; Maxwell to Thomson, 10 August 1872; Thomson to Maxwell 24 August 1872, Cambridge University Library MSS ADD 7655 / II / 59, 61A, 62.

35 R.J.Strutt, Life of Lord Rayleigh, 2nd ed. (Madison, 1968), pp. 49-50, 406-7; Cambridge University Reporter, 16 November 1870, pp. 93-97.

36 Maxwell to Rayleigh, 15 March 1871, in Life of Rayleigh, pp. 49-50.

37 Maxwell to Thomson 21 March 1871, Cambridge University Library MSS ADD 7655 / II / 45; Maxwell to Monro, 15 March 1871, in Lewis Campbell and William Garnett, Life of James Clerk Maxwell (London, 1884). For financing see Cambridge University Library MSS CUR 39.33 53 (3) and MSS ADD 7655 / II / 138.

38 Maxwell to Jenkin 17 March 1874; Hockin to Maxwell, 14 May 1874, Cambridge University Library MSS ADD 7655 / II / 240 and 78. For possible examples of Elliott brothers' produced resistance standards held in the Cavendish in the 1870s, see Kenneth Lyall, Catalogue of Electrical and Magnetic Instruments: Whipple Museum Catalogue 8 (Cambridge, 1991), nos. 278-82 (Wh: 1348, 1541-5).

39 On Pullin and Garnett see Life of Rayleigh, pp. 103-4, 412; Pullin to Maxwell, 5 July 1877, Cambridge University Library MSS ADD 7655 / II / 135.

40 William Garnett, "Maxwell's laboratory", in James Clerk Maxwell: a commemoration volume (Cambridge, 1931), 109-15, p.115; Arthur Schuster and Richard Glazebrook comment in History of the Cavendish Laboratory 1871-1910 (London, 1910), pp. 26, 36, 45. See also Arthur Schuster, The progress of physics during 33 years (Cambridge, 1911), pp. 30-31.

41 Todhunter attacked Tait in "Audi aliam partem", Macmillan's Magazine 26 (May 1872), 60-69; Tait riposted in "Todhunter on experimental illustrations", Nature 9 (February 1874), 323; Todhunter wrote against manufacture of experimenters in The Conflict of Studies (London, 1873), pp. 16-19. Compare Henry Sidgwick's attack on the professoriate in "Academic teachers", Cambridge University Reporter, February 22 - 8 March 1871, pp. 204-5, 221, 235-6.

42 Maxwell's reports of July 1873 and July 1878, in Cambridge University Library MSS CUR 55.2 ; Maxwell on Henry Cavendish in Maxwell to Chrystal, 21 September 1877, Cambridge University Library MSS ADD 8375 f. 14.

43 History of the Cavendish Laboratory, p.35 on Garnett; T.J.N.Hilken, Engineering at Cambridge University 1783-1965 (Cambridge, 1967), pp.63-74 and M.J.Cattermole and A.F.Wolfe, Horace Darwin's Shop: a history of the Cambridge Scientific Instrument Company (Bristol, 1987). See Lyall, Electrical and Magnetic Instruments, nos. 23-24, 451 (Wh: 1339-40, 4376).

44 For instrument design see Hopley, "Maxwell's determination", pp. 99-101 and "Maxwell's work on electrical resistance - proposals for the redetermination of the BA unit of 1863", Annals of science 13 (1958), 197-210, pp. 201-2. See Maxwell to Chrystal 7 August 1876, Cambridge University Library MSS ADD 8375 f. 4; Maxwell, Treatise, vol.2, p.396. The ballistic galvanometers are described in Lyall, Electric and magnetic instruments, nos. 14 and 17 (Wh: 1318, 1334).

45 For Horn see Hopley, "Redetermination of the BA unit", p. 208; see Hockin to Maxwell, 14 January 1878, Cambridge University Library, MSS ADD 7655 / II / 150; Ambrose Fleming, "Some memories", in James Clerk Maxwell: a commemoration volume (Cambridge, 1931), 116-24, p.119.

46 For Chrystal and Fulcher see Hopley, "Redetermination of the BA unit", p.209; Chrystal to Maxwell 9 July 1878, and Maxwell to Chrystal 9 July 1878, Cambridge University Library MSS ADD 7655 / II / 159 and MSS ADD 8375 f. 16. Maxwell's report, March 1878, is in MSS CUR 55.2 242a p. 11.

47 Maxwell, Treatise, vol. 1, pp. 326-7; programme notes in Maxwell, Scientific papers, vol. 2, pp. 505-27, discussed in Peter Galison, How experiments end (Chicago, 1987), pp. 23-27.

48 History of the Cavendish, pp.34-5.

49 R.T.Glazebrook, "Early days at the Cavendish Laboratory" in James Clerk Maxwell: a commemoration volume (Cambridge, 1931), 130-41, pp. 135-7.

50 Whipple to Maxwell, 6 May 1874 and Maxwell to Jenkin, 22 July and 18 November 1874, Cambridge University Library MSS ADD 7655 / II / 77, 241-2; Maxwell, Treatise, vol.2, pp. 382-91.

51 History of Cavendish Laboratory, p.19.

52 Strutt, Life of Rayleigh, pp. 75, 412; History of the Cavendish, pp. 41-42. Rayleigh's petition for an apparatus fund is in Cambridge University Library MSS ADD 7655 / IIIc / 1 ; Schuster to Rayleigh, 8 January 1880, Imperial College MSS Rayleigh Letters. Maxwell's advice to Rayleigh on the establishment of a private

laboratory is contained in Maxwell to Rayleigh, 8 July 1871, Imperial College MSS Rayleigh Letters: " the word [laboratory] denotes a place to work at experiments and connotes a place full of articles not wanted at present and liable to noxious fumes".

53 History of the Cavendish, pp. 44-46; Edward C. Pickering, Elements of physical manipulation, 2 vols. (London, 1878), vol.1, pp. vi-vii; Rayleigh to Chrystal, 22 August 1880, Cambridge University Library MSS ADD 8375 f. 28; Chrystal to Rayleigh 3 September 1880, Imperial College MSS Rayleigh Letters.

54 History of the Cavendish, p.254; Richard Glazebrook and W. Napier Shaw, Practical physics, 4th ed. (London, 1893), pp. ix, xii-xiii.

55 Maxwell to Chrystal, 17 July 1878, Cambridge University Library MSS ADD 8375 f. 18 on the despatch of units to Zurich; see also Hockin to Maxwell, 14 January 1878 and H.Weber to Maxwell, 19 May 1878, Cambridge University Library MSS ADD 7655 / II / 150, 154. See Rowland to Maxwell, March 1876, in Nathan Reingold, Science in Nineteenth-Century America (Chicago, 1964), p. 269 for the journey to Göttingen; Rowland to Maxwell, [1877], Cambridge University Library MSS ADD 7655 / II / 105 for the comparison of resistance coils.

56 Glazebrook and Shaw, Practical physics, pp. vii, 527; Maxwell, Treatise, vol.1, p. 465. For Kohlrausch's Leitfaden and German physics laboratories see David Cahan, "The institutional revolution in German physics 1865-1914", Historical studies in physical sciences 15 (1985), 1-65, pp. 48-50.

57 For Gordon see Glazebrook and Shaw, Practical physics, p. x; History of the Cavendish Laboratory, p.46; Strutt, Life of Rayleigh, p. 104.

58 Strutt, Life of Rayleigh, p. 427.

59 Lord Rayleigh, Scientific papers, vol. 2 (Cambridge, 1900), p. 5-12; Rayleigh laboratory notebooks, Imperial College MSS B / Rayleigh / 3 p. 268 (21 April 1882).

60 Rayleigh, Scientific papers, vol.2, pp. 38-40, 44; Lyall, Electrical and magnetic instruments, nos. 278-82.

61 Rayleigh, Scientific papers, vol.2, pp. 3, 47-50, 78-79; Rayleigh laboratory notebooks, Imperial College MSS B / Rayleigh / 3 pp. 242-5 (18 March 1881).

62 Rayleigh, Scientific papers, vol. 2, pp. 4-7; Strutt, Life of Rayleigh, p.115.

63 British Association Report (1881), pp. 424-5; G. Carey Foster, "Account of preliminary experiments on the determination of electrical resistance in absolute measure", British Association Report (1881), 426-31.

[64] Strutt, Life of Rayleigh, p.116; Rayleigh, Scientific Papers, vol.2, pp.135, 155-156.

[65] Strutt, Life of Rayleigh, p.125; Rayleigh, Scientific papers, vol.2, p.47.

[66] Rayleigh, Scientific papers, vol.2, p.181; Rayleigh to Eleanor Sidgwick on tuning forks and scale stability, 14-23 September 1881, Imperial College MSS Rayleigh Letters.

[67] Thomson to Rayleigh, 25 May 1883, Rayleigh MSS, Terling.

[68] For the 1881 show see "The international exhibition and congress of electricity at Paris", Nature, 24 (1881), p.512; "Congrès international des electriciens", Revue scientifique, third series, (1881) vol.1, part 2, pp. 382-4, 407-15. For publicity at the congress, see Robert Fox, "Edison et la presse française à l'exposition internationale d'électricité de 1881", in F. Cardot, ed., 1880-1980: Un Siècle de l'Electricité dans le Monde (Paris, 1987), 223-35.

[69] Strutt, Life of Rayeleigh, pp.124-5; History of the Cavendish, pp. 66-64; Rayleigh, Scientific Papers, vol.2, p.78; British Association Report (1889), pp. 42-43; Carey Foster to Rayleigh, 15 November 1882, and Henry Rowland to Rayleigh, 23 June 1883, Imperial College MSS Rayleigh Letters.

[70] William Siemens to Rayleigh, 3 March and 21 May 1882, Imperial College MSS Rayleigh Letters; Siemens, Inventor and entrepreneur, p.267; David Cahan, "Werner Siemens and the origins of the Physikalisch-Technische Reichsanstalt", Historical studies in the physical sciences, 12 (1982), 253-83.

[71] British Association Report (1891), pp. 154-60; Peter Alter, The reluctant patron: science and the state in Britain 1850-1920 (Oxford, 1987), pp. 139-40.

[72] W.E.Ayrton, Practical electricity, 6th ed. (London, 1894), pp. iv-v; Andrew Gray, Absolute measurements in electricity and magnetism (London, 1889), pp. 1-2. See Ayrton to Rayleigh, 10 March 1884, Imperial College MSS Rayleigh Letters; Graeme Gooday, "Teaching telegraphy and electrotechnics in the physics laboratory: William Ayrton and the creation of an academic space for electrical engineering in Britain 1873-1884", History of technology 13 (1991), 73-111.

[73] Thomson, Popular lectures and adresses, vol.2, pp. 106-7, 114, 120; Thomson to Rayleigh, 11 April 1883, in Silvanus Thompson, Life of William Thomson (London, 1910), p. 792.

Figure 1 The electrotechnology network in the 1880s: John Pender holds the ends of the cables, Charles Bright and Latimer Clark stand behind him. Following a cart carrying the staff of the post office march William Thomson, Sylvanus Thompson, Cromwell Varley, William Thomson, Sylvanus Thompson, Cromwell Varley, William Crookes, W.E. Ayrton, John Perry, Grylls Adams; behind them are the Siemens brothers. *Source:* *The City* *(5 May, 1883).*

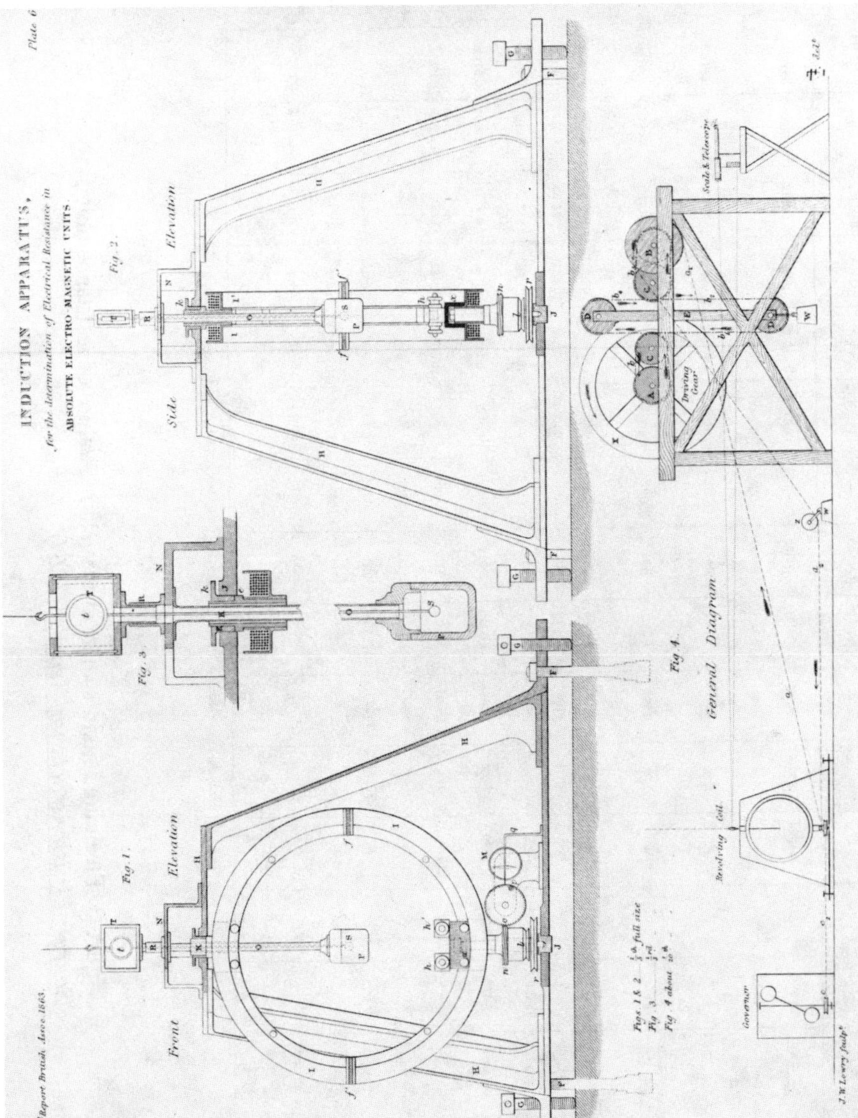

Figure 2 Maxwell and Jenkin's apparatus for the determination of a resistance standard, London 1863.
Source: *British Association Reports* (1863).

[To face page 349.]

FIGURE 3

Approximate Relative Values of various Units of Electrical Resistance.

Description.	Name.	Absolute $\frac{\text{foot}}{\text{second}} \times 10^7$.	Thomson's old unit.	Jacobi.	Weber's absolute $\frac{\text{metre}}{\text{second}} \times 10^7$.	Siemens 1864 issue.	Siemens (Berlin).	Siemens (London).	B. A. unit, or Ohmad.	Digney.	Bréquet.	Swiss.	Matthiessen.	Varley.	German Miles.	Observations.
Absolute $\frac{\text{foot}}{\text{second}} \times 10^7$ electro-magnetic units (new determination)	Absolute $\frac{\text{foot}}{\text{second}} \times 10^7$.	1·000	0·9520	0·4788	0·3316	0·3187	0·3168	0·3131	0·3048	0·03289	0·03123	0·02924	0·02243	0·01190	0·005307	Calculated from the B. A. unit.
Absolute $\frac{\text{foot}}{\text{second}} \times 10^7$ electro-magnetic units (old determination)	Thomson's unit	1·0505	1·000	0·5029	0·3483	0·3348	0·3328	0·3289	0·3202	0·03455	0·03279	0·03071	0·02357	0·01251	0·005574	From an old determination by Weber.
Twenty-five feet of a certain copper wire, weighing 345 grains...	Jacobi	2·088	1·988	1·000	0·6925	0·6655	0·6618	0·6540	0·6367	0·06869	0·06520	0·06106	0·04686	0·02486	9·01108	No measurement made; ratio between Siemens (Berlin) and Jacobi taken from "Weber's Galvanometrie."
Absolute $\frac{\text{metre}}{\text{second}} \times 10^7$ electro-magnetic units determined by Weber (1862)............................	Weber's absolute $\frac{\text{metre}}{\text{second}} \times 10^7$	3·015	2·871	1·444	1·000	0·9607	0·9556	0·9443	0·9191	0·00919	0·09416	0·08817	0·06767	0·03591	0·01655	Measurement taken from a determination in 1862 of a standard sent by Prof. Thomson; does not agree with Weber's own measurement of Siemens's units; by Weber 1 Siemens's unit $= 1·025 \times 10^7$ metres-second.
One metre of pure mercury, one square millimetre section at 0° C.	Siemens 1864 issue......	3·138	2·988	1·503	1·041	1·000	0·9950	0·9829	0·9563	0·1033	0·09799	0·09177	0·07047	0·03737	0·01666	Measurement taken from three coils issued by Messrs. Siemens.
One metre of pure mercury, one square millimetre section at 0° C.	Siemens (Berlin)	3·156	3·004	1·511	1·046	1·005	1·000	0·9881	0·9625	0·1038	0·09852	0·09227	0·07081	0·03757	0·01675	Measurement taken from coils exhibited in 1862 by Messrs. Siemens, Halske & Co. (well adjusted).
One metre of pure mercury, one square millimetre section at 0° C.	Siemens (London)	3·194	3·040	1·529	1·059	1·017	1·012	1·000	0·9742	0·1050	0·0997	0·09337	0·07166	0·03802	0·01695	Measurement taken from coils exhibited in 1862 by Messrs. Siemens, Halske & Co. (well adjusted).
British Association unit	B. A. unit, or Ohmad...	**3·821**	**3·123**	**1·570**	**1·088**	**1·0456**	**1·039**	**1·026**	**1·000**	**0·1079**	**0·1024**	**0·0959**	**0·0736**	**0·03905**	**0·01741**	Equal to 10,000,000 $\frac{\text{metres}}{\text{second}}$ according to experiments of Standard Committee.
One kilometre of iron wire, four millimetres in diameter (temperature not known)	Digney....................	30·40	28·94	14·56	10·08	0·0968	9·634	9·520	9·266	1·000	0·9491	0·8889	0·6822	0·3620	0·1613	From coils exhibited in 1862 (pretty well adjusted).
One kilometre of iron wire, four millimetres in diameter (temperature not known)..................	Bréquet	32·03	30·50	15·34	10·62	10·20	10·15	10·13	9·760	1·054	1·000	0·9365	0·7187	0·3814	0·1700	From coils exhibited in 1862 (indifferently adjusted).
One kilometre of iron wire, four millimetres in diameter (temperature not known).	Swiss	34·21	32·56	16·38	11·34	10·90	10·84	10·71	10·42	1·125	1·068	1·000	0·7675	0·4072	0·1815	From coils exhibited in 1862 (badly adjusted).
One English standard mile of pure annealed copper wire $\frac{1}{16}$ in. diameter at 15°·5 C.	Matthiessen..............	44·57	42·43	21·34	14·78	14·19	14·12	13·95	13·59	1·66	1·391	1·303	1·000	0·5306	0·2365	From a coil lent by Dr. Matthiessen (of German-silver wire).
One English standard mile of one special copper wire $\frac{1}{16}$ inch in diameter.............................	Varley	84·01	79·96	40·21	27·85	26·75	26·61	26·30	25·61	2·763	2·622	2·456	1·885	1·000	0·4457	From coils lent by Mr. Varley (well adjusted).
One German mile = 8238 yards of iron wire $\frac{1}{6}$ inch in diameter (temperature not known*)	German mile	188·4	179·4	90·22	62·48	60·03	59·71	59·00	57·44	6·198	5·882	5·500	4·228	2·243	1·000	From coils exhibited in 1862 by Messrs. Siemens, Halske & Co.

* Messrs. Siemens do not now manufacture coils with this unit, which has been abandoned by them in favour of the mercury unit given above.

Figure 3 Fleeming Jenkin's chart of the relative values of resistance units. *Source: British Association Reports (1864).*

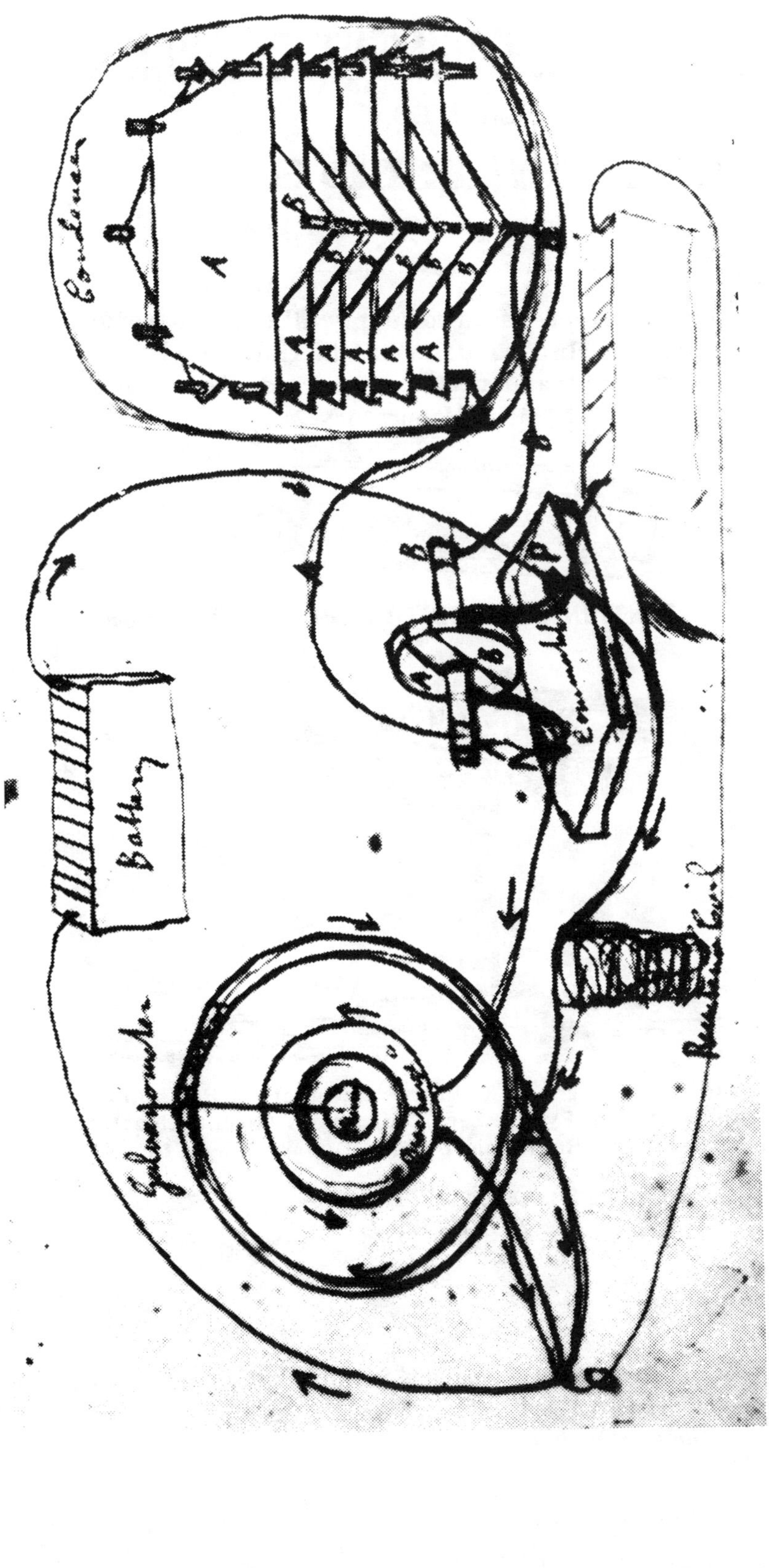

Figure 4 Maxwell's revised design of an experiment to measure the ratio of the electrostatic and electromagnetic units. The commutator is at the lower centre; the galvanometer is at the left. *Source: Whipple Library, Cambridge.*

SCIENTIFIC INSTRUMENT COMPANY,

LIMITED,

CAMBRIDGE.

(ADDRESS ALL COMMUNICATIONS AS ABOVE.)

The Company are prepared to supply all the Apparatus used in the experiments described in Glazebrook and Shaw's 'Physics,' and the necessary requisites for a Physical Laboratory.

SPECIAL ATTENTION DEVOTED TO EXPERIMENTAL WORK.

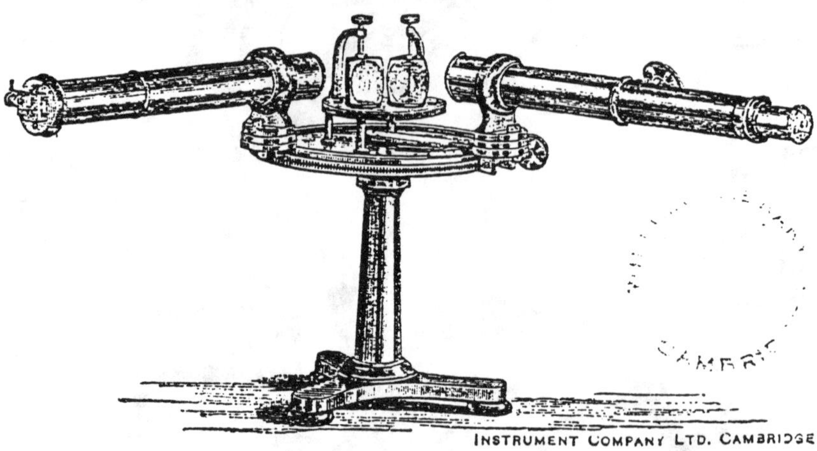

Makers of Callendar and Griffiths' Self-testing Resistance Boxes and Platinum Pyrometers, Callendar's Patent Electric Recorder, Boys' Radio-micrometer, Quartz Fibres, &c.

WRITE FOR CATALOGUES. SENT POST FREE ON APPLICATION.

Figure 5 An advertisement for the Cambridge Scientific Instrument Company printed in the endpapers of Glazebrook and Shaw, Practical Physics.
Source: Whipple Library, Cambridge.

Figure 6 Equipment for the standards project used at the Cavendish Laboratory in the early 1880s: the larger spinning coil stands at front left.
Source: Whipple Library, Cambridge.

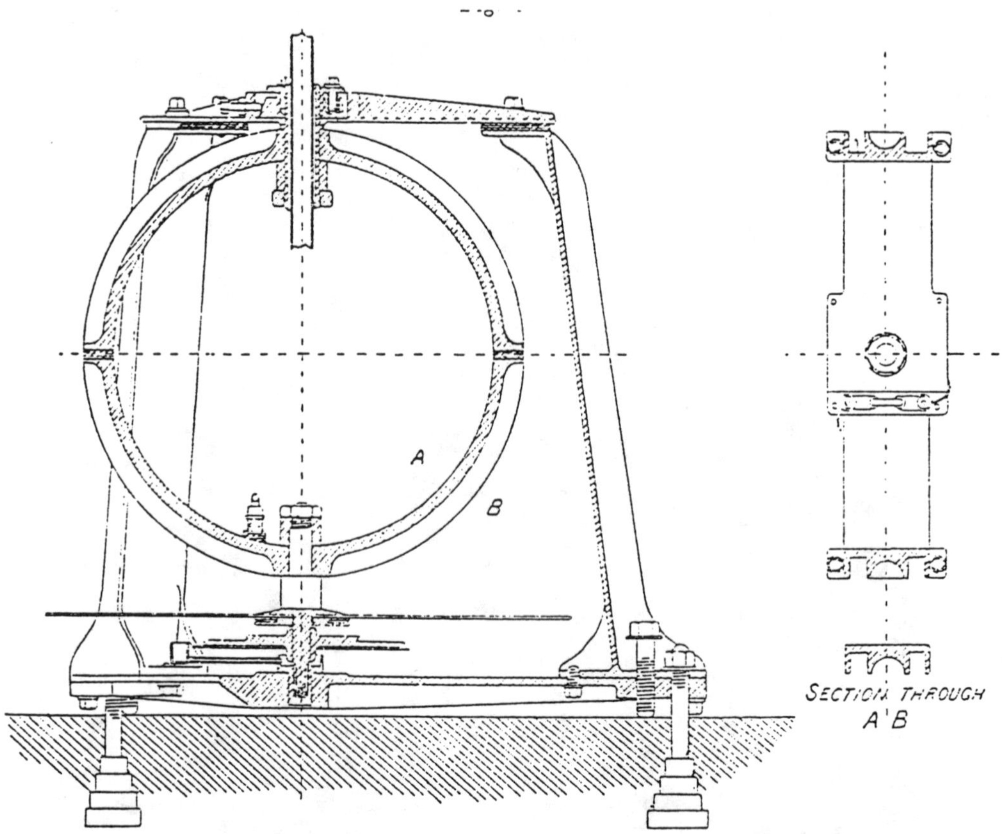

Figure 7 Schematic diagram of the spinning coil used in the Cavendish resistance experiments in 1881-2. *Source: Rayleigh, Scientific papers, Volume 2 (1900).*

Chapter 3

Rediscovering the instrument: research, industry, and education

Yakov M. Rabkin,
Université de Montréal (Canada)

1.INTRODUCTION

Physics, biology, geology, and chemistry all have had significant observational components in the course of their development. Chemistry projects the image of an experimental discipline *par excellence*. Its emergence depended so heavily on the use of scientific instruments that analytical chemistry was deemed "the mother of modern chemistry".[1] While this paper stems from a long-term interest in the history of chemistry, its scope has been broadened beyond the confines of one discipline.[2]

Historians of science have put much emphasis on theory as the motor of scientific development. Subsequently, the instrument has acquired the appearance of a tool manufactured expressly for the researcher in his quest for an ultimate theoretical breakthrough. This imagery is related to the commonplace subordination of technology to science.[3] Industrial history may provide a different perspective on the history of instruments in science[4]. It may show that instruments, often produced for industry, play a role whose confines and nature are more than instrumental. This may present a refreshing angle on science-technology relationship.[5]

Historians of science, most notably of physics and astronomy, have given greater prominence to the roles played by instruments in scientific research than historians of biology or geology. In fact, physics almost seems to be the only discipline whose rapport with instruments has been undergoing investigation in the history and sociology of science literature.[6] However, it would be unwise to extrapolate findings about physics, whatever the diversity within the discipline itself, on the history of other disciplines. No two scientific disciplines are alike, and chemistry, for example, because of its fundamental relationship with the instrument, must be understood in its own right. One therefore should distinguish between the history of instruments in different scientific disciplines. One may wonder, for example, about the relevance of a recent study of experimentation on elementary particles, perhaps the most elusive physical phenomenon, to the historical understanding of chemistry.[7] An instructive example concerns visual representations of Kekulé's formula of benzene. In 1865 he postulated, reportedly in the wake of a dream, that the molecule consisted of six carbon atoms in a ring. The field emission microscope showed many years later shadows of the molecule which proved Kekulé right. But the physical theory underlying the initial image had to be discarded eventually in favour of a totally different one. For chemists, however, the image was still true. "It made no whit of difference. People [more precisely, chemists] kept on regarding the micrographs of the molecules as genuinely correct representations."[8] This difference in the way chemists and physicists relate to the experiment and its interpretation should serve a note of caution to the historians of science about cross-disciplinary extrapolations. This episode seems to support the view that "the instrument presents a phenomenon; this insures the reliability and validity of the information supplied by the instrument, even when there is no way to know just what property the instrument is presenting to us."[9]

In another instance citation analysis reveals "a remarkable difference between the life sciences and the technical sciences with respect to the way they deal with the use of the electron microscope".[10] Their discursive style varies broadly with different aspects of experimental work deserving unequal attention from the authors. Consequently, unless the instrument is itself the object of an article, the instrument tends to disappear from public view as reflected in scientific publications.

There is another, perhaps, more mundane distinction between scientific instruments that reinforces the validity of the discipline-specific approach. Throughout the formative centuries of modern science, astronomical and physical instruments attracted the attention and interest of royalty, aristocracy and bourgeoisie whose houses they often adorned:

> De tous les objets de collection, les instruments scientifiques semblent avoir la faveur exclusive du sexe fort. Cela tient peut-être au fait que si la femme a la faculté de donner le jour à des mécaniques parfaites parce que vivantes et pensantes, l'homme a trouvé dans cette sorte de collectionomanie, un substitut quelque peu imparfait mais séduisant à ce besoin obsessionnel: Créer, créer sans cesse.[11]

But biological, chemical or geological instruments never acquired the same obsessive appeal. Consequently, the study of instruments in these disciplines cannot rely on the instruments kept in private and public collections to the extent that would the study of astronomical instruments.[12] Idiosyncrasies of individual collectors make even collections of physical or astronomical instruments a dubious foundation on which to build our understanding of the roles of instruments in the history of these scientific disciplines. This tends to limit the analogy that Jancz and le Pair suggest elsewhere in this volume with the integration of archeological data in scholarly knowledge of medieval history.

This work attempts to see how the diffusion of an instrument may affect the character and scope of scientific investigation, and to improve our understanding of instrumentation in the history of science, i.e. in the formulation of research objectives and their orientation. This objective is distinct from attempts to reveal the influence of instruments on "the political economy of experimental practice and the culture of laboratory".[13]

It will be argued that scientific instruments, even when these are produced at the express request of scientists and in accordance with their specifications, do not remain passive tools in the hands of the active researcher. Physics offers abundant examples of how a new measuring device can "boost" a field of research.[14] The acquisition of equipment, particularly when it constitutes a major investment, can shape the experiment and, therefore, the entire scientific process. While in some instances one may indeed wonder whether the scientist "uses" the instrument or, conversely, the instrument conditions the thinking of the scientist, the purpose is not to argue for any strong, and usually misleading, programme, whether of instrumental determinism or autonomous technology.[15]

A subsidiary purpose of this work is to analyze the impact of chemical industry as the user of instruments on their diffusion and subsequent contribution to research. The advent of serial, mass-produced scientific instrumentation increased the ease of exploitation. This led to certain alienation of the scientist from the actual design of the instrument, particularly in the 20th century. They report details on the use of instruments less frequently than, for example, on the preparation of specimens which, unlike the instrument, is seen as an individual contribution of the scientist, and therefore worth reporting. However, even in earlier centuries the production of instruments, mainly for astronomy and physics, was often affected by non-researchers, popularizers of science or instrument collectors. This phenomenon may not be quite so recent.

Another related point concerns the role of instrumentation in the emergence and development of the scientific profession. Since the use of instrumentation is indeed a major factor in the development of science, we shall also examine possible connections between the instrument-building capacity of a country and its relative standing in the world of science. This query may help us better understand factors of inequality in the world distribution of scientific activity.

2.ORIGINS AND ROLES OF INSTRUMENTATION

It has long been affirmed that chemistry "has been created chiefly by the careful use of the balance."[16] Lavoisier and Mendeleiev, to name just two classics of science, repeatedly emphasized the importance of quantification, only attainable through the use of measuring instruments, for the development of their science. This fundamental link between research and instruments has survived momentous changes in the history of science. While theoretical studies in chemistry or even biology may be at times hard to distinguish from physics, the diffusion of physical instruments into these disciplines has so far failed to produce a confusion with physics but rather reinforced the experimental, instrument-intensive character of chemical research.[17] The observation that "while theories may be conceived as as a species of belief, it is difficult to conceive of experiment as a species of belief"[18] seems to hold true for an entire range of physical sciences.

The birth of modern chemistry is usually associated with Antoine Laurent Lavoisier (1743-1794). His introduction of precision measurements offers a good perspective on the early interaction between the instrument and the conceptual development of chemistry.

"The prime instrument in Lavoisier's quantification was the balance", writes Trevor Levere. A significant leap in accuracy occurred around the middle of 18th century. Was the new precision balance developed ***for*** Lavoisier or were those advances of measuring technologies made for other reasons and other customers? To quote Levere again,

> ...there was a dynamic process in which the needs of science sometimes required the development of new apparatus - Lavoisier's use of the balance provides a prime example - but in which the development of new standards and tools of craftsmanship also indicated and sometimes dictated the development of science.[19]

While the former observation is unexceptional for the mainstream history of chemistry, the latter part of the quotation confirms that even in this most fundamental episode the history of chemistry the producers of instrument could actually **dictate** the development of science. Lavoisier was also shown to be "guided by his instruments in imposing logic on his results."[20] In other words, the instrument played a role which went well beyond the purely instrumental.

Lavoisier himself acknowledged this new role of instruments when he found the main reason for the delay in the reception of his results in the lack of affordable instrumentation needed for proper verification of his results. Nobody would "make precise instruments in order to test theories."[21] The instrument was essential for both the conceptualization and the acceptance of scientific data developed by Lavoisier. Moreover, even Lavoisier, a man of considerable wealth, found his instruments, particularly the gasometer, developed earlier in the century by Stephen Hales, to be very expensive. The credit Lavoisier accorded to his expensive instrumentation foreshadows the modern phenomenon of formulating the research programme in accordance with the expensive apparatus acquired earlier. Indeed, Lavoisier can be counted among the founders of modern science in more than one sense.

Unlike the balance, the microscope is primarily an observational instrument which literally opened new vistas in science. Microanalysis, i.e. identification of minute amounts of substances, is based on the investigation under the microscope of crystals formed with different reagents. The microscope came to be used mainly for medical and biological research, while also enabling scientists to investigate the nature and mechanisms of catalysis (Vanadium was thus ascertained as a catalyst for several organic reactions). The availability of the microscope oriented research of catalysis in the direction which no available theory would have otherwise suggested.[22] The microscope became a standard tool for approaching new substances and phenomena. Reasons for using the microscope, and later its electronic offspring, became basically akin to Galileo's reasons to use the telescope, namely to find out what a new instrument may reveal. Such use which is traditional in astronomy has become quite common in physics, biology and chemistry, in spite of eventual rationalizations of experiments that may appear in print.

Most advances in electrochemistry and spectroscopy which led to the discovery of dozens of new elements in the 19th century, came from outside the expanding bounds of chemistry. Physical phenomena which constitute a legitimate object of research in physics have often become instruments in the context of chemistry. An analogous relationship may also be observed between chemistry and biology.

An important exception to this pattern can be found in the work of Robert Wilhelm Bunsen and Gustav Robert Kirchhoff (1824-1887) who adapted the spectroscope for the express purpose of investigating chemical elements. In turn, their instrument which "embodied no new principles but brought together the necessary collimating and viewing tubes on a single stand for convenient operation,"[23] contributed not only to the development of chemical analysis but also to the discovery by the astronomers of the elementary composition of different planets. Reasons and functions of the use of the instrument turned out to be different, and they ought to be distinguished by the historian. The development of the classification of elements, a major contribution to the theoretical foundations of chemical research, ensued from instrumental advances in electrochemistry and spectroscopy. Theoretical and instrumental aspects of science are often held to be poles apart. However, a closer look at the biographies of prominent theoreticians of science would reveal a strong emphasis on the importance of instrumentation and precise measurement. For example, Mendeleiev whose main contribution to science is the Periodic Law of Elements, devoted significantly more attention to the question of measurement. In fact, the Institute of Metrology in St Petersburg not only bears his name, but his statue decorates the institute's front yard.

The very emergence of modern science is intimately linked with a new generation of instrument makers who were satisfying diverse customers: navies in need of precision chronometers to determine longitude at sea, amateurs of experimental demonstrations, and, ultimately, research scientists. But by concentrating on science and ignoring industry one misses important markets for instruments, and thereby essential stimuli for the development of instrumentation affecting both the nature and the direction of research.

The pivotal role of industrial uses in the progress of science is rarely reflected in the history of science. There are, however, two important exceptions: the microscope and the polarimeter, often dubbed as the sacharimeter. Needs of medical and biological research are credited for frequent improvements in the microscope while the development of manufacturing and international commerce in sugar is rightly considered the main motive force behind the refinement and the routinization of the polarimeter.[24]

My earlier work suggests that not only instruments, but in fact "extraneous" forces which regulate their development and production, affect the conduct of scientific research. This view was largely based on an analysis of infra-red spectroscopy, and later expanded to other instruments used in the 20th century chemistry.[25] The example of the infra-red clearly shows how the war effort, and more specifically the impact of military procurement on petroleum refining, affected the design and, later, the diffusion of the new instrument into chemical research. It challenges the traditional view of the instrument as a tool built in response to the needs of theory-driven scientists. The case of oceanography also discussed in this volume adds further emphasis to this challenge highlighting the importance of military uses for the diffusion of instruments into science.

Another example of an industry-driven instrumental technique widely used in chemistry is titrimetry. Efficiency of bleaching plants depends on the precision in determining the hypochlorite concentration. After a few years of largely industrial use the method of titrimetry spread to chemical research laboratories. This was done by Joseph Louis Gay-Lussac (1778- 1850) better known for his theoretical contributions to the science of chemistry. He modified earlier apparatus, introduced more finely calibrated burettes and other technical improvements, and was the first to use the term titrimetry in connection with this method of chemical analysis.[26] After him titrimetry became a standard analytical method which routinized determinations of the composition of substances and thereby contributed to a better understanding of reactions between them.

The history of refractometry offers a similar instance of diffusion of a new scientific instrument from industry to research laboratories. The phenomenon of optical refraction has been identified since the 17th century. This physical phenomenon finds itself at the basis of a new analytical instrument, since refractometry is a simple and reliable method of process control. It is produced serially by Zeiss Jena and its competitors elsewhere in Germany in the second half of the 19th century. Consequently, the instrument becomes routine in research laboratories doing work on correlations between properties and structures of the growing number of organic compounds.[27] A discussion of the origins of the ultramicroscope by D Cahan
suggests that the incidence of "industry-based science" is at least as high as that of the better recognized "science-based industry".

3.INSTRUMENTS AS A TOOL OF SCIENTIFIC EDUCATION

Didactic and social roles of instruments deserve particular attention. Laboratory training and access to scientific instruments are essential for the professionalization of science in the middle of the last century.[28] Practical work with scientific instruments comes to occupy an ever growing place in the curricula. At the same time instruments become a factor in scientific training. The sophistication stemming from industrial or military factors affects the definition of legitimacy and value in the initiation of future researchers. A measurement or an observation which at one time would earn a science student a doctoral degree, becomes a standard operation performed by a freshman according to printed instructions. In the popular image and in the more sophisticated perception of university peers, the road to professional life in science passes through extensive exposure to laboratory life and consequent socialization to instrumental techniques as a system of moral values.[29]

Such was the case of electrochemistry which owes its very existence to the invention of the electric pile by Alessandro Volta (1745-1827). The pile consisted of plates of different metals, and offered a convenient source of electricity which soon attracted the attention of a broad range of scientists. Many experimenters began to play with the voltaic pile to investigate the phenomenon of electrolysis and ascertain properties of various compounds. This exemplifies an observation, made by C Mukerji elsewhere in this volume, that scientific inquiry may become a process of creating a material, instrumental culture.

Much of this experimentation began without either a theory or even a clear purpose. This is how within a few years the voltaic pile enabled Humphry Davy (1778-1829) to discover the electric character of chemical bonding which had a profound effect of the development of science. It also led to the discovery of a number of new elements. Davy's protégé Michael Faraday (1791-1867) is reputed to be "content to uncover the secrets of nature, leaving it to others to develop the major theoretical ideas and to reap the practical benefits".[30]

The existence of this type of researcher agrees well with the role of scientific instruments as a motive force in the development of science. Or to quote Baird and Faust, "one aspect of scientific progress [is] the accumulation of new scientific instruments".[31] Experiments with electrolysis enabled Faraday to formulate his idea of electrochemical equivalents, i.e. a relationship between the amount of electricity and the quantity of substance decomposed. This discovery would greatly advance the atomic theories several decades after his death.[32]

Instrumentation becomes, perhaps, the main pillar of the professional identity of the scientist and of the general culture of instrumentation which distinguishes him from an amateur. Science is no longer done at home, however wealthy. Experimental science acquires proper housing which makes it both distinct and distinguished.[33] Moreover, it leads to a broad distribution of science graduates among government, industrial and university laboratories. For example, chemists become active in the public health movement, arguing for tighter controls on domestic fuels, foods and drugs. Their civic role is embedded in the morality of measurement, its reliance on the integrity and probity of the scientific practitioner.

Most prominent schools of science are grounded in experimental work, and excel in devising and refining new apparatus. First teaching laboratories, by definition the natural abode of scientific instruments, are often built through private initiative and financing of enthusiasts of chemistry. Central nations such as Germany and France in the mid-nineteenth century fare only slightly better than the then peripheral United States and Russia in building teaching laboratories. Attached to a university, a hospital or an industrial company, laboratories offer positions to technical staff some of whom use experimental work to become research scientists in their own right. The case of Jean-Baptiste Dumas[34] may well illustrate this point.

Hired as a *répititeur* at the Ecole Polytechnique, he has most rudimentary equipment at his disposal. It may have been sufficient for teaching future engineers but Dumas "built up a supply of apparatus and chemicals suitable for his own research. Then, following in the footsteps of those scientists who, both in Geneva and Paris, had been so generous in providing him with facilities to carry out his experiments, he too invited young men to work in his modest laboratory, to assist him in his research."[35] Subsequent improvement in the status of Dumas enabled him to acquire more sophisticated instruments and attract more students. While in Germany Justus Liebig's laboratory is a breeding ground for both research and industrial chemists, growing numbers of French chemists with advanced training join laboratories in chemical and related industries.[36]

Reliance on experiment and equipment combine in Dumas's laboratory with a strong motivation to advance chemical theory. In this sense Dumas can be compared to Lavoisier whose experimentalist bent is certainly not a goal in itself. By devising new laboratory procedures and techniques for research and analysis, and increasing considerably the data base of organic and analytical chemistry, foundations are thus laid for a rational science of organic chemistry.[37]

Professional scientists transfer techniques of experimentation acquired at the university to industry and government laboratories. They frequently choose the same equipment they have used in their formative years. The diffusion of professional scientists into industry creates not only a demand for experimental training but, no less importantly, a demand for instruments. Industry, alongside with medical and governmental institutions, constitutes an important market for scientific instrumentation. Thus begin the routinization of apparatus which, in turn, facilitates its acquisition and application for research.[38] While links between industry and university consist mainly of circulation of scientists and of contract research performed for industry at university research laboratories, both links enhance the similarity of equipment used at industrial and university laboratories. They also transfer the moral culture of science into the governmental sector, thus enhancing the image of the government as a neutral expert educated in the spirit of scholarly detachment and motivated solely by the search for truth.[39] Thus instruments play multiple roles in the professionalization of science in the last century.

4.INSTRUMENTATION AS A FACTOR OF INEQUALITY

Instrumentation is a major factor in the international distribution of research. Two brief examples may well illustrate this point and suggest avenues for further exploration. One has to do with the transfer of scientific centre form France to Germany in the 19th century. The other is the decline of Russian chemistry after the October revolution.

Germany emerges as the undisputed leader of world chemistry by mid-nineteenth century. The loss by France of the central position in chemical research profoundly injures national pride. It leads such a notable of French chemistry as Charles Adolphe Wurtz to begin his major compendium of chemical knowledge with the affirmation: "La chimie est une science francaise".[40] Several decades later, on the occasion of the Universal Exhibition in Paris in 1900, the decline of French chemistry is underlined by the obvious domination of the instrument-making scene by German companies.

French instrument-makers, following the lead of French science, are largely concentrated in the capital. Their fame is established in the early part of the 19th century. The disruption of commercial links with Britain, the then leader in scientific instrumentation, gives impetus to serious qualitative advances in the instrument-making in France.[41] French instrument-makers appear innovative. For example, they are the first to make use of aluminium anemometers, electrometers and sextants.[42] But their ingenuity alone cannot ensure that France retain its position. since the 1870s Germany takes the lead, and while the French may maintain the quality of craftsmanship, commercially France's instrument-making industry is badly lagging behind:

> Les nouvelles conditions du marché ne sont pas comprises, et l'artisan francais prend de plus en plus la mentalité de 'l'artiste', entendant par là le praticien de l'un des beaux-arts. Certes - et ceci contribue à expliquer cela - la consruction francaise donne encore à cette époque des réalisations spectaculaires, mais ce qui lui manque c'est la solidité en profondeur sur le plan commercial, acquise à la même époque par les constructeurs allemands.[43]

It is the industrialization of instrument-making that decisively accounts for the disparity. While some apologists try to demonstrate that French instrument-makers are on their way to triumph, this attitude is compared by contemporary historians "to the optimism feigned before terminally ill patients."[44] Praise for "the national temperament" of the French who can allegedly distinguish between "real" and "nominal" precision further illustrates the despair of the French instrument-makers faced, at the turn of the century, with an avalanche of mass-produced German instruments.

But the growth of German science alone cannot explain the spectacular advance of the country's instrument-makers. The emergence of science-intensive industries, chemical industries first among them, creates a new and substantially more significant market for chemical apparatus. Many of these instruments are first produced for, and used by the industry and only later reach the confines of scientific research. Unlike their French colleagues whom industrialists among their compatriots usually disdain as theoreticians incapable of offering practical expertise, German scientists are well integrated in the mass production of scientific instruments. Their advice is sought by such industrial giants as the optical concern Zeiss Jena, a leader in the production of both sophisticated scientific instruments and substantially more numerous optical devices for the industrial markets.

Globalization of markets for instrumentation may have lessened the importance of local instrument-making industries for the development of science. Industrialized countries may excel in science while purchasing instruments for science. But when autarkic ideologies or other reasons limit the effects of globalization, instrumentation becomes a crucial factor of inequality between national sciences.

For example, Russia is a major centre of chemical science in the 19th and the early 20th century.[45] The names of Mendeleiev, Butlerov, Chugaev illustrate the importance of Russian chemistry. The October revolution of 1917 causes relatively few Russian chemists to flee abroad. Moreover, the Soviet government puts particular emphasis on science as a key to "building a new society". However, Russian chemical research steadily declines in terms of its impact on world science. While restrictions of freedom and severe mismanagement impede normal development of Soviet science, inadequate supply in instrumentation is another major obstacle with profound consequences for the country's research potential.

Unable to mass-produce sophisticated instruments, advanced experimental chemistry in the Soviet Union uses occasional and badly maintained imports and local custom-made instruments. Low priority accorded to chemical research in the Soviet defense programme translates into poor instrumentation available to chemists, as distinguished from physicists, in the Soviet Union. Indeed, a comparison of Soviet and American participants in international scientific exchanges reveals a largely unilateral pattern of influence, from American onto Soviet chemists. Moreover, the most frequent factor identified by Soviet chemists as crucial for the success of their stay in an American laboratory was the level and availability of scientific instrumentation.[46] More theoretical, less technology-dependent disciplines in the USSR have by and large fared better than the experiment-bound disciplines like chemistry.[47] Even within chemistry more theoretical research proves more promising and yields significant results. For example, N N Semenov receives a Nobel prize for his work on chemical kinetics.[48] However, the decline of Russian chemistry in the Soviet period continues. The resulting marginalization of Soviet chemists, as distinguished from mathematicians or physicists, stems to a large extent from inadeqaute instrumentation.

The examples of France and the Soviet Union suggest that deficiencies in instrumentation constitute a decisive factor of marginalization in the world of science. They also suggest that mass production of scientific apparatus is an essential distinction between central and peripheral science.[49]

5.CONCLUSIONS

A blue-ribbon report published by the National Academy of Sciences in the United States,[50] suggests four stages in the development of a new instrument:

(1) discovery of suitable means of observing some phenomenon,

(2) exploration of this phenomenon with special, homemade instruments or commercial prototypes,

(3) widespread use of commercial instruments,

(4) routine applications of the instrument to control industrial production as well as research.

This scheme seems to be at variance with much of the evidence in the history of science. It has been shown that the integration of instruments has been rarely due to the demand on the part of the researcher. Rather it occurs through vigorous supply of advanced instruments on the part of the industry. The company that proposes these four stages in the report has itself had experience when stages 3 and 4 occur in the reverse order and, moreover, stage 4 is by far the most decisive factor in the development of new instrumentation.[51]

Beyond regular publicity, adult education campaigns are mounted, beginning with the infra-red in the 1940s, with the express purpose to introduce the researcher into the new world of instrumentation. An analogy would not be far-fetched with the general consumer market. Advances in stereo equipment or in personal computers rarely constitute a response to consumer demand but rather derive from industrial strategies of the manufacturers. Consequently, when the marketing approach is right, the habits of the consumer change, sometimes drastically, not because of the inner felt desire for a new product but because of its availability and prestige. For example, chemistry, traditionally linked with advances in instrumentation, naturally follows this path of development and undergoes changes in orientation and conceptualization which are often not of the chemist's own making.

Moreover, the scientist must be introduced and initiated to new instruments, often a novel domain whose fundamentals ought to be mastered. A telling sign of this trend is the fact that university books on instrumentation are frequently written by experts from instrument-making companies.[52] Indeed, it has been suggested that instrument makers constitute a third full-fledged category of scientists, alongside with experimenters and theoreticians.[53]

Another analogy between the consumer and the researcher consists in the growing acceptance of the new product or instrument as a black box whose inner working may be irrelevant, beyond one's technical proficiency or both. Consequently, while stages 1 and 2 may indeed occur in the development of a new instrument, this often happens beyond the confines of the laboratory. It is only if the instrument is meant mainly for the scientific market that stage 2 may involve laboratory tests. Otherwise, stage 4 may precede stage 2 in the above scheme.

Instruments may initiate and enhance interdisciplinary interactions. Thus, advances in chemistry of the 20th century result mainly from experiments carried out on instruments incarnating physical phenomena and later adapted to the needs of chemistry. While this circumstance does not blur borders between the two disciplines, the prominence of physical approaches to chemical realities represent a continuing trend which is at least a hundred years old. Physical phenomena become instruments in chemical research. They profoundly affect chemical practice introducing new ideas and, perhaps, more importantly, new, physical discourse which supersedes the conceptual apparatus inherited from the last century. Thus instruments, largely developed outside of the chemical laboratory, produce a massive impact on the way chemical research is conceptualized, conducted and reported. Similar interdisciplinary influences can be observed elsewhere.

Chemical science relies growingly on instrumentation which not only measures a variety of physical properties but directly reveals molecular and even atomic structures. For example, the passage from the optical to the electron microscope reinforces the observational element in the practice of chemistry to such an extent that it makes the tradition of analytical chemistry, based on intrusion and destruction, largely irrelevant. Advances in computers, optics and electronics regulate the progress of chemical analysis.

The transition to "the new analytical chemistry" represents a discontinuity no less striking than the transition from the horse to the locomotive. And, similarly, those responsible for the advent of the new instruments have no greater knowledge (or commitment to) of the traditional methods of chemical analysis than the railway engineers had with respect to horse breeding. Massive intrusion of new instruments may also subvert existing social structures of technology-based science by giving the young an advantage over the elders who may be less apt to master new technologies.[54]

Technologies do not necessarily blur the distinctiveness of scientific disciplines. While chemists are just beginning to see atoms and molecules on the screens of their apparatus, physicists face quite a different problem of distinguishing between real and artifactual phenomena. Concerns and agendas are not the same. Therefore, historical accounts of physical experimentation, by far the most numerous in the history of science literature, may not inform the history of other disciplines.

On another level, however, interdisciplinary inferences may well be legitimate. "Despite the slogan that science advances through experiments, virtually the entire literature of the history of science concerns theory."[55] Albeit borrowed from the history of physics, this paradox may be found in other disciplines. The habitual image of instrumentation in the mind of the scientist and layman alike remains that of a tool at the service of the researcher. This happens in spite of the fact that instrumentation has played a crucial role in the foundation of several modern scientific disciplines, their professionalization and international development. The standard literature on chemical instrumentation suffers from an even greater degree of amnesia that textbooks of science which make occasional references to the history of science. Our use of historical evidence in this work suggests that research should be viewed in the context of its instrumentation, that may often introduce approaches linked with its industrial or military origins. Instruments and their production profoundly affect the very identity of research, and remain crucial in the ongoing stratification of scientific nations. But these observations need further validation (or refutation) by historians paying specific attention to the instrument in a greater variety of scientific disciplines.

6.ACKNOWLEDGMENTS

The author acknowledges helpful comments from Vladimir Ganin, Spas Spasov, and participants at the Beckman Center conference "Chemistry in the Modern World".

7. REFERENCES

1 F. Szabadvary & A. Robinson, "The history of analytical chemistry", in: G. Svehla, ed., *Comprehensive Analytical Chemistry,* Volume X, Elsevier, Amsterdam, 1980, ch. 2, p. 63.

2 For a more focused treatment of the roles and images of instrumentation in chemistry see my paper in the forthcoming volume *Chemical Sciences in the Modern World* edited by Sy Mauskopf.

3 See: Derek De Solla Price, "Is Technology Historically Independent of Science...?" *Technology and Culture* 6(1965), pp. 550-568, and Yakov M Rabkin, "Science and Technology..." *Fundamenta Scientia* vol 2, nos. 3-4, pp. 413-423, 1981.

4 The term "scientific instrument" used in this paper does not presume the exclusive or even principal use of the instrument for scientific purposes. Rather it connotes an instrument which has been or can be used in science.

5 This view informed an earlier article about this subject: Yakov M. Rabkin, "Technological Innovation in Science", *Isis* vol. 78, pp. 31-54, 1987.

6 This impression is reinforced by a recent review of current sources on the subject of scientific instruments, even though it was written by an historian of chemistry: Jeffrey L Sturchio, "Artifact and Experiment", *Isis* 79, pp. 369-372, 1988.

7 Peter Galison, *How Experiments End,* The University of Chicago Press, Chicago, 1987.

8 Ian Hacking, *Representing and Intervening,* Cambridge University Press, Cambridge, p. 200, 1983.

9 Davis Baird, "Instruments on the Cusp of Science and Technology...", preprint.

10 W P Van Els et al., "The Citation Gap between printed and instrumental Output of Technological Research: the Case of the Electron Microscope", *Scientometrics*, vol. 17, Nos 5-6, pp. 415-425, 1989.

11 Maurice Rheims, préface, in: Henri Michel, *Instruments des sciences dans l'art et l'histoire,* Albert de Visscher, Bruxelles, , p. 8, 1973.

12 J.V. Field, "What is scientific about a scientific instrument", *Nuncius* vol. III, no.2, pp. 3- 26, 1988.

13 Sturchio, "Artifact and Experiment", p. 371.

14 See, e.g. a recent study of the thermopile, an instrument of consequence for the development of infrared spectroscopy: Edvige Schettino, "A New Instrument for Infrared Radiation Measurements" *Annals of Science* vol. 46, pp. 511-17, 1989.

15 The late Derek De Solla Price who time and again emphasized the importance of instruments for the development of scientific thinking, nonetheless stayed away from strong programmes.

16 W. Stanley Jevons, *The Principles of Science*, MacMillan & Co., London, p. vii, 1879, quoted from: Trevor H Levere, "Lavoisier: Language, Instruments and the Chemical revolution, in: T.H. Levere and W.R. Shea, eds., *Nature, Experiment and the Sciences*, Kluwer, Amsterdam, pp. 207-223, 1990.

17 More on the early relations between chemistry and physics can be found in : A. Thackray, *Atoms and Powers*, Harvard University Press, Cambridge, MA, 1970.

18 David Baird and Thomas Faust, "Scientific Instruments, Scientific Progress and the Cyclotron", preprint.

19 Levere,"Lavoisier...", p. 209 (emphasis added).

20 Levere,"Lavoisier...", p. 217.

21 Hacking, *Representing and Intervening*, p.242.

22 Szabadvary & Robinson, "The history of analytical chemistry", p. 163.

23 Aaron J. Ihde, *The Development of Modern Chemistry*, Harper & Row, New York, p. 235, 1964.

24 Aaron J. Ihde, *The Development of Modern Chemistry*, p. 295.

25 This was developed in the *Isis*, 1987 article and a presentation at the annual HSS meeting in Cincinatti in December 1988.

26 Szabadvary & Robinson, "The history of analytical chemistry", pp. 177-80

27 Boris V Ioffé, *Refraktometrichiskie metody v khimii*, GNTIKhL, Leningrad, p. 5-6, 1960.

28 J.B. Morrell, "The Chemist Breeders..." *Ambix* , vol. 19, pp. 1-46, 1972.

29 See the paper by S Schafer in this volume.

30 Szabadvary & Robinson, "The history of analytical chemistry", p. 136.

31 Baird and Faust, "Scientific Instruments, Scientific Progress and the Cyclotron", p. 3.

32 A similar example of Einstein as a "parasite, feeding haphazard on long dead experiments" is brought up with respect to Armand-H. Fizeau and his experiment on the velocity of light (see: Hacking, *Representing and Intervening*, p. 237.

33 S Shapin, "The House of Experiment in Seventeenth-Century England", *Isis*, vol. 79, no. 296, pp. 373-404, 1988.

34 Leo J. Klosterman, "A Research School in Chemistry in the Nineteenth Century..." *Annals of Science* vol. 42, pp. 1-80, 1985.

35 Klosterman, "A Research School in Chemistry in the Nineteenth Century...", p. 8.

36 Dumas married a daughter of the director of the Sèvres porcelain factory. This was another kind of connection between academic and industrial worlds which remains to be explored.

37 Klosterman, "A Research School in Chemistry in the Nineteenth Century..." , p. 80

38 This idea was developed more fully for a later period in the United States in: Yakov M. Rabkin, "Technological Innovation in Science".

39 This diffusion of scientific values into society, which has taken the appellation of scientism, reaches its peak in the middle of this century when several governments rationalize as "science-based" most abhorrent acts of violence.

40 C A Wurtz, *Dictionnaire de chimie pure et appliquée*, Hachette, Paris, 1868. This provokes a major controversy among German chemists analyzed by Alan Brocke in a paper presented at the International Symposium "Interaction between Jewish and Scientific Cultures" in Montreal in June 1990.

41 Jacques Payen, "Les constructeurs d'instruments scientifiques en France au XIXe siècle", *Archives internationales d'histoire des sciences* , vol. 36, pp. 84-161, 1986.

42 Anita McConnell, "Aluminium and Its Alloys for Scientific Instruments, 1855-1900", *Annals of Science* vol. 46, pp. 611-620, 1989.

43 Payen, "Les constructeurs d'instruments scientifiques en France au XIXe siècle", p. 86.

44 Payen, "Les constructeurs d'instruments scientifiques en France au XIXe siècle", p. 152.

45 Alexandre Vucinich, *Science in Russian Culture*, Stanford University Press, Stanford, CA, 1970.

46 Yakov M Rabkin, *Science between the Superpowers*, Priority Press, New York, ch. 2, 1988.

47 see, e.g.: Thane Gustafson, "Why Doesn't Soviet Science Do Better than It Does", in: *The Social Context of Soviet Science*, Westview press, Boulder, Co, 1980.

48 N N Semenov, *Tsepnye reaktsii*, Leningrad, 1934. See also an historical essay on the subject: V A Kritsman, *Razvitie kinetiki organicheskikh reaktsii*, Nauka, Moskva, 1980.

49 An outline of the centre and periphery in science can be found in: Shils, E., "Center and Periphery", in: *Center and Periphery: Essays in Macrosociology*, University of Chicago Press, Chicago, p.5, 1975. See also: Schott, T., "International Influence in Science: Beyond Center and Periphery", *Social Science Research* vol. 17, , pp. 219-238, 1988.

50 *Chemistry: Opportunity and Needs*, NAS/NRC, Washington, p. 92, 1965.

51 This point is documented in my *Isis* article of 1987.

52 Howard A Strobel and William R Heineman, *Chemical Instrumentation*, John Wiley & Sons, New York, p. vi, 1989.

53 Baird and Faust, "Scientific Instruments, Scientific Progress and the Cyclotron", p. 36. A related idea that architects also impose their values on science through the laboratory design may be harder to prove.

54 This observation was made by Professor S Blume of the University of Amsterdam.

55 Galison, *How Experiments End*, p. ix.

Chapter 4

Information in instrumentation

C.N.M. Jansz
C. le Pair
STW - Technology Foundation (Netherlands)

1. INTRODUCTION

History is a respected branch of learning, as is archaeology. However, although both deal with the past, they approach it from different angles and, thus, see different things. Unfortunately, the interaction between the two disciplines is weak. One of the most famous historians of our times, Lynn White, observed this phenomenon and won renown by integrating archaeological knowledge of mediaeval artefacts with academic knowledge[1]. Following White other historians rewrote parts of the economic and political history of the Middle Ages. The picture that emerged was quite different from the one that existed before[2]. Not only did the 'dark ages' become less dark, in some aspects they even seemed to be more enlightened than the early Renaissance. It became apparent that, in these so-called dark ages consistent use was made of ancient knowledge. Gimpel showed, for example, that mediaeval engineers had a fairly good knowledge of the works and writings of their counterparts in ancient Rome. The scholarly writers of the period, who are the main sources for students of history, were not aware of this. The fate of Vitruvius' work illustrates this point. The *Ten books on Architecture* by this Roman engineer who lived at the time of Emperor Augustus were 'rediscovered' in 1414 by the humanist Poggio. Renaissance scholars believed them to have been lost during the Middle Ages. They were wrong, however, for 55 examples of the book still exist today, which were copied between the 10th and 15th century. In other words, mediaeval engineers knew and used Vitruvius' work, but their scholarly contemporaries were unaware of its existence.

Some years ago a summer school was organised in The Netherlands on the subject of the interaction of science and technology with national governments. Among the lecturers were some of the most outstanding historians of science in the country[3]. They were fascinated by the topic. They admitted that, with regard to this area there was a real gap in our historical knowledge[4]. The role of technology in society had been overlooked. One of the participants suggested that this was an extension or rather a generalization of the concept of C.P. Snow's 'two cultures'[5]. Examples were given of some far-reaching historical descriptions, given by famous writers who had an unprecedented influence on political thinking, but who had clearly not made a proper assessment of the impact of technological developments[6]. Some years before H. Mark[7] had described how the growth of Portuguese maritime dominance in the 15th century had come about as a result of advanced navigational and geographical technology, promoted and cherished by the Portuguese government.
Similar developments could be traced in The Republic of the Netherlands but they seem to have gone unnoticed by the chroniclers. There is evidence that the growth of centralised power in Europe in the 15th century, eventually leading to the

building of the Habsburg empire under Charles V, and, ultimately, to the supremacy of Spain, was rooted in the advanced state of weapon technology in the Low Countries[8]. The starting capital came from the same region: more than one third of the imperial income came from the Low Countries. And, since that was precisely the area that had hardly any natural resources, the strength was based on manpower and skills, in short: on technology. At that time and in that area there was but little emphasis on scholarly knowledge. This is illustrated by the distribution of Universities in Europe in that period, as shown in fig. 1: the north-western corner of Europe is quite empty[9].

Fig. 1. The distribution of universities and law schools in 15th century Europe.

Nevertheless, the Netherlands defeated Spain and grew to be a world-power in the 17th century. In most traditional history books we read that the Dutch defeated Spain and that subsequently the arts and the sciences prospered. Only recently has it become clear that one of the major causes of the Dutch success was their technological leadership, which was revealed for example in their skill in building faster and cheaper ships[10].

2. INSTRUMENTATION IN SCIENCE

It is not only in the humanities that the role of technology is neglected. Even in the natural sciences there is a tendency to overlook the crucial role that technology and instrumentation play in the development of our understanding of nature. Mukerji[11] also finds a persistent cultural subordination of technology to science.

Kuhn formulated the theory, that scientific revolutions, the milestones in the development of science, are marked by the downfall of a paradigm caused by internal crises. His theory has itself been a paradigm for years. However, in a number of cases the development of a new instrument has led to a break-through in science. Casimir[12] has pointed out that there is a continuous flow of development from science to technology to science etc., which he calls the science-technology spiral.

When considered from this viewpoint, the Copernican Revolution becomes a triumph of technology. Price[13] gives a very clear account of the developments at the time. At the roots of the Revolution we find an improvement in the craft of making eye-glasses. This technological innovation led to the invention of the telescope. News of this reached Italy when Sacharias Janssen and Hans Lipperhey from the Dutch town of Middelburg tried to sell their invention to the Medici family for military purposes. They did not succeed, but Galileo heard of the device and was quick to replicate the invention. He was the first to turn the instrument skywards and he published his results. His *Siderius Nuncius* of 1610 had a great impact and created tremendous enthusiasm for the instrument among a large group of people. It was this, rather than the work of Copernicus, which created the Copernican Revolution.

The same technological advance led to the invention of the microscope and the discoveries made with it by Antoni van Leeuwenhoek and others. This instrument and its successors had a great impact on the development of science. David Cahan also stresses the importance of the microscope and the improvements made in it around the beginning of this century[14]. And these are by no means the only examples of technological innovations opening up whole new fields of science.

Rabkin's[15] outline of the history of chemistry shows the importance of instruments in the development of that discipline. Similar views on the role of instruments in the history of other fields are expressed in several contributions in this volume (e.g. Shaffer, Mukerji, Georghiou and, to a lesser extent Feeney).

Although there have been periods during which instruments had greater social standing[16], those who initiate such innovations rarely receive proper recognition. Recently some Nobel prizes have been awarded for technological achievements: e.g. to Van der Meer at CERN for accelerator development and to Ruska for his work on the electron microscope. However, these exceptions are so rare that they do not really indicate an increase in the appreciation of the influence of technology on science, or of the influence of instruments on society in general.

During the last few decades, tight budgets have caused policy-makers to look for more and more refined evaluation methods by which to judge scientists and their work. In line with the historical picture drawn above, these methods mainly concentrate on the paper outputs. Unfortunately, not all those who apply bibliometric methods are aware of the possible traps and pitfalls. The recent article in *Science* on 'uncitedness'[17] and the ensuing flood of letters and comments provide a clear illustration of the difficulties encountered in bibliometric analyses. In most evaluative studies, the role of instruments as the output of research is ignored. For an engineer, however, the instrument itself is often the primary 'publication' of the results of his research. Le Pair postulated the existence of a 'citation gap' in such cases[18]. Subsequently, several studies have been undertaken to estimate the size of that gap and the implications of its presence.

3. THE CASE OF THE ELECTRON MICROSCOPE

The work of the technologists who introduced the study of the electron microscope in The Netherlands is known only to others working in the field. The impact of their work has been tremendous, both economically (here lie the roots of the ca. 70% market share which the company that utilised their results was able to achieve at a certain time) and scientifically (research done with the instruments has led to important discoveries in various fields). But, with today's emphasis on publication and citation counts, these people are almost invisible.

We considered the electron microscope to be a very suitable instrument on which to base a study of the citation gap since it is directly comparable to a paper publication: it makes the results of research available to other scientists and it is used primarily for further scientific research, i.e. knowledge is used to produce more knowledge. In addition, much of the research work done with the instrument is in fields where paper publications are the main output. An instrument does not have a title page mentioning the authors, so it will not be found in the list of references. However, it may be mentioned in the text; we counted these references in a random sample of the literature in order to estimate the number of 'invisible' citations.

Author	SCI	Instr.
W.H.J. Andersen	9	2275
S.L. van den Broek	0	1470
A.C. van Dorsten	2	1470
J.B. le Poole	11	1470
C.J. Rakels	2	6335
J.C. Tiemeijer	0	6335
K.W. Witteveen	0	4830

Table 1. Citation counts for the main 'authors' of the electron microscope in The Netherlands for the period 1981 - 1985; comparison of normal citations taken from the Science Citation Index (SCI) and textual citations of the instruments (Instr.), showing the citation gap.

The results have been published[19] and they clearly confirm the citation gap. Table 1 shows a comparison of the citation counts for the main 'authors' for the period 1981 to 1985, taken from the Science Citation Index (SCI), with the estimate from our study.

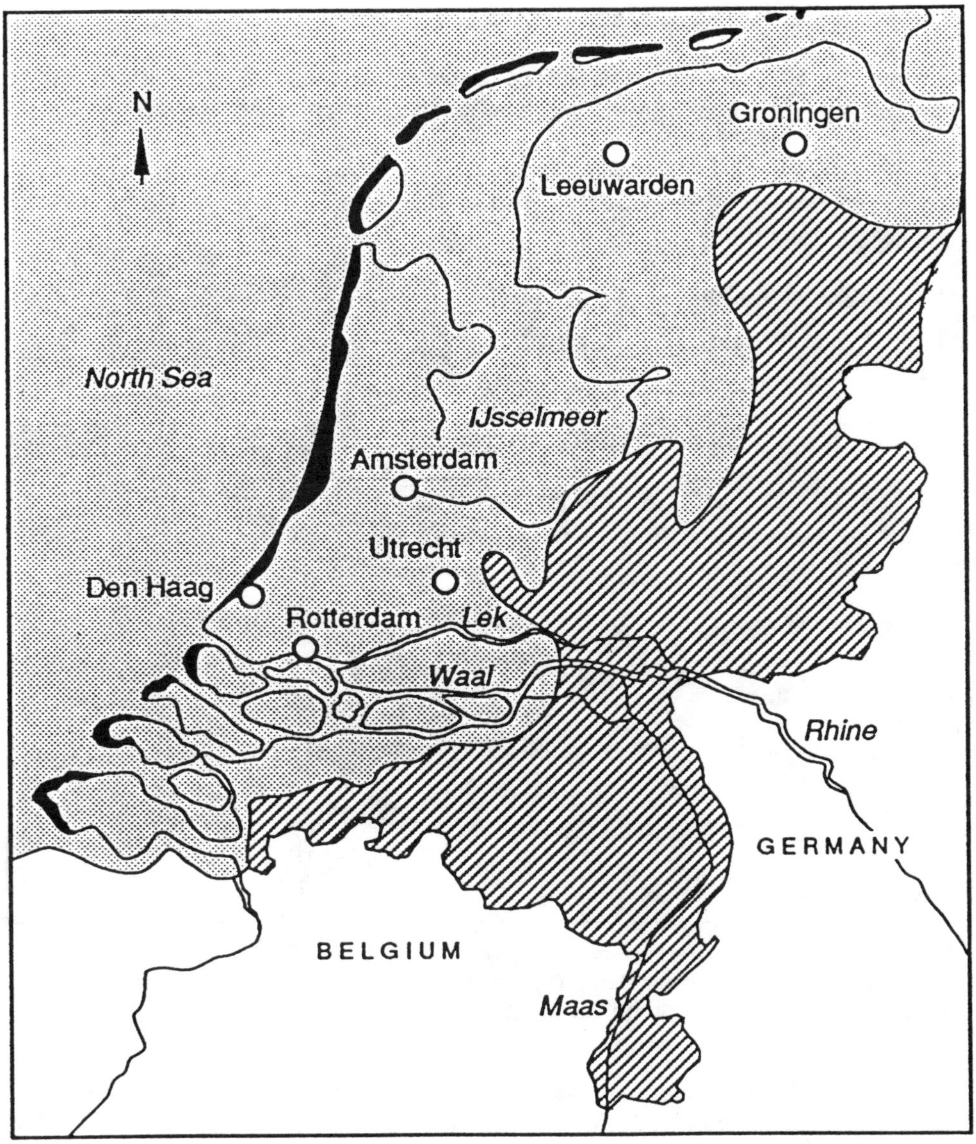

Fig. 2. The Netherlands. Without dikes, the Western part, i.e. more than half the country would be under water.

The SCI numbers refer to the entire work of these authors, whereas we, in our literature search, considered only three instruments. Moreover, the authors with the lowest instrumental counts were involved only with the earliest instrument, developed in 1958. For these reasons our estimate must be a conservative one. The actual citation gap is much larger.

4. THE CASE OF THE STORM-SURGE-BARRIER

Another case that illustrates the lack of proper paper output for technology and the resulting citation gap is that of the Storm-Surge Barrier in the Eastern

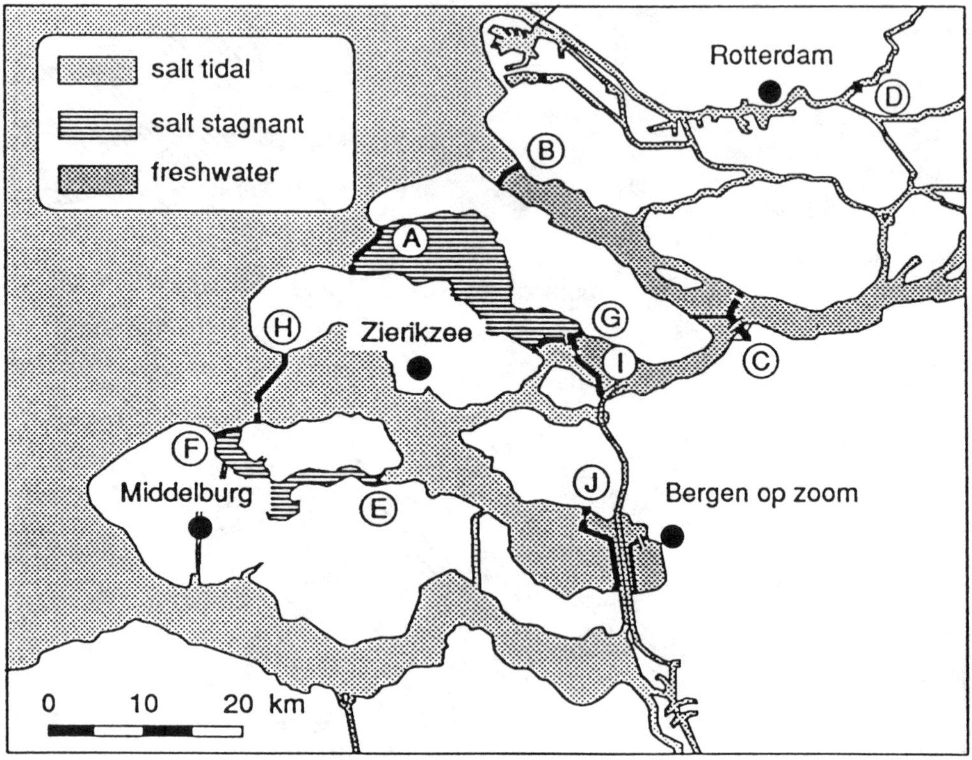

(A) Brouwersdam; (B) Haringvliet Dam; (C) Volkerak Dam; (D) Hollandse IJssel Storm Surge Barrier; (E) Zandkreek Dam; (F) Veerse Gat Dam; (G) Grevelingen Dam; (H) Eastern Scheldt Storm Surge Barrier; (I) Philips Dam; (J) Oester Dam.

Fig. 3. The South-western corner of The Netherlands with the estuary of three large European rivers (the Rhine, the Maas, and the Scheldt), showing the location of the main waterworks constructed to shorten the coastline.

Scheldt, the most innovative part of the impressive waterworks constructed in the southwestern part of The Netherlands; figures 2 and 3 show the location. This project seemed of particular interest because many different disciplines played a role in the design of the barrier, ranging from civil and electrical engineering to ecology and marine biology.

We studied the bibliometric visibility of the innovations[20]. To this end we conducted a number of interviews with people involved in the project, using snowball sampling to produce the list of interviewees. Our spokesmen agreed readily about the innovations but were more reluctant to attach names of 'authors'. A search in various literature databases, including the Science Citation Index, confirmed our fears that the technologists involved are almost invisible bibliometrically. Table 2 shows some of the results of these searches. Evidently, there are some fields where the situation is different. The authors with higher publication and citation counts all work in more fundamental areas.

Table II

Name	Fld.	1	2	3	4	5
Agema	pd	2	0	0	1	0
d'Angremond	co	6	0	0	5	3
Awater	sh	0	1	0	0	3
Bijker	co	66	2	0	6	10
van Duivendijk	ci	0	0	0	0	2
Engel	pm	0	0	0	0	0
de Groot	sm	2	2	0	1	1
Heijnen	sm	0	0	0	5	4
Huis in 't Veld	co	1	1	0	0	1
Leenaarts	ee	0	0	1	0	0
Lok	me	2	0	0	0	0
Nienhuis	mb	168	25	0	4	18
van Oorschot	co	8	1	0	2	4
Saeijs	ec	29	5	0	3	7
Spaargaren	gt	1	0	0	1	0
Stelling	he	12	5	0	4	1
Verruijt	sm	109	5	5	12	17
Vos	ci	2	1	0	4	2
Vrijling	pd	5	0	0	2	2

Table 2. Citation and publication counts taken from the Science Citation Index (1 and 2) and publication counts taken from the bibliographic databases INSPEC (3), COMPENDEX (4) and PASCAL (5), for some of the contributors to the Storm-Surge Barrier in the Eastern Scheldt. Names of principal contributors are underlined. Fld. indicates the fields in which the contributions were made: pd = probabilistic design, co = coastal engineering, sh = soil hydraulics, ci = civil engineering, pm = program management, sm = soil mechanics, ee = electrical engineering, me = mechanical engineering, mb = marine biology, ec = ecology, gt = geotextiles, he = hydraulic engineering.

The reason for the bibliometric invisibility of the technologists cannot be attributed to lack of written material, but is to be sought in the nature of the documents. Numerous reports were produced in-house in many of the organisations involved in the project. Although such reports are accessible, they are not part of the open literature. Many publications are written in Dutch, which severely limits the number of potential readers. And last but not least, many of the written documents we traced, did not even bear the names of the authors. Here we come up against important cultural differences between scientists and technologists.

5. SCIENCE AND TECHNOLOGY

Scientists and technologists often work on the same subjects, but they do so in different cultural domains. In technology, credit and reward do not derive from publications. Thus, there is very little inducement to write papers for scientific journals. There may be in-house reports, large numbers of them in some cases. In one of the institutes which contributed its expertise in soil research to the Eastern Scheldt project, we found a basement-room full of such reports. Krige[21] told us that at CERN they keep about five metres of documentation on the development of one particular detector.

Technologists, then, have other communication patterns. Although much remains to be done, several points have emerged from our interviews about the methods which technologists use to assess their fellow researchers. Conferences play an important role in the exchange of information; to be an invited speaker is considered to be a sign of recognition as is an invitation to become a member of an international professional association. And, of course, to have taken part in a successful innovative project is also considered to be an asset. This suggests that sources other than paper publications must be used to evaluate technologists and their work. In some fields, patents may be used. Or we can listen to technologists talking about their work; if we do, we recognise other forms of (citation-like) referencing; they refer to other instruments and say 'we built it like ..". Other sources of information were suggested during the discussion: inventories of instruments (Rocher) and a comparison of (references in) the curricula vitae of scientists and engineers (Kruytbosch) might yield interesting results.

6. CONCLUSIONS

Information is fashionable. We read about information management, information planning, information systems, etc.; complaints are often heard about the flood of information. Indeed, information plays such an overpowering role that it has been suggested that our society should be called the information-society.

In all these expressions, information consists of words and numbers, or is in computer readable form, i.e. reduced to bits, to zeros and ones. However, everything we see contains information, not only what we can read. The description of an instrument, if the description is complete, contains the same information as the instrument itself; both are representations of the same idea. Likewise, the same information is to be found in a plant and in the DNA in one of its cells. But, different skills are needed to extract the information. Reading is only one such skill (to understand a scientific article, one needs more skills than just reading; background knowledge is also essential). While studying a subject, one automatically also acquires the skill needed to 'read' the information as it is coded in that field. Thus, an engineer will learn to 'read' instruments. He will also come to consider an instrument the proper form to publicize the results of his work.

Such communication channels deserve greater appreciation. If we were to include them in our study of the history of science, we would find a confirmation of Casimir's science-technology spiral[22] as well as of Price's theory that advances in instrumentation and experimental techniques (what he calls instrumentalities) have been of major importance in stimulating and permitting both radical theoretical advances in fundamental science and radical innovations in practical application[23]. We would thereby obtain a better understanding of the role played by technology in the course of history.

Unfortunately, the increased use of 'science indicators' has led to renewed emphasis on written documents as pointers to developments in science. As a result, other important evidence of advances in knowledge is ignored. Instrumentation is an important source of information.

7. NOTES

1. White, Jr., Lynn. Medieval Technology and Social Change. Oxford: Clarendon Press, 1962. White, Jr., Lynn. Technology assessment from the stance of a medieval historian. The American Historical Review 79-1 (1974) 1. White, Jr., Lynn. Medieval Religion and Technology. Collected Essays. Berkeley: University of California Press, 1978.

2. Gimpel, Jean. The Mediaeval Machine. The Industrial Revolution of the Middle Ages. Harmondsworth: Penguin, 1976 (La Révolution Industrielle du Moyen Age).

3. In Dutch/German the term 'science' (wetenschap) also includes arts and humanities.

4. Broeder. J.J. (ed.) Zomerschool: De wetenschappen in relatie tot de overheid in het verleden en nu. Utrecht: STW en FOM, 1988.

5. Snow, C.P. The Two Cultures and a Second Look. Cambridge University Press, 1974.

6. See also Pair, C. le. Limits of science through erosion of information. Paper presented at a colloquium of the Gesellschaft für Wissenschaftsforschung of the Academy of Sciences, Berlin, January 1991.

7. Mark, H. The aim of national research facilities, past and future. In: The management of Science. Relation to Industrial and National Needs. North-Holland, 1982.

8. Devries, Kelly R. A 1445 reference to shipboard artillery. Technology and Culture 31 (1990) 818.

9. Kinder, H. and W. Hilgemann (eds.) Sesam Atlas bij de Wereldgeschiedenis. Kaarten en Chronologisch Overzicht. Deel 1. Van Prehistorie tot Franse Revolutie. Apeldoorn: Van Walraven, 1989. (Cartography by H. and R. Bukor. Translation from German by J.M. Vreugedenhil).

10. Marx, Robert F. De verovering van de Zilvervloot: het verhaal over Piet Hein. Baarn: Hollandia, 1986. ISBN 90-6045-468-5. (The capture of the treasure fleet: the story of Piet Heyn. New York: McKay, 1977.).

11. Mukerji, C. Scientific techniques and learning: Laboratory "signatures" and the practice of oceanography. This volume.

12. Casimir, H.B.G. Industries and academic freedom. Research Policy 1 (1971/1972) 3.

13. Price, Derek de S. The science/technology relationship, the craft of experimental science, and policy for the improvement of high technology innovation. Research Policy 13 (1984) 3.

14. Cahan, D. The origins and early use of the ultramicroscope: between theoretical science and industry. Paper presented at the SPIE Institute on 'Instruments and Institutions: Making History Today', 12-14 April 1991, London, England.

15. Rabkin, Y.M. Uses and images of instruments in chemistry. This volume.

16. See, e.g., Schaffer, S. Late Victorian metrology and its instrumentation; this volume. Y. Rabkin points to the role of popular science in the dissemination of instruments; in addition, he refers to instrument collectors.

17. Hamilton, D.P. Publishing by - and for? - the numbers. Science 250 (1990) 1331.

18. Pair, C. le. The citation gap of applicable science. In: A.J.F. van Raan (ed.) Handbook of Quantitative Studies of Science and Technology. Elsevier, Amsterdam, 1988.

19. Els, W.P. van, C.N.M. Jansz, and C. le Pair. The citation gap between printed and instrumental output of technological research: the case of the electron microscope. Scientometrics 17 (1989) 415.

20. Jansz, C.N.M., and C. le Pair. Bibliometric invisibility of technological advances. Paper presented at the International Conference on 'Science and Technology Indicators', Bielefeld (Germany), 10 - 12 June 1990. To be published.

21. Krige, J. private communication.

22. Casimir, 1971/72

23. Price, 1984

Part II
Modern Science

Introduction to Part II

Moving firmly into the era of post-War science, the papers in this section begin to reveal patterns in the contemporary origins, diffusion, and consequences of instrumentation. These patterns appear despite the quite different goals of the authors in the individual articles, some of which postulate broad pattern changes, some of which make theoretical points for the analysis of science, and some of which reflect the more technical orientation of instrument innovators.

The patterns which appeared in the last section are again reflected in some of these papers. For example, the importance of a material and technical culture shared between research laboratories and industrial practice appears in the arguments surrounding the construction of detectors to be used at CERN, the European high energy physics facility, as described by Krige. In this case, national communities are aware that the activity of detector construction will both draw on and build a local base of know-how, and they therefore argue for construction "at home" in member countries, rather than centralized at CERN. The shared technical base appears again in a form which is only slightly modified in Mukerji's discussion of ocean science and in Edge's presentation of mosaic array astronomy. In both these cases, research benefits from advanced technologies developed for the military; but there are also drawbacks. Scientists have only limited access to those technologies, Mukerji points out, with the rules set by the military sponsors. And in the mosaic array case, the specifications of the technology itself are more or less set by the military market, and only minor modifications are affordable for the few, marginal research users.

The two military-related cases thus draw attention, as Rabkin's paper in the last section did, to one of origins of contemporary research instruments, namely technologies used in other areas of practice. In their papers on nuclear magnetic resonance techniques, Feeney and Morris provide glimpses into a different but related route for instrument innovation. In this instance, fundamental knowledge of physical processes sparked the idea for a measurement technique, which then coevolved with increasing understanding of the basic physics involved and tinkering with the other parts of this complex, system technology, in particular computers and magnets. In the NMR case, a unique sort of research network developed around the technology, a network which encouraged interplay between fundamental analytical research, tinkering, and systems integration, and which allowed for interaction among the specialized forms of the instrument which were developing in response to differentiation in possible applications. The network knew no institutional boundaries: research laboratories, hospitals, manufacturing firms, and instrument companies were all sites of participation. The dynamics of this kind of network have received remarkably little attention in either science studies or science policy research, despite its obvious and crucial role in broad areas of innovation.

More typical for science studies is the focus on the ways laboratories incorporate new instruments. The social analysts represented in this section describe a wide range of forms of adoption, from creative tinkering to perfunctory consumption. Mukerji makes the creativity point most strongly in her description of ocean scientists as using instruments to explore at the edges of their cultural knowledge, as children use toys to create meaning in the world around them. The active element of experimental knowledge—its intervention in the research material—is another mark of the process of constructing meaning. The x-ray astronomers Edge describes can likewise be seen as deriving intrinsic satisfaction from pushing the edges of both their techniques and their knowledge of the universe. Each of these authors also makes a point about the individuation of laboratories through instruments. Competition in science stays in balance with the communal task of knowledge production through the formation of research niches, or in Mukerji's term, "laboratory signatures." Instrumental configurations play perhaps the most important role in that differentiation.

Gökalp et al. and Blume describe more predictable diffusion processes for more standardized equipment. Gökalp et al. show how quickly laser Doppler anemometry replaced earlier hot-wire techniques for measuring velocity in turbulent reacting flows, because the clear technical superiority of the new method made the investments in skill worthwhile, especially for new researchers. In contrast, in pure turbulence studies (where no combustion was occurring), hot wire retained some advantages, and the adoption was slower and more prone to obstruction by factors such as the existing skills of technicians who could not be replaced. Blume attributes a large part of the diffusion of a new instrument, at least after the initial stages of its introduction, to a dynamic of conspicuous consumption: "Though the first organizations...to introduce a particular innovation may be concerned with the pursuit of formative goals...followers may be much more concerned with maintaining their standing in the organizational field." Hospitals, for instance, "need to be able to display the most glittering possible array of instrumentation." These patterns of simple adoption, rather than delicate adaptation to a carefully-chosen research niche, supplement rather than supplant Mukerji's and Edge's findings. Many diffusion dynamics are underway at once in the complicated world of research.

The whole differentiated pattern of diffusion and adoption, however, may shift in character over time as the characteristics of the technology base itself change, and as social forces work to reinforce one or another part of the distribution. Several of the authors, for instance, mention the tremendous impact of cheap computing power on the instruments they describe. With little imagination, one can read between their lines to find the major changes in laboratory organization that went with the shift from mainframe and mini-computers to today's powerful micros. Gökalp et al. also mention the laser revolution, which opened new vistas in access and imaging power. Such changes sweep across experimental science, altering adoption costs and social interactions in their wake. Many of the shifts in instrumentation are more abstract, however, like the factor of scale in Krige's account of detectors at CERN. Most abstract, but also perhaps most intriguing, is Blume's suggestion that "standardization" is on the increase, in science as in clinical practice. Standardized research is more planable, fundable, overseeable; thus standardized, "black box" instruments may be gradually replacing "sealing wax and string" as the dominant mode of laboratory equipment. Instrument manufacturers would surely welcome this trend, but its implications for innovation networks like the ones reflected in Feeney's and Morris's papers on NMR should not be welcome to policy makers.

Finally, while changes in the resource environment of research can shape instrument choices (as Blume argues), shifts in instrument form can also be linked to changes in the organization of science itself. As Krige shows, scale settled the organizational issues surrounding detectors at CERN by closing off some of the options. Gökalp et al. describe both macro and micro reorganization in combustion research that follows from the introduction of laser Doppler anemometry: the establishment of an autonomous domain of turbulent combustion research at the macro level, and model-unifying, place-unifying effects at the micro-level. Indeed, the increasing and changing use of research instruments may even be linked to the appearance and prominence, but surely at least size, of the kinds of instrument-based research networks reflected in Feeney's and Morris's discussions of NMR.

The science studies literature on instruments has focused almost exclusively on the role of instruments in individual laboratories and in the knowledge-production process. The combined message of the case studies in this section is that there is just as much to be learned about contemporary science by examining the connections between instruments, institutions, and broader-scale patterns of social organization in research.

-SEC

Chapter 5

Whatever happened to the string and sealing wax?

Stuart Blume
University of Amsterdam (Netherlands)

ABSTRACT

Looking back over their professional lives, eminent scientists often contrast current laboratory practices with those of their own early years. 'In those days,' they tell us, 'it was usual to build one's own equipment. Today no on starts work without the most advanced equipment money can buy.' This essay takes this observation as its starting point, and suggests that it results from two rather different kinds of phenomena.

2. EXPERIMENTATION AND THE LIFE OF SCIENCE

What do scientists have to tell us when they reflect upon the passage of their lives? To write an autobiography is typically to try to explain one's development, whether inward or outward: to show how one has become what one believes oneself to be. On the whole scientists do not do this very well, as S.E. Luria reminds us in the introduction to his own book[1]. The difference between the autobiographies which come to be regarded as literary classics and those soon forgotten is not a consequence of the momentousness of the events through which the writer has lived. It has a great deal to do with the writer's ability to interpret the mass of detail in terms of a "master idea" with its own "momentous dignity"[2]. Luria believes that the difficulty for scientists is partly that the science of the past does not make for a powerful narrative: only science in its creation, the process of discovery, does that. What is this process of discovery? What has to be described?

> "Did the scientist use a screwdriver, or break a test tube at a critical moment, or find crystals in a solution? And how did these events lead to discovery?"[3]

The scientific autobiographer necessarily works under certain particular constraints. Most scientists who write their autobiographies assume that public interest in what they have to say derives from the discoveries with which they have made their names. These discoveries are embedded in the histories of the disciplines in which their authors worked. The histories are collective achievements, for science is a collective endeavour. What one did has to be representative of what others did, whilst in some way worthy of unique attention

> "By going from fundamental research into industrial research management, I exemplified as a person what happens to ideas all the time. Nearly all the innovations of modern technology came into being thanks to preceding fundamental research. On the other hand modern experimental science could not exist without the tools made available by technology. I believe that I have been in a good position to observe this powerful science-technology spiral"[4]

One possible reading of any scientific autobiography is as a participant's contribution to the historiography of the discipline. This imposes a certain constraint on the autobiographer, for what he or she has to say must fit the facts as generally acknowledged, and can only attempt marginally to 'set the record straight'.

There is a second important constraint. What is written must accord with the notion of science as <u>progressive</u>. This is unlikely to be experienced as a constraint. Unless the writer has been scientifically ruined, his or her achievements dramatically consigned to oblivion, the conventional view of science as inherently progressive will almost inevitably permeate the account[5]. Not only can one not say that 'we knew more in those days', or that 'the good old theories were better', but one may very well try to show the part one has oneself played in making progress happen. Progress, needless to say, implies theoretical progress...

The ideology of science, powerfully enunciated within the dominant philosophy of science tradition, specifically concerns scientific <u>theory</u>. It is in reflecting on the theoretical development of science which they have witnessed, or to which they have contributed, that scientists are most particularly constrained in what they can say. This is much less the case with regard to those elements of the practice of science which philosophers have tended to ignore... Luria's screwdrivers or broken test tubes. The material culture of science, precisely because it was ignored by philosophers, has escaped incorporation into this dominant ideology. Indeed it is only recently that historians and sociologists of science have begun to try to develop an analytic vocabulary for discussing this material culture. So when they come to reflect on the practice of science, on their own experimental activities, we might well imagine that scientists use a rather different kind of language.

Experimental scientists are well aware of the definitive significance for their work of instrumentation, and the ways in which it is used[6]. The new interest of historians and sociologists of science in studying the material culture finds constant resonance in natural scientists' own reflections and musings. In this essay I want to argue that these reflections attest also to something else. Consider the eminent British physiologist Sir Alan Hodgkin, for example, reflecting on changing experimental practice.

> "In those days laboratory life was rather informal, at any rate in Cambridge. I never worked for a Ph.D. and didn't have a research supervisor. You might easily start in a bare room and have to build most of your equipment yourself, apart

> from a few standard bits like smoked drums, Palmer stands and kymographs. This sounds depressing but it actually wasn't. Nowadays, scientists are apt to become neurotic and give up work if they know their equipment is markedly inferior to other people's. Indeed it is regarded as somewhat unscientific to carry out experiments with anything but the best equipment. This certainly wasn't my feeling when I started research and, to begin with, all that mattered was that one would have enough equipment to do something new...the general attitude to equipment was certainly very different from that existing today."[7]

If one were systematically to peruse memoirs, or were to ask other experimental scientists, certainly British ones and of a certain age, about the changes in science which they had remarked, I believe that many would refer similarly to a change in **experimental culture**. It might take the form of a rather wistful reference to pride in traditional British experimental **ingenuity**...in putting together a simple but elegant experiment with modest components. Accounts are typically suffused with nostalgia for some golden age of science, now apparently passed. But why? Why should scientists write nostalgically about the way in which they used to conduct experiments? I am going to suggest that there are two kinds of reason.

When we are being nostalgic, we are not trying objectively to compare two states of the world, even if it might feel like that. It's much more a symbolic expression, loaded with affective value: call it 'a celebration of lost values' if you like. What we are doing is generating a contrast between events from our own lived pasts and the "events, moods, descriptions" of our present circumstances[8]. Nostalgia is typically associated with reflexion upon the course of one's life: with "the subjective appreciation of transitions" associated with the life cycle

Perhaps then this sense of a richer, more authentically demanding way of doing science - now lost - is to do with the progression of the scientific life cycle. Reflecting upon the early stages of a scientific career from the vantage point of later stages, this 'loss' is one of which scientists might be acutely aware. There is something to be said for this view, and it is noteworthy that nostalgia for a lost age of experimentation is not a purely contemporary phenomenon.

Here are two examples from the early 1950s

> "What is important, and a change in kind, is that users of instruments are increasingly not their masters. Once, any physiologist could tinker a kymograph into good behavior and even make or have one made in the shop in the basement. Few today dare open the crinkle-finish black boxes purchased from some "radio" firm, and even of those who do, a small number indeed could carry on without the services of an expert electronics engineer. This may be unfortunate, but it is certainly inevitable"[9]

> "It was therefore perhaps natural that one began to attempt the more detailed analysis of more or less known functions. For this purpose refined, specific methods frequently were needed and developed. Many of them were extremely complicated and long training was necessary to become familiar with them. It appears that these circumstances, together with an apparent general tendency of our time to worship methods and gadgets, has quite frequently placed emphasis on technical procedures, and diverted interest away from physiological problems. This has gone so far that sometimes one has the impression that in contrast with former times, when one searched for methods in order to solve a problem, frequently nowadays workers look for problems with which they can exploit some special technique"[10]

If we pursue this line of thought then the question which immediately follows is why these nostalgic feelings should express themselves in relation to experimentation? There seem to be two necessary circumstances for this to happen. First, experimentation, doing things, building things, fiddling, would have to be in some sense constitutive of the period of their careers which elderly successful scientists looking back see as formative, as having made them what they are[11].

A few years ago this would have surely seemed a pretty strange idea to scholars who think about science. If anything could be viewed as 'constitutive' of the scientific career, surely it would have had to have been the theoretical perspective within which one was reared. What other significant measure of change could there be than the contrast between the outdated theories of one's youth and those now current? Curiously enough, when scientists do reflect on the outdated theories with which they might have been brought up, their reflections don't seem to result in nostalgia.

Learning to do things with instruments, to manipulate them, to judge the significance of the data they yield, to deploy them in experiment...these things are of the essence of the scientific apprenticeship. The mature scientist, advanced in his career, spends his time doing very different things... perhaps for many the nearest they get is vicarious experimentation. They design experiments, interpret results, possibly theorize, but with rare exceptions do not themselves engage in 'hands on' experimental practice[12]. Though little work has been done on the subjective side of the scientific career[13], it seems to me that this disengagement from the material world of science, this transition inherent in the kind of scientific careers generally deemed successful, must certainly contribute to the feelings of nostalgia.

3. THE CHANGING MATERIAL PRACTICE OF SCIENCE

Scientists' references to the changing material culture of science are not a consequence of life cycle effects alone. Mixed with nostalgia is an awareness of changes which can also be identified in other ways. Instruments are becoming more powerful.

New techniques are introduced into the laboratory, many of which spread rapidly. For example, the first demonstration of the nuclear magnetic resonance effect took place at the end of 1945, performed simultaneously by Bloch at Stanford and Purcell at Harvard. Subsequently, a number of Bloch's collaborators joined Varian Associates, which had close ties with Stanford. Varian built its first NMR spectrometer for sale in 1952. Between 1958 and 1964 the growth in laboratory use of NMR was estimated at 40% per annum. This is not unique. At the same time, use of infra-red spectroscopy was estimated to be growing by 33% annually and of gas chromatography by 40-45%[14]. To have spent those years in the chemistry laboratory was surely to have witnessed a major transformation in the material practice of the science.

How can we capture these changes analytically? How can we describe what is happening other than by listing innovations or plotting increases in available energy levels, resolving power or field strengths?

One concept which was intended to do something like this, and which was introduced into the science policy discussion of the late 1960s, was the concept of 'sophistication'. This sought to express the changing instrumental requirements of research in terms of cost.

Two British studies of the late '60s examined the per capita expenditure on scientific equipment at the level of the research performing unit. "...it was hoped that a study of the 'sophistication factor' would produce evidence relevant to the widely held notion that the cost of scientific research keeps growing in real terms, because the equipment required is increasingly complex and expensive"... The first study examined the budgets of 13 government research establishments and 3 university departments, covering almost all fields of S&T, over a 10 year period around 1960. The question posed was: does the real cost (net of inflation) of maintaining a scientist at work change over time? The study found that it did: that the rate of increase in expenditure on equipment in real terms varied between 0 and 20% for the various institutions. A subsequent study looked at 69 different university deparmtnets classified now by groups of science (biological/engineering...), corrected for equipment allocated for teaching, and tried various 'per capita' measures. The conclusions of the study were that although fields differed considerably in their unit costs, growth rates in real terms were similar and large : around 9-10% per annum.[15]

Through the 1960s scientists had become accustomed to rapid rates of growth in the funds available for research: a life of luxury. By the end of the 1960s these growth rates were levelling off, and spokesmen for science viewed the future with some alarm. If it could be demonstrated that for research to remain of high quality a certain level of investment was absolutely essential a powerful argument could be deployed in negotation with Ministries of Finance. It was for this reason that the Council for Scientific Policy, which advised the British Secretary of State for Education and Science on the allocation of resources for research, commissioned these two studies of sophistication.

However, even as the second study was being completed its utility had been undermined. As Sir Frederick Dainton rather regretfully pointed out in his introduction to the second study, the trouble with taking past expenditure as a guide is that what is measured is just that, the amount of money which happened to have been made available for instrumentation. The concept hardly stood up as a measure of anything fundamental in the material practice of science[16]. Moreover, as Galison has emphasized, charting growth in the scale of instrumentation (whether or not expressed in money terms) tells us little about the consequent changes in the practice of science[17]. That, after all, is what scientific recollections like that of Hodgkin refer to. In order to get to grips with the relationship between instrumental change and changes in 'doing research' we need some means of analyzing what experimenting entails. We need a theoretically grounded conceptual vocabulary. Here Pickering's work in the sociology of science seems useful[18].

Roughly summarized, his argument is as follows. Scientific work demands all sorts of resources for its performance; different scientists have different access to the necessary resources; "resources to hand are opportunistically assembled as contexts for constructive work are perceived". Research groups and individual scientists formulate the goals of their research in relation to their perceptions of the opportunities open to them. Moving on in his later work to a finer grained analysis of what scientific practice oriented to goals so-formulated looks like, Pickering emphasizes the attempt to produce coherence between material practices, instrumental theories and phenomenological theories... trial and error

In experimenting, in trying to generate what colleagues will regard as plausible knowledge claims (Pickering speaks simply of 'facts'), the scientist is seeking coherence between his material practice (what he does: setting up the apparatus, running it...), his instrumental model (his understanding of how the apparatus works: the understanding that goes into the design of the experiment); and his phenomenal model (his understanding of the part of the world under investigation: the relevant theory in the sense the word is generally used, shared with his professional community). Pickering charts the conceptualisation of an experimental procedure (to do with the hunt for quarks) (instrumental model/material procedure) on the basis of particular theory...its realization...and response to failure: to 'resistance' of the material world. The scientist can respond in various ways when his experiment doesn't work out quite as he had anticipated. He may modify his instrumental model (for example by introducing extra terms in the equations), he may modify his material practice (for example by varying some dimension which had hitherto been fixed)...sometimes dramatically through the design of an wholly different experimental set up. Modification to the phenomenal model would be attempted rarely: Pickering refers to such relatively stable elements in scientific work as 'fixed points'.

Resistances can arise not only in the material world, but in the conceptual and social worlds too.

> "Scientists' engagement with the material world is never unmediated and direct: it has always to be understood as conditioned by the culture in which it takes place, and as react ing back upon the culture. Morpurgo's repeated revisisons of his interpretative model of his apparatus - conceptual rather than material tunings - can stand as an emblem of the latter...one has to think of a complex process of reciprocal and interdependent tunings and refigurings of material procedures, interpretations and theories...Further, one needs to add to the picutre a concomitant tuning and refiguring of social entities and their relations...work-styles, institutional forms, and more are all progressively implicated in this process as one examines higher and higher levels of aggregation..."

It is important to see the connexions between the broader context of opportunities to which Pickering alludes on the one hand, and the details of experimental practice on the other. Consider what is involved in assembling the space, the research team, the regular supplies of electricity, liquid helium, genetically pure rats, time on the supercomputer or accelerator beam, volunteer patients, which research may involve, not to mention assembling or building the necessary apparatus.

Experimental scientists themselves are acutely aware of these kinds of problems, inherent in their activities. The point I want to stress is that the material practice of science (establishing the scope and nature of a series of experiments, choosing or designing the instrumentation to be used, making the observations) and the kind of negotiations which experimentation involves (with professional peers, laboratory heads, funding bodies of various kinds) are inter-dependent[19]. Mukerji's work on oceanography shows this with particular clarity. In this field the most advanced technologies by means of which scientists can identify and approach their research sites, collect precise data, process the data are typically developed by the military for military purposes and in secret. Civilian scientists have to negotiate access to the equipment they covet...in so far as they are informed of its existence. This is an extreme case, but the phenomenon to which it points is a general one. In the research that I have been studying - research related to innovative techniques of diagnostic imaging - the experiments carried out can only be understood in relation to negotiations involving funding bodies, industry, the medical profession[20].

Consider first research done on the medical applications of nuclear magnetic resonance: a technique to which I referred earlier. In the early 1970s it was shown that, in principle, it ought to be possible to modify the standard NMR instrument in such a way that new kinds of 'representations' (images) of 3 dimensional structures could be formed. Very soon, on the basis of earlier biochemical work, a hope emerged that cancerous and non-cancerous tissue, thought to have different NMR parameters, would show up differently. By 1976 there had been considerable successes. One of the groups, that of Raymond Damadian in New York, published an NMR image of the thorax of a live mouse in Science. Many were starting to think of clinical applications: in effect, of finding a way of imaging the whole human body. The problems involved were substantial. They included constructing an apparatus around a magnet with a hole in it large enough to accomodate a human being. In Britain the academic groups involved at this stage (two in physics, one in medical physics) were able to convince the Medical Research Council to support their work. Each was able to commission construction of the very large magnets they needed from a firm specialised in magnet technology. Damadian, a physician and biophysicist, whose ideas partly inspired the whole enterprise, could not raise the money. The only possibility, he decided, was to build a magnet themselves. One of Damadian's collaborators has described how the laboratory's programme of work had to be transformed.

> "Shortage of funds as well as the desire to make the strongest magnet possible in the light of the claims of theoretical impossibility, he decided to build his own magnet, a five thousand gauss superconducting magnet with a 53" room temperature bore...Damadian had never built a magnet before, let alone a 53" liquid helium cooled magnet and dewar to be built from scratch inside a medical hospital [near] a medical ward...Designing the magnet and performing the computer calculations for the Niobium Titanium solenoid himself, he laid out the basic design for the dewar and we got expert help...Dr Damadian and I spent the better part of a year, mostly through the day and late into the night (typically 4.00 A.M.) welding and constructing the dewar and ferreting out leaks. My job was to supervise the machining of all magnet and dewar parts in the machine shop that Dr.Damadian put into his laboratory and to become proficient at the heliarc welding fast so we could build the dewar. He designed the magnet to provide the shaped magnetic field needed to perform the focused scan of his 1972 invention"[21].

The second example is more complex. In the early 1950s various doctors and some engineers and physicists began pointing ultrasonic devices into the human body, to see if they could learn something of the internal structure from the pattern of returned echoes. Ultrasonic devices were widely available by then: they had been developed in the war, both for locating submarines ('sonar') and for detecting flaws in large metal structures. The kinds of experiments the various physicians conducted depended upon the concerns of their specialist practices, and these differed greatly from one to the other. A brain surgeon at the University of Lund, who wanted to see which way the the mid-line of the brain had been shifted by a severe blow to the head, obtained extremely useful information. He acted on it, operated accordingly, and published his account. The professor of obstetrics and gynaecology in Glasgow managed to get hold of a similar device

> "By good fortune, the wife of an eminent engineer at this time gratefully survived a hysterectomy at my hands and introduced me to her husband. Following a lunch

> party with their research directors a session was arranged in the research department of Messrs Babcock & Wilcox. I shall always remember that hot sunny afternoon of 21 July 1955 when we took to the factory some selections of the last few days operating, in the boots of two cars, large ovarian cysts and uterine fibroids...The firm very thoughtfully provided a truly massive piece of prime steak as a control material.
>
> All I wanted to know, and this was surely not asking too much, was whether a metal flaw detector could show me... the difference between a cyst and a myoma and so forth. To my surprise and delight the differences were exactly as my reading had led me to expect" [22]

Later, using the same apparatus to see if he could characterize the contents of the pregnant uterus, the results were disappointing. The echoes were far too complex to interpret. Further work was possible only thanks to a collaboration with a research engineer who worked for the firm manufacturing the flaw detector. They set about redesigning the apparatus. A cancer surgeon, John Wild, working in Minneapolis, used a somewhat different device (obtained from the Air Force) to try to distinguish cancerous from normal tissue in living patients. Although his early work was rather successful Wild's applications for further National Institute of Health (NIH) funding soon began to be turned down. Koch has recently explored the conflict which developed, and her conclusions are very relevant to the argument I am trying to make. Koch found that Wild's pragmatic style of experimentation - moving back and forth between the patients and the prototypes his engineer-collaborators were developing - clashed with then-dominant beliefs about how biomedical research should be done

> "Wild's insistence that research on clinical applications could proceed in advance of an understanding of the phyical characteristics of ultrasound went completely contrary to the tenets held by his biomedically and physically trained reviewers. When Wild dismissed the value of working on animals in the laboratory, in advance of clinical testing on humans, he attacked the very foundation of American medical research and education" [23].

Koch shows how NIH and its reviewers tried - without success - to get Wild to change his approach: to do different kinds of studies. (He wouldn't, and his funding was cut off.)

I hope that these examples show that the experimental practice of the scientists has to be understood in terms of the priorities of their medical specialisms, the collaborations they were able to negotiate with skilled engineers, their success at persuading funding bodies of the likely medical utility of the devices they were working on, and so on.

If we try to characterize this experimental practice, we can see that what is entailed certainly <u>includes</u> a number of distinct dimensions

* Demarcating the project

What is going to be done, and what bracketed, ignored or held constant? For example, what parts of the necessary instrumentation will be purchased or borrowed, and what built? Will the building be viewed as part of the research (in the sense of reflected upon, parameters allowed to vary), or will it be viewed as a preparation for the research? Are safety studies to be done, or are they to be deemed outside the group's expertise and left to others?

* Defining the internal structure of the project

How are sub projects and their end-points defined? For example, on the basis of what considerations do shifts from studies of phantoms - possibly via in vitro tissue studies, possibly via animal studies - to the study of live humans take place?

* Assembling the necessary skills

How are the skills which the planned experiments need acquired: through trial and error/apprenticeship/practice; through collaboration with another group possessing the necessary skills; or through recruitment?

With the aid of these categories we could begin to analyse changes in the material culture of science, which was of course our starting point. In particular, we can return to the suggestion made by Professor Hodgkin that scientists are increasingly unwilling to start from scratch, that they prefer to begin research only when they have assembled -ready made - all possible apparatus in the form of gleaming black boxes. Why might this be? In the final part of this essay I am going to suggest what I think lies behind Professor Hodgkin's observation.

4. THE COMMODIFICATION OF SCIENCE

This term was inspired by a book published a few years ago called Science as a Commodity[24]. That book was concerned with whether or not the rules governing the interactions between scientists had been so changed by things like military secrecy, industrial sponsorship and the like that one could no longer speak of the existence of a community of scientists in the way one used to. I want to use the term slightly differently to suggest that the acquisition, and to some degree the significance of instrumentation has also changed. In essence my point is that the significance of advanced research equipment for a research group is partly social. In other words the purely cognitive terminology experimentalists use to justify their need to replace their MkI instrument by the Mk II - in terms of the need to get to higher energy levels, higher resolution, and so on - will not suffice. The outline of an analysis which I want to give is very different from the 'sophistication' concept, for that rests ultimately upon the 'cognitive needs' argument.

First, I believe that the pressures under which research laboratories today operate, with an emphasis on efficiency, selling services, and so on make increasing recourse to standardized equipment a wise strategy. A related point has been made by Fujimura, who studied a line of cancer research. She found that scientists were very well aware of advantages attaching to the construction of what she called 'standard packages'. These advantages were of various kinds (even apart from the profit which the originators could obtain). For example, it opens up problem areas to laboratories which did not have experimental skills previously required

> "For example, the development of gel density gradient electrophoresis systems (a physical apparatus and a series of tasks) in molecular biology made it possible for researchers in many laboratories to participate in biological problems which required skills in DNA separation and measurement. Slow and difficult DNA measurement techniques were simplified, standardized and packaged. These packages were then easily disseminated to many laboratories...Problems whcih required such techniques were thus more doable in those laboratories"[25]

At the same time, by pre-structuring as much of the research as possible, more discrete and overseeable sub-projects can be formulated. Under present day circumstances this has much to recommend it. For example, it makes research more easily plannable; the different parts can be submitted seperately for funding if necessary; partial projects make for a stream of publications; PhD work can be better and apparently more efficiently organized.

Standardization also facilitates the deployment of research skills in the provision of services. We are all under pressure to provide services (problem-solving, consultancy, policy analysis) for economic and social institutions. It seems to me that the distinction between research and services is going to be harder and harder to sustain. To understand the implications for the performance of research, think of the field of medicine.

In the modern university-affiliated hospital research and patient care are intermingled: there is often no clear boundary between the two[26]. We can see this in many ways: the way in which the structure of care shapes clinical investigation; the way in which concerns of local practice set research agendas; the content of the literature of clinical research; the ethical dilemmas which arise when the dictates of one and of the other point in opposite directions...Patients want to be reassured that the care they are receiving is the best possible: today this has typically come to mean based on the most advanced possible techniques

In medicine, because hospitals are competing in their provision of care, and because of the intermingling of research and service, imperatives of the care system impact on research. Let us take this reasoning further by considering its implications for the acquisition of new instrumentation.

Evidence drawn from a wide variety of organisations suggests that when organisations innovate - for example, introduce a new procedure or technology - they may not only be concerned with the organisation's overt goals. There is a convincing argument in the sociological literature[27] to the effect that though the first organizations in a given area of activity to introduce a particular innovation may be concerned with pursuit of formal goals (of efficiency, productivity, quality of service...) followers may be much more concerned with maintaining their standing in the organisational field. I can illustrate the argument with an example from the area of acquisition of new technology by hospitals. The first hospitals to acquire a fancy new piece of equipment - the CT scanner, or MRI for example - or to introduce a new technique - like transplanation surgery - seem to base their decisions on some assessment of the quality of the care they provide, and the possibilities which the new technique seems to offer[28]. Gradually, as increasing numbers of hospitals acquire the technology, assessments of these kinds become less important in their decision making. Instead, hospital decision makers (including both administrators and senior specialists) focus on the fact that other hospitals already have the technique in question. The social logic is simple "if they, then us"...in order to maintain their status (or in the competitive American system <u>compete</u> for the best physicians, patients, resaerch grants) hospitals need to be able to display the most glittering possible array of instrumentation.

I am arguing two things. First, that universities increasingly find themselves in a quasi-market similar to that in which hospitals (increasingly) operate: ie that the competitive selling of education and research services is coming to dominate. Second, that as a result the possession of sophisticated instrumentation becomes important not only because of the scientific problems to which it can be applied, but by reference to the competitive rules of this quasi-market. Of course the two are not wholly independent, but I want to stress the importance of possession - not use - for status within the group of similar but competing institutions. The significance of acquiring the most advanced and powerful instrumentation is to be understood in relation to the need to enhance status within the 'organizational field' and thereby to maintain one's competitive position in the market. I can't prove all this. Anecdotal evidence from Eastern Europe, where lack of modern instrumentation is legion, does attest to its plausibility. For example, I have heard of good scientists whose work has apparently been rejected by journals not by virtue of any inherent lack of interest, or flaw, but because the instrumentation used was regarded as outdated.

One scientist in his memoirs has reflected, with some bitterness, precisely on this broader transformation in the character of scientific practice. Edwin Chargaff, a major figure in 20th century biochemistry, wrote

"When I was young, I was required - and it was easy - to go back to the origins of our science. The bibliographies of chemical and biological papers often included references to work done forty or fifty years earlier. One felt oneself part of a gently growing tradition, growing at a rate that the human mind could encompass, vanishing at a rate that it could apprehend. Now, however, in our miserable mass scientific society, nearly all discoveries are born dead; papers are tokens in a power game, evanescent reflections on the screen of spectator sport, news items that do not outlive the day on which they appeared. Our sciences have become forcing houses for a market that in realtiy does not exist, creating, with the concomitant break in tradition, a truly Babylonian confusion of mind and language"[29].

Chargaff, romantically and unrealistically, pleads for a return to the science he knew in the 1930s and the 1940s, "in which one man, perhaps together with two or three younger ones, can pursue his search in a quiet and dignified manner. I should like to see the noises and the crowds of the market place...kept away"[30]. In this essay I have tried to show the connexion between Chargaff's harking back to the science of an earlier time and the more specific reflexion on past methods with which I began.

5. REFERENCES

1. S.E.Luria, *A Slot Machine, A Broken Test Tube* Harper & Row, New York, 1984

2. A.O.J.Cockshut *The Art of Autobiography* Yale University Press, New Haven, 1984, p.216

3. Luria, p.5

4. H Casimir *Haphazard Reality* Harper & Row, New York, 1983 p.3

5. Graham Lepenies and Weingart 1983 *The Functions and Uses of Disciplinary Histories* and P.Galison 1983

6. Chandra Mukerji *A Fragile Power*, Princeton University Press, Princeton N.J. 1989

7.A L Hodgkin "Chance and design in electrophysiology: an informal account of certain experiments on nerve carried out between 1934 and 1952" in *The Pursuit of Nature* Cambridge University Press, Cambridge, 1977, pp1-22

8. Fred Davis *Yearning for Yesterday: A Sociology of Nostalgia* (N.Y. Free Press 1979)

9. R W Gerard "The organization of science" in *Annual Reviews of Physiology 14* (1952) p 152. I am grateful to Olga Amsterdamska for this and the following illustration.

10. Otto Loewi "Reflections on the study of physiology" *Annual Review of Physiology 16* (1954) p272

11. The eminent Dutch endocrinologist and physician Andries Querido begins his recent autobiography with the following paragraph

> "I sat in the cellar of the physiological-chemical laboratory on the Jonas Daniel Meyerplein in Amsterdam and built a pH meter... I had to buid the pH meter around the so-called electrometerlamp which Philips, shortly before in 1930, had invented. I had not understood half the explanation and my knowledge of physics was purely theoretical. I thus had to think out and test every step in the construction myself. Mistakes occurred regularly, and these I had to analyse and find solutions...

A. Querido *De Binnenkant van de Geneeskunde: Een Autobiografie* Meulenhoff, Amsterdam, 1990

12. Relationships between position in formal and informal laboratory structures and the nature of engagement in research work are being studied in current work by Terry Shinn. Clearly the feelings surrounding this shift are complex, and can provoke certain ironic notice. The Dutch physicist Hendrik Casimir - one-time assistant to Wolfgang Pauli - reflects on this in dealing with his own move from research to administering a major industrial laboratory. Casimir writes

> "At some point in my career I felt that I had more or less mastered the most advanced branches of physical theory and that in some areas of specialization I might occasionally even be slightly ahead of others. But I did not do all the running I might have been able to do, and after I had become a co-director of the Philips Laboratory I began to lose ground rapidly."

Pauli was well aware of this. Whenever I met him in later years he always addressed me as "Herr Direktor".
H.B.G. Casimir *Haphazard Reality* Harper & Row, New York,1983 p.244

13. The relation between subjective perception of career and the career 'structures' available in different sciences is the subject of a doctoral dissertation being written by Hans Postel in the Department of Science Dynamics.

14. These figures are taken from D. Shimshoni "The mobile scientist in the American instrument industry" *Minerva* 8 (1970) 59-89

15. A study carried out in the USA in the late 1970s reached a similar conclusion.

16. Deprived of the notion of some basic and essential sense of sophistication, the line of argument had to be worked out in other ways. An alternative which was developed was through comparison. What instrumentation was necessary - which translated into what level of investment was necessary - for a laboratory to be comparable with - or competitive with - another? Out of this line of reasoning emerged the notion of the equipment inventory (and related to it that of obsolescence) which has received considerable attention in the USA. Here perceived obsolescence (% equipment viewed as obsolete) or comparison with some other institution becomes the basis for the assessment.

17. P. Galison, *How Experiments End* Chicago University Press, Chicago, 1987 p 14

18. A. Pickering *Constructing Quarks* (Edinburgh University Press, Edinburgh and Chicago University Press, Chicago, 1984; "Living in the material world" in D.Gooding, T.Pinch and S.Schaffer (eds) *The Uses of Experiment* Cambridge University Press, Cambridge, 1989; "Knowledge, practice and mere construction" *Social Studies of Science*

19. This point has been worked out nicely and fruitfully by Joan Fujimura, who has analysed cancer research in terms of what she calls the 'articulation' of 'alignment' between the levels of experiment, laboratory and social world (composed of laboratories, collegues, sponsors...). See J. H.Fujimura "Constructing 'do-able' problems in cancer resarch: articulating alignment" *Social Studies of Science 17* (1987) 257

20. See S S Blume <u>Insight and Industry</u> The M.I.T. Press, Cambridge Mass. 1991

21. L.A. Minkoff "On the history of NMR scanning in medicine" *Physiological Chemistry and Physics 17* (1985)

22. I. Donald "Medical sonar: the first 25 years" in A. Kurjak (ed) *Recent Advances in Ultrasound Diagnosis 2* Excerpta Medica, Amsterdam, 1980

23. For a study of innovation ultrasound in relation to other diagnostic technologies see S S Blume *Insight and Industry* Chapter 3. For a detailed study of Wild's dispute with NIH see Ellen Koch *American Research on Ultrasound 1947-1962* Philadelphia, unpublished Ph.D. dissertation, University of Pennsylvania 1990 (p88)

24. Michael Gibbons and Bjorn Wittrock (eds) *Science as a Commodity: Threats to the Open Scientific Community* Longman, London, 1985

25. Joan H Fujimura "Constructing 'do-able' problems in cancer research: articulating alignment" *Social Studies of Science 17* (1987) 257

26. Still the best study of how this works is, in my opinion, Renee C Fox *Experiment Perilous* Free Press, New York, 1959, republished by Pennsylvania University Press in 1974. See also A.A.M. Prins "Behind the scenes of performance: Performance practice and management in medical research" *Research Policy 19* (1990) 517

27. P J DiMaggio and W W Powell "The iron cage revisited: institutional isomorphism and collective rationality in organizational fields" *Americal Sociological Review 48* (1983) 147

28. It isn't likely to be so simple because the first hospitals to acquire this sort of equipment, or introduce these sorts of procedures, are typically academic/research hospitals. Research opportunities, or opportunities for making reputations in research, are likely to be as much or more important than considerations deriving purely from patient care. See *Insight and Industry* Chapter 1. See also L.B.Russell *Technology in Hospitals* Brookings Institution, Washington DC, 1979

29. E. Chargaff *Heraclitean Fire: Sketches from a Life Before Nature* The Rockefeller Universityi Press, New York, 1978, pp.81/2

30. Chargaff, *loc cit* p.122

Chapter 6

Scientific techniques and learning: laboratory "signatures" and the practice of oceanography

Chandra Mukerji
University of California, San Diego

We tend to think of techniques as practical means for getting things done in the world, moving objects from here to there, looking more deeply into some sample of tissue, or measuring the salinity of water samples. However, I want to argue in this chapter that techniques are also and importantly means for projecting meaning into the world by acting upon it. When a scientist studies a sample of seawater for chemical components, water in the area where the sample was collected is given qualities that it did not have before. It gains a scientific identity based upon quantitative and qualitative measures that come from science. The chemicals were there already, but the meaning of the chemicals was not. In the acts of gathering and measuring, nature is laden with the culture of science.

This way of conceptualizing technique is useful for thinking about equipment and its use in the practice of science. It differs from most approaches to the subject because it treats scientific research not as a system for determining truths (as it is seen in the philosophy of science and sociology of knowledge), but rather as a means of learning about the world. (Learning in this context is a way of making the environment meaningful, hence part of one's culture). Since scientists are, by their own admission, trying to learn about nature when they do research, one ought to be able to apply theories of learning to the practice of science.

The kind of learning theory that I use here is derived from the sociocultural school of psychology, based on the ideas of Vygotsky.[1] It begins from the assumption that children learn about their world not by studying it in some detached or contemplative way, but rather by acting in it. By using tools and language (toys and talk) to make the environment serve their purposes, children fill their world with meaning. They develop skills in using the cultural resources at their disposal to order physically and to construct symbolically a meaningful world.

Vygotsky argues that young children master their environment first and primarily with tools, just like apes. Babies hit the ground (and themselves) with spoons and toys, suck on everything they can grasp, and drop things off high chairs, seeing how they fall. They act in the world and watch what happens. They do similar things with language. They play with sounds and learn from the results. When they know some words, they experiment with using them.

According to Vygotsky, children begin to create higher psychological functions when they combine the use of tools and speech. Language allows reflection and planning, imagining what is beyond perception, and this opens up new possibilities for action. Young children begin a second stage of exploration of the world when they can combine tool-use with language. They begin to produce continuous verbal commentaries on their activities and initiate social interactions to search for any culturally scripted solutions to their learning problems. By adulthood, Vygotsky argues, thought and planning by far overcome manipulation with tools as the primary way of shaping action in the world. Adults know so many cultural scripts that they can predict more outcomes of their behaviors, and can join scripts in complex ways. There is still a merging of action and language, but the role of communication is the most obvious.[2]

The kind of simultaneous talking and tool manipulation that Vygotsky sees in young children as ways of learning about the world is suggestively typical of scientific researchers in the laboratory. I do not want to argue here that scientists are childlike; that is a cliché that deserves to be forgotten. But I do want to suggest that the merging of tools and talk that Vygotsky identifies in young children is less a characteristic of an age than a relationship to culture and learning. Children and scientists experiment at the edge of their cultural knowledge. Most of the time adults act out known cultural scripts that require less active learning (or they try to hide mistakes while learning). Scientists faced with large unknowns do not have this luxury. Even when they try to plan experiments, they make many miscalculations. They would not need to experiment if this were not the case. What they must do when they start a new project is not drawn out in any cultural map that they know. Hence scientists, like children, use tools and forms of talking that they already know (their culture) to try to explore new objects in their environments. This is not childlike behavior; it is learning through active experiment and the production of cultural accounts for the results.

Repetitious tool-use is central to this kind of learning. Just as children play with the same toys in similar fashions over and over again (just as they listen to the same story books for months at a time), so scientists use repetition to learn. A single run of an experiment is not enough to establish a pattern that a scientist will happily rely upon. There may be unique characteristics of a trial that could only be discovered in contrast to other trials. Moreover, significant details of an experiment, revealing crucial elements of its success or failure, might only come to the attention of the researcher over time. So for scientists and children alike, making elements of the world relied-upon parts of the cultural universe requires carefully scrutinized repetitious actions upon that world with one's cultural tools.

Scientific practice, conceptualized from this point of view, is a system for placing meaning in parts of the world not yet explained by scientific culture. The layers of meaning may be hard to see as "meanings" since they are designed (by cultural prescription) to represent the world in the most accurate possible fashion. Science, by the tenets of our culture, is something extracted from the world, a part of the world captured for culture (not vice versa). It is supposed to help people to see better what is already out there in the world. But the practice of science is a form of activity in the world, not just a way of seeing farther into the world. And since research technologies and techniques are central to the activities necessary for making science, we cannot understand them well until we see how researchers use them to intervene in nature to give it shape and meaning.[3]

The action-in-the-world that is scientific practice has cognitive significance for individuals that might profitably be explored to develop further Vygotskian cultural psychology, but what concerns me in this chapter is how it works for groups in laboratories. Different laboratories have distinctive sets of techniques that they apply to a known range of scientific problems that define the lab's approach to science. I call these techniques "signatures," since they are known within the social world of scientists (or at least members of a scientific discipline or sub-discipline) as the distinguishing features of that lab. Applying signatory techniques to defined scientific problems is what a lab is funded to do, and what labs use for learning about and laying cultural claims on nature.

In the application of signatory techniques, labs do many kind of collective things at once. These techniques constitute the basis for the selection and allocation of research activities within the lab. They also provide an identity for the chief scientist whose "mind" is supposed to be represented in the "signature."[4] It is also a public face of the lab, since this is primarily how the lab is distinguished in the world of science. But most importantly for this chapter, the signature is a set of techniques by which lab members enter the world and mark out cultural territories, first for the lab, and then for science.

Let us take two extreme examples of how this can be done. One is the case of a submarine on the sea floor mapping hydrothermal vent sites. The submarine is used not simply to move over this area and take pictures of it, but to put down experimental devices and pick up pieces of rocks and specimens of animals there. Nature is not just observed, but manipulated and marked with transponders to help the scientists return to the sites they have studied. The vent field when the scientists go home is not the same as it was when they found it. It has bits of scientific apparatus now in it, and samples of rock, animal and seawater are no longer there. In its transformation, this part of the world is marked for science and infused with meaning as a research site.

The cultural stereotype of research technologies takes them to be extensions of the senses. Like any stereotype, this image is based in enough truth to be sustained by casual observation. One could argue, for example, that the research submarine extends human vision to the sea floor, and allows photographs and videotapes to be taken that that "hold" vision and make it both transportable and easier to study. But researchers on the submarine concentrate much more on and depend upon the manipulations of the area with the submarine and additional equipment in order to do their science. If they just went down there, took a look around, took a few pictures, and went home, they would not have a very good sense of the chemical and biological processes going on there that make it a scientifically interesting site.

Let's take another example. If you watch people use an electron microscope and look at the pictures that result from it, it would be easy indeed (in fact common sense) to believe that this machine is a means for extending the senses to very small spaces and not a means for manipulating the world. But the microscope has to have some sample to photograph, and it is taken from the world and prepared before it ever gets to the electron microscope. Interactions with bits of nature are central to any image-making and image analysis that goes on in science. Researchers do not simply look at the world but transform it to make it meaningful when they do research.

Bruno Latour has made this case. He argues convincingly that scientists do not study nature, per se, but experimental objects -- ones that are marked, tagged, sliced, ground, heated, pressurized, or otherwise manipulated before being an appropriate object for scientific study. The analysis of science in Critical Theory also stems from the similar observation that scientific practice involves engineering, manipulation and domination of nature, in the experimental process itself. Experiments, by this analysis, are not detached contemplation, but a way of acting in the world to learn about it. I would add that the engineering involved in scientific practice is also a reason why science has been so valuable to the military and to medicine where control of nature is at their heart of the enterprise, albeit in quite different fashions for these two realms of action.

1. THE STUDY

It is conventional in analyzing the role of science in modern societies to study well-established physicists or biologists (i.e., elite members of traditional and prestigious sciences), and use them as models for the rest of science, even though they are exceptional cases. Instead I have studied members of laboratories (from all the social ranks in those labs) of one of the less prestigious sciences, ocean science, one whose status as a unified discipline as well as its status within the scientific community as a whole is questioned even by ocean scientists. But major recent research on the deep ocean has been powerful in fueling revolutions that have shaken many sciences-- for example, the revitalization and reformulations of plate tectonics and (on a smaller scale) the challenges to marine biology and seawater chemistry presented by the discovery of the hydrothermal vents. [5]

Ocean researchers may also seem an odd group of scientists to study for another reason. They work on submarines as well as labs. They go to sea much of the time and work in laboratories on board ship that cannot compare to the more sophisticated ones on shore. But in needing expensive and sophisticated equipment, they are quite representative of large numbers of researchers in the greater scientific community. They are like space researchers who are dependent on rockets to get themselves or their experiments into space, and physicists and astronomers who find themselves only able to do their work when they can get "onto the machines" (telescopes, particle accelerators, or whatever) that they need for their experiments or observations.[6] In addition, their technical dependence helps to highlight the centrality of technique to their science, so it makes the analysis of laboratory signatures easier than it might have been in a less machine-based area of research.

In addition, oceanography is a particularly good discipline to study in order to understand the learning process in research. More than most parts of the world, the deep ocean has been relatively understudied. Some recent breakthroughs in ocean research have had far-reaching results because they have simply replaced complete ignorance (or bad theory) with some empirical analyses. The shape of the mountain ranges running through the major oceans of the world in itself has been a relatively recent discovery. Clearly, the level of ignorance has been great enough to allow rather great empirical and conceptual leaps.[7] Work on spreading centers in the ocean has turned out to be crucial to geology as a whole and to the development of plate tectonics. So ocean science has presented researchers with particularly daunting learning problems coupled with opportunities to make major contributions to cultural representations of the ocean.

In the course of my research I conducted 74 interviews with 63 ocean researchers and/or science policy experts located in nine institutions around the United States. I used snowball sampling to generate a list of the scientists, engineers, technicians and graduate students to interview. Because I focused on deep ocean research, many of those I interviewed (18) were in marine biology or marine geology or geophysics (19). Many (21) had been centrally involved in the development and use of new research technologies, some sampling systems, some <u>in situ</u> measurement systems, and a number of imaging systems used for producing indirect observations (recordings) of deep ocean phenomena. I also talked to 14 scientists who had been involved in science policy and/or worked at different levels of the major funding agencies that support oceanographic research, eliciting from many of them the clichés about science that simultaneously legitimate its funding and provide the foundation for public perceptions of scientists and science.[8] In addition, I hung around and asked questions in four laboratories and on one research vessel where research equipment was being set up for a cruise. During these periods of observation, I acquired most of my data on the work process in laboratories, but many of the characteristics I noticed were also described by researchers during interviews or were apparent to me as I briefly toured labs or watched technicians while waiting for an interview.

In analyzing the data, I found two patterns of interaction in labs joined in the delineation of a distinctive laboratory signature. If a laboratory signature consists of a set of technical protocols designed to address a particular scientific problem, then it is realized, not in the chief scientist's quiet office, but in work stations scattered around the laboratory. The most noticeable aspect of the work is that it is located where machines are manipulated. Thinking consists not of abstract contemplation but structured patterns of interaction with equipment and on-going interpretation of the results of those interactions. Researchers at one work station pay attention to what is going on at other stations in order to assess what they are seeing in their part of the experiment. They look and they talk. They manipulate and compare. The analysis of an experiment goes on intermittently and socially among those responsible for different parts of the signatory technique. It is the combination of the talk integrating signatory techniques and the manipulation of samples with machinery that gives laboratory work a distinctive character.[9]

2. TECHNOLOGY AND SIGNATURES

There are a variety of ways in which technologies enter into the work of labs. Some laboratories take strictly conventional techniques, using equipment bought off the shelf and simply assembling it in some slightly new and ingenious fashion to answer some familiar but intractable scientific problem. At the other extreme, some laboratories are known primarily for their technological innovations. They make research equipment to solve fundamental problems in acquiring data from the ocean that then gets adopted by many researchers in the field. They make flexible machines with lots of uses rather than equipment customized to fit one particular research process. In the middle are the bulk of laboratories that adopt a standard genre of research, using extant techniques and equipment, customizing parts of it to constitute a signature.

Journal articles by scientists are a good source for evidence of the use of conventional research techniques by laboratories because, interestingly, articles tend to describe not so much the innovations used in labs as the conventions on which experiments are based. Most articles do little more than list the set of named techniques that form the conventional bedrock of the work. Very little detail is given about how these stock methods are tailored to the task at hand and next to nothing about how customized equipment is used either to link or add to the conventional methods.[10] The culture of science rather than its margins is the stuff of journal articles, so the activity on which it is built is given its most conventional face.

Data about the techniques that make laboratory signatures distinctive are easier to find in research proposals than journal articles. Proposals give prominence to the activity of research and must stress the innovative qualities in a laboratory's approach to scientific problems. The proposals for "big science" projects are particularly revealing of signatures and their uses in science. Large group projects depend to a great extent on assembling groups of researchers with a combination of signatures relevant to a collective enterprise, and the success of the projects depends on the appropriateness and effectiveness of the techniques brought to the task at hand.

> [I was] a early, very early user of *Alvin* [research submarine] and one who concentrated very early in the game in turning a submersible into a field mapping device. And that then began after I did my thesis which ended the day before I sailed for Project FAMOUS. Yeah, just turned in my

> thesis and sailed... on Project FAMOUS....My reasons for being involved in Project FAMOUS initially was not that I was a guru -- I mean I just got my PhD; I was wet behind the ears. [It] was that I knew how to map. Because I was put into Project FAMOUS as a, almost as a techn, technician.... I can remember just being terrified...I'm gonna go down and look out of the window but what does it mean? Now how in the hell am I gonna look out the window and make a statement that's earthshaking to plate tectonicists? The point of using the submarine was to show its value for producing major scientific results. The only was to justify the expense of building and using it was to do so.
>
> ------
>
> That was, that was *Alvin*'s, the very first time that *Alvin* did major science. And of course, as well we know, it was a huge success. Partially because of being along on that, I was offered a post-doctorate, uh, a sabbatical at [the university] here, immediately after this. And I spent a whole year here with all the geological data, everything except the hard rocks, and I had a seminar with geophysics students. And [a colleague] came over for a couple of months. And between those students and the two of us, we published it all in one year.

The success of group projects and of the individual members of groups engaged in large projects depends on the match of signatory techniques to a cruise. In this kind of work more than others, research groups are constrained to live up to their signatory reputations to help coordinate these vast forays into the ocean.

What kinds of reputations do groups cultivate? The vast majority of research groups rely on a unique combination of standard (and accepted) research techniques combined with a small number of customized instruments and process with which to act on the world in distinctive ways. These arrays of techniques are tailored to scientific problems --tears in the fabric of scientific explanation that scientists want to repair. One laboratory I studied, for example, developed a deep ocean free-vehicle camera system that was designed to sink to the sea floor, take pictures of fish attracted to bait, drop ballast, and return to the surface where it could be picked up by a ship. This machine was used to distinguish between areas of the sea with live organisms and those without them. Another laboratory I visited built fish traps and respirometers to study fish populations in situ to determine (among other things) to what extent specimens brought to the surface in pressurized traps differed in their behavior from those on the sea floor.

For the purposes of studying instruments in science, it is important to point out that a small number of laboratories develop "signatures" as designers of new research vehicles or analytical systems. These groups identify limits in old machinery and possibilities in current engineering that could be used to fashion new systems for research. A scientist, who developed a robot-based system for deep ocean geology, described his impetus for the making the innovation in these terms:

> I had spent up to that point, a decade of my life in a submarine on the bottom of the ocean, going to sea four months a year, for a decade, forty months at sea. And I'd seen less than one tenth of one percent of the mid-ocean ridge. I mean, what a way to go. And the mid-ocean ridge is forty thousand miles long, seventy-two thousand kilometers, covers twenty-eight percent of the planet. It's the biggest feature of the planet. Am I really gonna spend the rest of my life on my hands and knees in the dark with a flashlight, crawling around? ...I had now at that time, had maybe two hundred deep sea dives, and the thrill of getting up at six in the morning and not drinking anything and crawling in a submarine and making a two hour descent in a cold sphere, starting to freeze, cramped up and legs going to sleep, getting a migraine headache from lack of oxygen, getting a wet shoulder from the hull, to get down on the bottom to, to stick my head beneath my legs and step on my feet and crack my head and talk for hours after hours into a microphone and become a machinery and digi... I was a human digitizer. I must admit the first time a saw a black smoker [a hydrothermal vent where the water comes spewing out in a great black cloud of super-heated water through a chimney of rock], it blew me out of the seat. The first time I saw a hydrothermal vent I couldn't believe it. That represented about five percent of my bottom time, if that, perhaps less. Then after four hours, to return to the surface, two hours back up, writing out little clichés for little kids that have sent in their stamped envelopes. Eating a lousy sandwich. Walking out with a migraine headache that just made your eyeballs jump out of your head, cold, tired, to sit down then for two days before you dove again and transcribe all your tapes.

Researchers like this are not simply ocean engineers since they develop and use their equipment in order to do basic research, but they make engineering innovations more than scientific ones the focus of their labs' activities. Their successes in science "sell" their machines more than vice versa. So, while they pride themselves on being scientists, many of their colleagues do not see them this way.

Many scientists with reputations as first-class researchers who know how to design technological innovations to further their research see themselves as entirely different. They produce original equipment to serve their particular research needs, without designing for the field as a whole (or thinking about finding large markets for their products). The best of these groups are praised for their technological originality, but they do not take equipment development as a major purpose of the lab. In fact, one member of such a group showed an astonishing contempt for labs that have reputations for engineering.

> I have very little use for people that create new machinery and then look for a use for it. I mean, the only time I've ever made a new piece of machinery is because I had to have something. I mean I...it was...been the combination of seeing some kind of real interesting problem and seeing a way to do it. But I, you know...there's certainly a lot of people in the field who...you know, will develop, you know, very fancy pieces of machinery and...and become very committed to that piece of machinery. Uh...and...the..that's.. that's not ever been my approach really..now. If anything I decided I really don't particularly like making new..[equipment]. I mean, I do enjoy doing it, but if at all possible, I'd rather buy it off the shelf. Ha ha ha ha. Because it's just a huge amount of work and effort and it's a lot of just creative energy...of just sitting there and just picturing you know, how's this thing gonna work? And then.. translating that into the.. the hard.. things.

This level of disdain for ocean engineering is not unusual in science. The Vannevar Bush proposal for the American science system set up after World War II stated explicitly that engineering depended on science, not vice versa.[11] Bush justified the science funding system in large part on the ground that engineering would become weak if not guided by innovations in science. Not many scientists, even those who have high regard for research engineering in science, have wanted to undermine this assumption. Any gratitude for those who have produced the vehicles, sonar systems and analytic equipment that make deep ocean work possible today has not affected the cultural subordination of technology to science. Labs with signatures as engineering centers have what Bourdieu calls low cultural capital,[12] in spite of the centrality of their machines to the research signatures of other labs, and the importance of their work to the field of oceanography. Making and manipulating tools is as culturally subordinated as it is central to learning.

In spite of the prejudices against engineering, it is clear in the history of oceanography that many innovations in science have only been possible because of engineering activities. This is certainly true in the early post-war period in oceanography where an intellectual revolution in geology was created in large part with new technologies for deep ocean research.

> With the Midpac expedition of 1950 the new marine geology was fairly launched and the coming tectonic revolution became inevitable. ...In 1950 the first generation of new marine tools was assembled: echo sounders, magnetometers, temperature probes, explosion seismometers, piston corers, and dredges. Within five years they would be adequate to discover the magnetic anomalies that take the measure of all the world, the great fracture zones that offset them, and the median rift where the sea floor is born. Meanwhile the cadres and tools for the other phases of the revolution were being assembled. ...paleomagnetic studies were about the begin. So was the

> new seismology --triggered by efforts to detect clandestine tests of nuclear weapons-- that would verify transform faults and track the plunging lithosphere. All was ready to begin the revolution, but the campaign would be long.[13]

The seafloor studies of plate tectonics were not just a sidelight to a major land-based geological movement. Understanding what was happening in the spreading centers in the ocean was essential for analyzing what was more visible on land. The intellectual revolution that shook all of geology, then, depended on generating readings of areas of the earth crust covered with miles of water. This was a difficult task, but possible because of the massive increases in ocean instrumentation. In no sense did technological innovations <u>force</u> scientists see the earth's crust in new ways. The technology gave scientists new ways to probe the earth's crust in more inaccessible regions of the deep ocean. Scientists drew new maps of the seabed filled with hills and mountains, bringing to the surface complex geological images for one another to explain. At the same time, measurement systems were sent down the water column so that more scientific testimony about spreading centers and plate tectonics could be retrieved from the deep. This movement through the water column was achieved not with vehicles but research instruments.

Recent instrument systems have given scientists access to even newer realms (like vent fields) for science to contain within its culture. An article on exploration of the East Pacific Rise began with the following (partially edited) explanation of the project's research design:

> In April 1976 a group of ocean-oriented earth scientists gathered in La Jolla to discuss ways in which newly available capabilities for detailed studies of the sea floor, particularly manned small submersibles, might best be further developed and used in the Pacific Ocean. ...The plate tectonic model of the earth's crust has focused the attention of marine geologists on the oceanic edges of the plates -- the rise crests, transform faults, and trenches -- as the regions in which investigations would be most likely to lead to an improved understanding of fundamental geological processes. The existence of exposed, small-scale phenomena at depths easily reached by today's submersibles led us to concentrate on the rise crests as particularly informative sites. ...The principle technical aspect in which the RISE program differed from the FAMOUS and Galapagos spreading center projects was a thrust to develop and exploit geophysical measurement capabilities which the submersibles and unmanned vehicles at that time did not possess. The result was a plan that emphasized geophysical measurements with complementary geological observations and sampling.[14]

Research sites for science are marked and characterized with customized measurement systems, defining and setting up distinctive programs of research. The "new" instruments in this case were not engineering breakthroughs but their use in combination with deep ocean vehicles was innovative and hence useful. (This balance of convention and innovation in instrumentation corresponds to the mix of innovation and convention that marks creative work in the arts.)[15] For this cruise, familiar bits of scientific material culture (instruments and vehicles) were sent to the sea floor, allowing scientists to act upon the vents sites in ways that would yield useful science.

Looking at another cruise, the one where the Sea Beam sonar system was first used by scientists, we can see how important a technology can be in destroying extant assumptions about both what the natural world looks like and how one should study it.

> We discovered that the Navy had a secret sonar that was like to us, a James Bond incredible piece of machinery. And [it] was called the SASS system. And the Navy had been mapping the oceans for their own purposes with this sonar. And we didn't...know it. And we got the Navy to go into the Project Famous area and map the area with their sonars and the resulting maps were numbing, just numbing to the scientists who.... I can remember the first time they gathered around a table and looked at them. Almost in just utter disbelief. It was like someone had turned on a light.... It was just, you could hear a pin drop; people went, you know (gasp) my God! Look at those maps! The detail was just incredibly high detail.... [And] they do it (gshhh), rapidly, twenty knots, wummmp and there was the map. You know, it wasn't like they spent a lot of time. Boom, you know, they have a map to keep you busy for a year.... [It was like] the first time anyone had a periscope, I mean a telescope or a microscope. Just my God, look, look, look! You sat there, watching the machine.

Because the images on this machine were so much more accurate than anything these scientists were used to seeing, they simply exploded old assumptions about how to do deep ocean research. Old signatory systems were immediately out of date. Scientists had to think about how to take advantage of the more accurate maps to mark useful research sites.

Most signature changes stimulated by the introduction of new technologies are not so dramatic.

> What I've been doing to count the colonies [of bacteria] and the distributions of them in tubes is to count them with an image analyzer. Part of that was just to develop the technology to see what the image analyzer was good for and it's a really sophisticated piece of equipment.

> It's fantastic....[The chief scientist in the lab] has this device which is a temperature gradient block; it sets up a temperature gradient from like zero to fifty degrees in pressure chambers. So you put a bunch of bacterial cells in a long glass tube in agar which is a solidifying agent and so they're distributed along the whole length of the tube. You shove them into a temperature gradient from zero to fifty degrees and pressurize them. And I had like eight of these chambers and you put them at all different pressures at the same gradients, then you leave them for a period of time, take them out and then you look to see in what temperature range the colonies formed. Well that's fine and [the chief scientist] had used this device for a long time. You'd mark the extremes and then you'd find that at higher and higher pressures the extremes get narrower and narrower. That's fine but there were some subtleties going on that were really hard to resolve. Like there were areas where the numbers of colonies were different in the tubes and so I wanted to be able to quantify the numbers of colonies and it was really hard. I mean [if] you sit at a microscope and try to count these things you go nuts. So I sat down with [the lab technician] who is our programmer and I said, you know, we have this image analyzer and I..and the two of us worked out a program. Basically I was dictating the criteria for the program and he was writing the code and it works great. It's really a nice program. In fact I just finished the paper on it....The image analyzer looks through a microscope and it photographs an area..a small area of the tube; then it transforms it to high contrast and then it circles and counts each of the colonies....You have to have something that's fairly uniform and the world isn't uniform.... The only reason I had any success I think is because the system I was dealing with was fairly uniform and the processing that I was doing were relative measurements so that because it was uniform I could say, "Well, even though I'm not counting a hundred percent of what's in this tube, I'm doing the same thing all along the tube so that the relative numbers are going to be valid even if the absolute values aren't." And the image analyzer was great for doing that. It generated some data that it would have taken me ages to be able to get by eye.

This post-doc uses the image analyzer to elaborate the signatory technique of the lab he works in, adding to the temperature gradient/pressure studies of bacteria growth. By using the image analyzer, he replaces the measures of distances between extremes of colonies at different pressures with a measure of changing densities of colonies within the temperature gradient. The image analyzer in this case was mobilized to expand the uses of the temperature gradient block, the instrument that was at the heart of the signature of this particular lab. As he designed a protocol for the image analyzer, he gave the lab as a whole a new way to think about possible uses of the temperature gradient block itself.

Experimentation with new techniques, elaborating those stock techniques characterizing a lab's signature, is a means scientists use to expand the world of science from knowns into the unknown. The old and new techniques provide links for connecting a lab's practices from one historical moment to the next and for delineating what is being learned from the world. The gains made by single labs from any one change in its signatory technique may be modest. They may only marginally improve what the lab can tell other scientists about their research subject, but the cumulative effects of such changes are more dramatic. The search for new ways to act on nature to learn from it, the repetitious experimentation with promising techniques, gives scientists the ability to make the world meaningful within the cultural system of science.

3. THE DISCOVERY AND EXPLORATION OF HYDROTHERMAL VENTS

We can get a better picture of how signatory techniques are mobilized to learn about a new area of the deep ocean and claim it for science by studying the early history of research on the hydrothermal vents. While geologists had suspected for some time that hot water ought to be seeping out of the rocks along the mountain ridges in mid-ocean and went to find them to confirm their theories, they discovered something much more dramatic than they expected: dense animal populations in streams of hot water that were spewing out large quantities of toxic chemicals into the water. The fascinating puzzles and opportunities presented by the vents made them important research sites for oceanographers.[16]

> It was a long development in this hot spring program, starting fifteen years ago. Then there were various chemical and geophysical anomalies in the ocean and we found a problem that couldn't be explained by any obvious process. And then the consensus began to develop among theorists at [Big Lab]. Almost everybody involved in the original hot spring work was [from Big Lab].... It was a fantastic crop of graduate students who went through [Big Lab] in the '60s and early '70s in all fields. And..we, you know, we knew, of course, plate tectonics.... And the consensus was [that] there had to be hydrothermal activity on the sea floor, the ridge axis. But the question is, was, how do you find it? .... And then a big consortium of us dove over that thing and sort of sat on it, & discovered the Galapagos [hydrothermal vents].

When scientists went down to study the hydrothermal vents, they dove to the sea floor on the deep submergence vehicle, *Alvin*. Videotapes made automatically on these dives recorded much of the conversation and many of the activities of the scientists. They constitute a rich source of data on this particular piece of territorial expansion for science. The stationery camera on the front of the submarine did not always record the movements of the vehicle's robot arms, particularly when the scientists were setting up experiments on the ground close to the *Alvin*. In those cases, the activities went below the camera's line of sight. But the movement of the *Alvin* itself is well-documented along with most of the talk and many of the activities of the scientists.[17]

The vents studied during these dives are essentially hot springs on the sea floor. What surprised scientists first about the vents when they discovered them was the abundance of life. There were lots of clams, mussels, and large worms. Nothing in the literature predicted such a thing; there appeared to be too little food in the deep to support this amount and variety of life. What were the animals living on? How did they get there?[18]

In addition, researchers found that the water escaping from these vents carried high concentrations of substances dissolved from the rocks under the ocean. Circulation of the hot water through the rocks was bringing minerals into the seawater in surprisingly high concentrations. How did the character of the seawater emerging from these vents relate to the metal deposits on the deep sea floor? And would this water finally explain the chemical composition of sea water?[19]

To begin to answer these questions, scientist had to think about how to manage the practical problems of trying to learn something new about natural processes at these remote and perplexing locations. How should they get to the vents? What could they do once they got there? What were the hazards of heat and chemicals in the water? What equipment could work in the inhospitable environment? And how could they put together effective experiments?

They faced some daunting learning problems because the vents were so full of surprises. What exactly should they consider a vent field? Did vents always have hot water coming from them, or could they sustain life just with cold water expelled from the rocks? What about "dead" vents, where there were lots of clam shells but no living ones? How did live vents come and go? What were the salient features of vents? Was heat more important than chemicals spewing from the vents or vice versa? And what constituted an important discovery for science that they might be able to find there? Only if they could answer these and other questions could they begin to make this kind of natural phenomenon part of the domain of science, a source of its empowerment as a body of knowledge and a territory known to the world in scientific terms. And the only resources they had to bring to the learning task were their signatory techniques: the skills they already had developed for exploring some other parts of the ocean. The practical thing to do was to take known equipment and ways of using it to the site (adapting it to higher temperatures and pressures, as necessary), and to use them to structure their activities at the new research site. They already knew how to do something. They would see what they could learn by doing it at the vent fields.

Studying dive tapes from the *Alvin* reveals how researchers focus on research tools and their manipulation to do new science. When they mobilize their signatory techniques at a new research site, they play out a familiar cultural script and protect themselves from the overwhelming novelty around them. They concentrate on their machinery, and use it as a translator from the unfamiliar to the familiar; the readings they get, the measurements they make, the samples they collect, and the pictures they take back to their labs all capture novelty in familiar nets, allowing comparisons with parts of the ocean these scientists already think they can explain in scientific terms. The techniques provide a means for assimilating the novelty within the categories of science.[20] At the same time, by carrying their equipment to the far corners of the planet, these researchers change the environments they enter. They take bits and pieces of their culture into the unknown, helping to claim it for science as a site for research.

Most importantly, different research groups pose specific and diverse scientific questions when they apply their signatory techniques. (Is temperature or water chemistry more important to the vent's ecology? Measuring one or the other is a signatory "bet" about the answer, and the equipment taken by a laboratory down to the vents will define the "bet" laid down by that group.) Putting down a simple piece of equipment, then, is a culturally complex act, one in which learning and exercising power are deeply intertwined. It is also a risky act, an act of faith, because no one knows what the experimental results will be. One can see on the *Alvin* dive tapes that researchers are often not quite sure even how to use their equipment in a vent site; they are not clear how to identify the proper placement for a given experiment.

> #2: I haven't seen a good place where water comes out yet. If you see any we can drop the pump down there.
> #3: Okay look to those clams right over here. You've gotta look real hard and you can see real shimmering water. See?
> #2: Uh huh.
> #3: I'm not sure you'll see any more than that.
> #2: Okay, I'm all in agreement to use an area like this one.
> #1: To put the pump out.
> #3: Okay, this is the best area that I know of. I mean, this is the area that we were all talking about coming to. This is the......
> #2: Okay. Let's use that big hole right next to it. [The nearer one] is too dangerous to use.... It could be great trouble for us xxxxxx xxxxx in the hole. But that's.....that [other one] would be fine with me if we......put the pump out here and.....and leave it for an hour.
> #1: Okay, do you want...? Do you need a temperature reading.. when I set the probe or...?
> #2: Yeah, it would be nice.
> #1: I know it would be nice but..
> #2: If it can't be done, it can't be done, you know.
> #3: I'm gonna move over here. I think I see more, more clams, and a little more hot water.
> #2: That one's good.
> #3: Now you may, pretty soon, be seeing the white smoker.
> #2: Is that...close?

#1: There's a clump of tube worms and it looks like, just on the other side of the tube worms, is something...artificial kicking up...it might be.. [inaudible]. I don't know where....
#2: Okay, here I see warm water almost everywhere.
#1: Here's some of the clams on the mantel stuck WAY out.
PAUSE.
#3: You see that pit there, Arlen, the hole? There are a couple of tube worms and a lot of warm water coming up out of it.
#2: Okay.
#3: Is that sufficient for me to just throw the nozzle in there?
#2: Yes, yes. That's fine.
#3: There's nothing to stick it in around here.
#2: That is fine.
#3: Okay, I think we can try to get a temperature [reading] when we come back over here, if we have time.
#2: Yeah, right. I can see now...coming out of the clams, like a bag, hanging out.
#1: Yeah. Yeah. That's some xxxxxxx coming in.
#2: Right. Yeah. I can see it here, looks right too.
#3: Okay, so I'll just throw the thing down that hole.
#2: That would be fine.

The scientists recorded here are engaged in <u>acts</u> of learning. They are not like the careful cultural planners that Vygotsky describes as adults. They are more like children trying to learn about their environment. They learn by moving, by trying to do something and seeing what happens. In some sense, it matters less where they finally put down their pump than the fact that they put the pump out at all. By acting on the environment with the submersible, placing their equipment in it, and talking about it as they go, the scientists transform the site, making it into a part of their cultural world. They move along the vent site, not as disembodied minds surveying and comprehending the hot springs, but as tool manipulators, defining a structured way to learn through action: collecting water samples for analysis. When these researchers look out from the windows of the submarine, they may indeed see a world of wonder and awe, but as scientists who must learn from it, they also must find a place to put down a pump to measure the vents against their extant conceptions of the deep ocean. They apply their signatory techniques to yield new ostensive models of the vents for explaining the characteristics of hydrothermal processes and the life-forms they support.[21]

There are a number of other examples from other kinds of deep ocean work in which we can see scientists using signatory techniques to try to learn about the animals and rocks there. One woman used "wood islands" to determine what wood-boring animals existed on the sea floor, and to test theories about how quickly they could destroy a known quantity of wood. First she put onto the abyssal floor the "islands" made of timbers strapped together. Then the borers were left to do their work. Later, the islands were retrieved, and the changes in their weight and density were taken as measures of the borers' activities.

It is easy to think that the borers who ate into these islands were doing something perfectly natural, but what they did depended on the artificial introduction of great hunks of wood. The results were partially natural and partially artificial. Boring was natural enough for these creatures, but boring into these islands was not ordinarily an option for them. Getting them to do it made the animals participants in a modeling process intended to imitate their "normal" behavior in a more controlled fashion. One could argue that, in this kind of experiment, scientists manipulate organisms, getting them to act abnormally so that they make good data; they are coaxed into entering the culture of science for a while, where their behavior can be measured in scientific terms.[22]

Another signatory type of research used for vent work was a photographic survey. A laboratory of marine biologists did surveys of animals near the vent sites to study the composition and distribution of this population. What scientists brought back from the sea floor was not animals, but rather pictures of animals. The whole population could not be brought to the surface. Only a model based on photographs like this could be. So employing these photographic techniques, scientists made an analysis of animals populations possible. In doing the work, the researchers faced some difficulties. They had trouble, for example, distinguishing between related species of crabs from the photographs because, when the images were magnified, the grain obscured relevant details. On some occasions, the animals would climb on top of one another, and a picture taken from overhead could not record their presence or might make the crab underneath too obscure to be identified. Obviously the model of nature drawn from these pictures had some problems due to the limitations of the photographs, but the surveys still provided a model of animal distribution that could then be compared to models of temperature variations and maps of chemical changes in the water to test theories about how the animals derived nutrients from the environment.

Other systems measured the vents in other ways. Animals were caught in cages where their metabolic rates could be studied in one fashion or another. These "readings" rather than animals themselves were then brought from the deep. Video images of geological formations were recorded during cruises, with images being coded according to a "geological video atlas" so that their labels as well as their images would be available for analysis. Bits of animals, water samples, and chips from rocks were subjected to chemical analysis. In all these ways, signatory techniques were applied to the vents, yielding ostensive models of them and addressing theories about their formation and ecosystems.

As they use their signatory techniques in the deep, researchers begin to claim territories. Symbolically, scientists make claims to areas of the sea floor with their machines while they make conceptual claims with their own peculiar research designs. Scientists on *Alvin* routinely put down markers and transponders to tag research areas and make precise navigation to those sites easier. The vent fields are changed by the presence of these machines. They begin to become less natural environments than areas that have characteristics of both the natural world and the social world of science.

Some of the signatory techniques that affect the research site the least might seem to delineate no particular territories for research, but the opposite can be the case. The group mentioned above that did photographic surveys of the vent areas were on one dive trying to correlate heat with animal populations in the vent fields. While other research groups marked their research sites with machinery strewn about or defined research sites as the places they picked up samples of animals and rock, this group wanted to work where there was no evidence of science. They sought a pristine "wilderness" to describe in ecological terms. Ironically, this territorial claim was made apparent to me from a dive tape made when some researchers were taking samples of clams. They took some from this area, and said in a troubled voice that the chief scientist who was doing the ecological survey would kill them if he knew they had removed some animals from this area. By violating the territorial claim of the ecological research group, they made clear the nature of that claim.

When these sites are claimed for science, what does that mean? Like a playground, the research site is a place where socially shared definitions of the situation can be projected and tried out by a group. Local resources and externally acquired skills are joined in culturally scripted ways. Tools and talk are mobilized to see what can be done there. Will temperature correlate with population densities? Will levels of sulphur make more of a difference? Scientists measure and test, not just to make models of the world but to find ways to act in understudied parts of the environment in order to learn something new about it. Tools and speech are joined for these researchers as they are for children not to repeat cultural scripts but to make them up. Through this process scientists can expand the domain of science. From this point of view, the research site is not just a location, but a social space where people can learn through structured action.

4. CONSTRAINTS ON SCIENCE FROM TECHNOLOGICAL DEPENDENCE

Understanding the centrality of tools to the learning process in science is important for drawing analysts away from temptingly simple models of scientists acquiring scientific knowledge through independent evaluation of arguments and evidence without any visible leg-work. Learning about the world for science is better described as a way of manipulating the world with tools, as a form of culturally shaped social action, dependent on an elaborate material culture.

This being said, it becomes important to know more about the material culture with which science it built. The question of where and how scientists get their tools emerges as primary for understanding how they get their ideas. In American deep ocean science, the question of where equipment comes from is simple to answer: scientists get most of their tools (ships and instruments) from the U.S. Navy.

> ...MPL is an offshoot from World War II, The University of California Division of War Research, which was set up to support the Navy for submarine and anti-submarine warfare. And it's a group that developed a lot of sonar equipment, acoustic equipment. So that's how MPL started doing that kind of thing. But that led into learning more about the sea floor and it's effects on acoustics. ...And then, when you do work on acoustics in the ocean, you start collecting a lot of data; then you've got a lot of data processing, signal processing. So that led into a whole chain of technology development in signal processing which has been very productive for the Navy. And then of course when you want to go to sea and do things, you have to devise new gadgets and pretty soon you find yourself in the world of ocean technology. So, it all stemmed from the ocean acoustics.

The culture of knowledge and practice that scientists have been elaborating and projecting into the far reaches of the world's oceans since the second World War has been made, therefore, using tools designed to serve strategic interests.[23] Ocean science has grown dramatically in the Cold War period because it has been a direct beneficiary of some of the technological competition surrounding the arms race. The U.S. Navy built new ships and instruments to gain strategic advantage at sea. Some of this technological innovation has been designed directly to support ocean research. The Navy had obvious general interest in keeping itself on top of any new ideas about the ocean and its use. Moreover, when the U.S. Navy put nuclear missiles on submarines, it needed to know how to launch missiles from under water and how to hide these strategically valuable vessels in the geological features of the sea floor or the heat layers of the upper ocean.

> This dates back to 1946. [If The Navy wants] to get information out of the ocean, [it has] to go to sea to do it. They can't buy the stuff, the instruments. You have to find some engineers that can help you design the things and technical people to help you build it and operational people that can go to sea and use it. So you get a whole little community that all stems from some individual that wants to go out and get some information. And the technology [is developed that way] and so that's what happens. Year after year at our [Navy instrumentation lab] we stimulate new technology and then there are offshoots of that. When you gather something new and good then you look for other ways it can be used [including use for science].

The usefulness of the new equipment to science has been dramatic. Most of the techniques used to study plate tectonics in the 1950s (described earlier) came from

Navy or Navy-affiliated labs where military needs led to rapid proliferation of ocean technology.

While Navy efforts at ocean engineering has been the primary source of new equipment used for research by scientists, much of the resulting machinery has had limited applications for science. For example, the Navy has built a number of small submersibles that can dive into the deep ocean, but most of them are not particularly appealing or useful for scientific cruises.

> Part of the submersible mission is to support scientific research --Navy approved scientific research. And that's done through the Office of Naval Research, that every year goes out to the scientific community and says, "Who wants to do what with the DSVs [deep submergence vehicles]?" ...And they all get together with the individual in ONR who's the coordinator of all this and then come up with a prioritized list that goes back to the Chief of Naval Operations and it goes to the Fleet. ...Then what they try to do is fit that into a schedule of operations, which is difficult because they have Navy tasks to do and when something like, for instance, oh, last year a Marine Corps helicopter crashed off Point...the San Clemente Islands. And the *Seacliff* went out, and gosh, I think she was out there for like a couple of months. So anything that was scheduled to be done for science in those two months, you know, we quit doing it. And it's hard on the scientist because he's got his programs laid out and his funding is limited, so making that thing match is difficult but we're working on it. So the *Seacliff* is the one that has everybody's attention now because it was converted about a year or so ago to a 20,000 foot depth capability. And there are only two in the world that can do that. ...*Seacliff* has been enlarged, now it's bigger. The *Turtle* I think is still basically the same size as *Alvin* but it has a...a tougher sphere which is good to 10,000 feet. Last year they had an electrical fire which did a lot of damage so they had to take the whole thing apart. We have it, and when it gets finished, which shouldn't be much longer, should be probably this summer, the *Turtle* will be a good operating boat...and available for scientific purposes down to 10,000 feet. The DSRV's, of course, were not designed for scientists and they're limited....in terms of observational capabilities....The deep submergence vehicles basically have portholes. You can go down and look out the porthole and you, then you can take pictures; you get cameras, and you have external cameras, things like that. The DSRV's are built to do everything with optics. ...Most of [the research scientists] prefer to use the *Alvin*...because the *Alvin* was built for that purpose and is organized and run for that purpose with ONR support. And there's a steady cadre of people that are available back in Woods Hole [Oceanographic Institution]: the pilots, you know. And they all know how to work with the scientists and there aren't any other Navy missions that get in the way. Well, not usually. There could be but...not

> usually. It's, oh you know, science. So that's much the preferred platform but there's two problems with the *Alvin* --at least two. The first problem is you can't go to 20,000 feet, and of course everybody wants to go to 20,000 feet. And the other is that everybody wants to use it, and therefore you've got a long waiting line for that.

To the extent that scientists can build some of the instruments to use on submersibles for collecting data or can design the analytic equipment that they use back in their labs to experiment with their samples, they gain some autonomy from Navy-based systems for acting in and upon the ocean. But even in the area of instrumentation, many scientists find that they cannot produce the equipment they need and they end up relying on unique configurations of commercial systems for their signatory analyses.

> The days when the operator of a machine could actually build it really ended [in ocean research] about ten years ago. There are few exceptions that still do it, but they're very, very few. ...It's almost all bought commercially. And that's the, part of the problem is the, the under-capitalization which has been chronic for a long time. ...There's a big gap between what's possible and what's available and what you can actually get through academic research funding. That's what we mean by undercapitalized. We can stagger on. It won't be the end of the world if I don't get my two hundred thousand dollars [for a new instrument] but it, it will certainly make life very [difficult]. I would lose a substantial competitive advantage if some other guy at some university had his university give him the money. ...There's equipment out there that cannot conceivably be in common use in, in a field like ours. ...One or two of these things will come into the field. And the field has no way of stopping them. And often it's not the guy who's, who's.... Often the guy who gets the thing is not the guy who should. ...Complicated deals get done and so the, the relationship to technology is very complex. It's not just simply a question of NSF deciding to give people money; it's deals that people do. Some guys are in a better place to do deals than others.

Learning through tool-use in science, then, embeds scientific practice in a world of power and politics that shapes the design of and access to complex technologies. As much as being able to manipulate tools can help scientists learn about the world, being deprived of tools can adversely affect individual careers or the development of a science as a whole (where an instrument is withheld by the military for security reasons or kept from labs because of its expense). The result of military and commercial experimentation with instruments has been a cascade of new tools for interacting with the ocean that have allowed scientists to claim more aspects of the sea as known to science and important to scientific theory, but the price of this development has been high. Scientists have lost some control over the signatory choices they can make in designing their own research. Scientists doing deep ocean research have become particularly dependent on outsiders to develop the tools that will shape the intellectual growth of their field. This does not mean that their work has been or will be determined by Navy interests. Still, some of the negative consequences of scientists' dependence on the Navy are clear enough.

> Basically what happened was that the US Navy, for totally unconnected reasons, mainly pressures from the suppliers, commercialized many technologies that it had developed for its own purposes like the transponder navigation systems and the big cameras, mainly the navigation system and, oh yea, the Sea Beam, the bathymetric mapping capability.... The US Navy has developed acoustic transponders which were basically smart pingers [for seafloor navigation]. You could send a package down that would emit a ping that would be received by these, these what we call transponders which would ping at it again in a different frequency. So if you have accurate clocks, you can clock travel time, use triangulation, and there you are. So they [the Navy] obviously had those for positioning their own instruments to listen for submarines and all that stuff. And the companies who manufacture them, you see, were seeing other applications. This always happens. And so they were demanding that the Navy declassify. You see what the Navy does then is they truncate; what you get is a degraded version of the technology. You never get state of the art.... [A famous inventor and researcher] developed the transponders. But it was a classified thing, well, classified basically, but it was a Deep Tow development. And then the Navy commercialized it, and let the big contract with AMI, American Manufacturing. And then they got greedy and said "shit we could sell a couple of hundred more." Which they did. And the Sea Beam [mapping system] was the same thing. The Navy for so many reasons decided it wanted extremely accurate... it wanted to map the sea floor with the same resolution that we map the land....What happened was then that they started to commercialize a degraded version of this system that wasn't quite so accurate, but accurate enough....We managed to get our hands on a Sea Beam map, first one that was ever released. And then of course after the expedition they reclassified them so you couldn't get them. It was unclassified and then some admiral, somewhere, obviously of greater seniority, found out that we had this map and it was at the walls of [Big Lab] and we had to take it down and hand it back in. But we had it for long enough. Then the institutes purchased their own systems through Sperry so now it is all on the up and up.

The Navy's decision to keep the SASS (prototype for the Sea Beam) system away from the general population of researchers (for security reasons) deprived scientists of opportunities to map great swaths of the sea floor efficiently, even in the areas where scientific expeditions were going on. Once the new technologies were released, scientists saw how impoverished their earlier signatory techniques had been. Even when they had the Sea Beam to use, there was some reason to be discontented because the commercially available system did not have the same quality as the Navy system. In these situations, the relationship between science and the military is both strained and a bit clearer. The Navy may want sophisticated American scientists to work on the ocean (and has funded much research for this purpose),[24] but the military is nonetheless fundamentally in competition with scientists over the definition of the oceans and their importance. While scientists mutter about plate tectonics, admirals grumble about how to hide and monitor the movement of missile-laden submarines. To the former, the deep ocean is a place where science can expand its domain; to the latter it is a place where strategic interests are defended. To the extent that strategic interests have coincided with scientific ones --where the need for information, skills, and information-gathering machinery has made these definitions of the ocean compatible-- the post-war period has been particularly productive for U.S. ocean scientists. But the fact remains that the science done in this affluent and technique-rich period has necessarily been limited by a pattern of technological innovation designed by the Navy to realize a cultural vision of the ocean as a site of strategic contest.

The diverging interests of science and the military are not always so visible and insistent in part because scientists have vested interests in minimizing their sense of dependence, and in part because the military and science often use the ocean in similar ways. Researchers trying to move into and explain the ocean are not so unlike Navy personnel trying to make the deep sea a place where the U.S. military can move and track others' movements with efficiency and effect. The technological interests of the two are often not that different. The Navy needs surveillance equipment to track ships on and in the ocean, and scientists can easily use the sonars and magnetometers developed for this purpose. Surveillance is, after all, a means for gathering information. Even some of the equipment developed by scientists turns out to be strategically useful. To the extent that water chemistry affects the transmission of sound in water, for example, then instruments for measuring it can be useful to the Navy. These coincidences of interest explain why the Navy has intermittently supported ocean engineering efforts designed to serve science. In the end, science and the military both seek to control territory through experiments with technology and seek to learn from those experiments. These compatibilities mean that the extent to which science has been shaped by the technological agendas of the military can be easily minimized and the results in scientific productivity applauded as a fundamentally independent effort. But the resources scientists have available for shaping their signatory techniques remain circumscribed in profound ways by strategic interests. Learning continues, but not as freely as the lore of science would like us to believe because learning is not a matter of discriminating between true and false assertions through careful assessment of arguments and evidence. It is first and foremost a way of acting in the world with tools and language, and the tools at the disposal of American ocean scientists are part of the material culture of the U.S. Navy.

5. CONCLUSIONS

Thinking about scientific learning as a way of acting in the world with tools and language highlights the importance of research equipment and social relations to science. It points to the complex material culture of science that impresses itself on most observers when they enter scientific laboratories. It suggests why the "shop talk" that so many sociologists of science have noticed in labs is so important to scientists as they work.[25] It also provides analysts with a new way to look at the equipment they encounter within a laboratory. They can appreciate its role in the definition and coordination of "signatures" in laboratories. Students of science can think about how researchers may be trying to alter their laboratory's scientific identity when they develop special instrument packages to put on standard equipment or they find new ways to adapt existing technical systems to new kinds of scientific problems. Observers of laboratory life can also begin to recognize how the learning processes in the lab, while not determined by outside interests, are nonetheless tied to extrinsic political and economic relations through the material culture that enters the laboratory (and the funds that bring it there).

Most of all, looking at research as a site for learning through action, play and experimentation invites analysts to consider what is and is not "childlike" in scientific practice. Scientists, like children, work at the margins of their cultural system, weaving a reality from their actions and from the responses to them by the social and natural environment. But scientists play games with quite different consequences. The stakes are very high in the game played out by scientists. They may be engaged in acts of cultural naming, but their purpose is not simply to master a world known culturally by the adults around them; they look for intellectual, economic and political advantage to be gained through the elaboration of the cultural control of nature. There is play and learning going on in the laboratory, but we can only appreciate its importance to the development of science and the social world in which it occurs by realizing that a research site is no school yard and the game of generating science in the U.S. today is a way of harnessing nature for human service, and that is quite clearly a game of power.

NOTES

[1]Vygotsky, Mind in Society. Cole et al, eds. Cambridge: Harvard University Press, 1978. The philosophical source of this theory is explored in David Bakhurst, "In the Beginning Was the Deed: Ilyenkov on Activity and the Ideal," paper presented at the Conference on Work and Communication, La Jolla, 1989.

[2]Vygotsky, chs. 1 & 2.

[3]I have argued elsewhere (Chandra Mukerji, A Fragile Power. Princeton, N.J.: Princeton University Press, 1989) that when scientists do research, they expand the domain of science, claiming parts of the world and its processes as "known" to science. This process has political dimensions to it. Groups like fishermen who were formerly the reigning experts on the ocean are displaced by fisheries experts. The domain of science expands as domains of folk knowledge are displaced. The expansion of science also has territorial dimensions. By intervening in the natural world, by engineering experiments in which nature is manipulated and transformed to make it knowable to scientists, researchers physically mark animal bodies with dyes, take cores of geological formations, create reactions with chemical compounds, and smash atoms to study the results so that these located physical phenomena can be "known" (at least in probabilistic terms) by science. New physical locations in animal bodies, geological features, chemical compounds and atoms are claimed as they undergo manipulation in the name of science.

[4]Chandra Mukerji, "Scientific Genius" paper presented at the University of Bielefeld, Bielefeld, Germany 1988.

[5]Susan Schlee, The Edge of an Unfamiliar World. New York: Dutton, 1973, ch. 9; Mosaic 11:2-28; Peter Rona et al., Hydrothermal Processes at Seafloor Spreading Centers. Plenum Publishing Co., 1984; J. Corliss et al., "Submarine Thermal Springs on the Galapagos Rift." Science, 203:1073-83, 1979; J. F. Grassle, "The Biology of Hydrothermal Vents" Marine Tech. Soc. Jr. 16:33-38, 1982; F. N. Spiess, "East Pacific Rise," Science, 207:1421-33, 1980; R. Hayman and K. Macdonald, "The Geology of Deep-Sea Hot Springs" American Scientist 73:441-449, 1985; J. F. Grassle, "Hydrothermal Vent Animals" Science, 229:713-717, 1985; H. Jannasch and M. Mottl, "Geomicrobiology of Deep-Sea Hydrothermal Vents. "Science, 229:717-725, 1985; J. M Edmund, et al. "Chemistry of hot springs on the East Pacific Rise and their effluent dispersal. Nature 297:187-91, 1982; L. D. Kulm et al., "Oregon Subduction Zone." Science 231:561-66, 1986; K. Johnson et al., "In Situ Measurements of Chemical Distributions in a Deep Sea Hydrothermal Vent Field." Science 231: 1139-41, 1986; William Cromie, "The New Oceanography." Mosaic 11:2-7.

[6]To highlight the technological dependence of the scientists I studied, I focused on deep ocean researchers. Oceanographers distinguish between blue water and white

water researchers, those who work far offshore and those who work close to land. I interviewed exclusively blue water scientists, but I was more selective than that. With a few exceptions, I talked to scientists who did work on or near the sea floor off the continental shelves. Just getting to their research sites was necessarily a difficult task; using equipment there to conduct experiments or collect samples was also a technological challenge.

[7]See Schlee, particularly ch. 9.

[8]I used a variety of archival materials for this study as well, but the primary source of information came from interviews that I conducted with a tape recorder and in person (with occasional telephone supplements). Most of these interviews lasted for an hour or more; many lasted two. Some were followed with second and third interviews; one poor man has been interviewed numerous times, answered myriad questions over the phone and even given me a guest lecture. In my questions I tried to follow the line of questioning most appropriate for the subject involved. In many cases, I wanted to understand how they developed and deployed equipment to do their research. In many cases, I wanted to see how their work related to other work on seabed disposal of wastes or on hydrothermal vents. I tried to elicit details of the research process in my questioning as well as information about funding patterns. I wanted to find out what made this work seem valuable to the scientists; what they would tell an outsider (like me) about its value; and how they translated its conceptual value into its funding value. I also wanted to know details about the use of technology, since I found technology a major reason for the recent flowering of deep ocean research as well as an essential link between science and funders, defining their mutual dependence. So I pushed in these directions until I reached a sense of clarity about how the activities of the scientist were organized and executed. I did this with varying degrees of success. Some scientists gave me formulaic explanations of their activities that seemed to have been developed long ago and reused for many audiences on many other occasions. Others were simply reluctant to tell me anything. They did not quite understand what I was doing and were not interested in jumping into my research under those circumstance. But the vast majority of those I studied took the project at face value and told me whatever I wanted to know about in a fairly ingenuous fashion.

[9]Jean Lave, Cognition in Practice. Cambridge: Cambridge University Press, 1988; Karin Knorr-Cetina, The Manufacture of Knowledge. New York: Pergamon, 1981; Michael Lynch, Art and Artifact in Laboratory Science. London: Routledge, 1985.

[10]This level of description allows readers just enough information about research designs to deduce for themselves their appropriateness for the problem, but, interestingly, not enough to really allow readers to question the particular techniques that make the lab's work distinctive. In this fashion, the solidity of the work is generally tauted more than its innovativeness. The great exception to this is when the lab's signature is a new technology or technique rather than the science generated

by it. In the latter case, great amounts of technical information about what new machinery is capable of doing is placed in the journals.

[11]Vannevar Bush, Science: The Endless Frontier. Reprint of 1945 edition. Washington: National Science Foundation, 1960.

[12]See Bourdieu, "The Aristocracy of Culture." Media, Culture and Society. 2:225-254. July 1980.

[13]H.W. Menard, The Ocean of Truth: A Personal History of Global Tectonics. Princeton, N.J.: Princeton University Press, 1986, p. 3.

[14]The RISE Project Group: F.N. Spiess et al., "East Pacific Rise: Hot Springs and Geophysical Experiments." Science. 207: 1421. 1980.

[15]H.S. Becker, Artworlds. Berkeley and Los Angeles: University of California Press, 1982.

[16]See John Edmund and Karen von Damm, "Hot Springs on the Ocean Floor." Scientific American, 248:78-93, 1983 particularly pp. 78-81; John Lupton and Harmon Craig, "A Major Helium-3 Source at 15o S on the East Pacific Rise" Science , 214:13-18, 1981; and Ken Macdonald and Bruce Luyendyk, "The Crest of the East Pacific Rise." Scientific American 244:100-116, 1981, esp. 100-102.

[17]These tapes are available for study in the archives at the Woods Hole Oceanographic Institution.

[18]J. Corliss et al., "Submarine Thermal Springs on the Galapagos Rift." Science, 203:1073-83, 1979; J. F. Grassle, "The Biology of Hydrothermal Vents" Marine Tech. Soc. Jr. 16:33-38, 1982; J. F. Grassle, "Hydrothermal Vent Animals" Science, 229:713-717, 1985; H. Jannasch and M. Mottl, "Geomicrobiology of Deep-Sea Hydrothermal Vents. "Science, 229:717-725, 1985; J. J. Childress, A. J. Arp, and C. R. Fisher, "Metabolic and Blood Characteristics of the Hydrothermal Vent Tube-Worm *Riftia pachyptila*, Marine Biology, 83:109-24, 1984; R. Hessler and W. Smithey, "The Distribution and Community Structure of Megafauna at the Galapagos Rift Hydrothermal Vents." in Hydrothermal Processes at Seafloor Spreading Centers. Peter Rona, K. Bostrom, L. Laubier, and K.L. Smith. Plenum, 1984, pp. 735-70.

[19]Edmund and von Damm.

[20]See Piaget for a theory of assimilation and accommodation in the process of learning. Jean Piaget, Play Dreams and Imitation in Childhood. W. W. Norton, 1962.

[21]The structuring of scientists' observations with signatures and routine measurements is all the more useful for scientists who are trying to make sense of a novel site like the hot springs. Much of the deep ocean biology done prior to the discovery of the vents, for example, focused on organisms living in the silt layers along the vast plains on the sea floor. The vent areas were mountainous, some of the creatures they

found there were often unknown to them, and all the animals were huddled around scalding water seeping or soaring out of the rocks, carrying massive amounts of toxic chemicals. Scientists faced with explaining what they saw there began using their signatory techniques. Ecological biologists spent a great deal of time simply labeling and counting animals, doing a census to see who lived there. They applied routine scientific approaches to biological problems so they could answer basic scientific questions. Other scientists did more specialized experiments on metabolic rates and the processing of toxic chemicals by the vent animals, using signatory instruments and techniques developed from earlier work with "plains creatures." Some researchers found their existing signatures inadequate or inappropriate for this work, and revised them as they learned more about the vents. Through the application of techniques, researchers carried themselves through the three stages for the alienation of nature: framing (simple ostensive modeling), making data (advanced ostensive modeling), and theoretical modeling. In this way, they made the vents into images in the "big picture," and used them to make new and bigger pictures, thereby claiming once again a part of the natural world for science.

[22]Ruth Turner, "Wood, Mollusks, and Deap-Sea Food Chains." Bulletin of the American Malacological Union, pp. 13-19, 1977.

[23]Mukerji 1989, ch. 6.

[24]Mukerji 1989, particularly chs.1 & 4.

[25]Knorr-Cetina 1981; Lynch 1985.

Chapter 7

Mosaic array cameras in infrared astronomy

David Edge
Edinburgh University (United Kingdom)

1. INTRODUCTION

This paper explores an episode in the development of instrumentation in modern astronomy. It describes how five groups of researchers embarked on the process of designing and building infrared cameras incorporating 58 x 62 element InSb mosaic arrays, and reports on the situation reached by mid-1986. These detector arrays were originally manufactured for military purposes, but were made available to a wider market in the early 1980s. Several astronomers realised their potential, and engaged in something of a 'race' to exploit it. However, their aims were different, and the resulting 'competition' was complex. A range of factors led to a variety of designs, with each group giving its own instrument a distinctive 'signature'. The paper investigates the origins of this variety.

The paper highlights the importance, in the development of this instrumentation, of two sorts of relationship:-

(a) between scientists and industrial suppliers: mutual interests had a significant influence on instrument design; and

(b) between 'academic' scientists (astronomers) and their associated technicians and engineers.

The process of designing an infrared camera involves the 'translation' of scientific aims and strategies into *technical goals*. Scientists and technicians have different criteria of what constitutes 'satisfactory performance': how this influences the design, production and commissioning of the instruments will be discussed, and the question is raised as to whether the resulting drive towards excellence by purely *technical* criteria gives the process a semblance of 'technological determinism'.

First, two sections outline the theoretical background to this study within the sociology of technology, and the technical background within astronomy. The five groups are then described. The camera designs of the two major groups are then compared in detail. Three later sections focus on general issues: the relationship between astronomers and suppliers; competition; and the relationship between scientific aims and technical goals.

2. THE THEORETICAL BACKGROUND

This research was intended to add to the small number of empirical case studies inspired by Pinch and Bijker, in their advocacy of research on the Social Construction of Technology.[2] In particular, Pinch and Bijker stress the role of 'relevant social groups' in moulding the precise form of new technical artefacts. They write:-

> The use of the concept 'relevant social group' is quite straightforward. The term is used to denote institutions and organizations (such as the military or some specific industrial company), as well as organized or unorganized groups of individuals. The key requirement is that all members of a social group share the same set of meanings, attached to a specific artefact.[3]

Pinch and Bijker use the notion of 'interpretative flexibility' to refer to the fact that a *range* of meanings can be attached by actors to concepts or artefacts. Initially, when concepts or artefacts are in a half-formed, emerging stage, this interpretative flexibility may be large: but, as time passes, a process of 'closure' occurs, and some consensus of meanings is achieved. Thus 'scientific facts' are established, and artefacts reach their 'normal' state, with the associated aura of inevitability. Pinch and Bijker see technological artefacts as the solution to *problems*, and 'closure' is then the emergence of consensus that the problem has been 'solved':-

> The key point is whether the relevant social groups *see* the problem as being solved.[4]

Given the wide range of social groups which may be associated with the evolution of any given artefact, they comment:-

> It can be shown that different social groups have radically different interpretations of one technological artefact. We call these differences 'radical' because the *content* of the artefact seems to be involved. It is something more than what Mulkay rightly claims to be rather easy - 'to show that the social meaning of television varies with and depends upon the social context in which it is employed'. As Mulkay notes, 'it is much more difficult to show [that] what is to count as a "working television set" is similarly context-dependent in any significant respect'.[5]

It is agreement on what constitutes a 'satisfactorily working' artefact that establishes what one can call the 'content' of any new technology. From that point on, both the ability of the artefact to carry out its function successfully, and the definition of what it is to 'carry out its function successfully', are not at issue. They have become reified in the new artefact, in its specific *content*. Drawing their illustrative material from a study of the development of the 'safety bicycle', they show the range of options once thought viable, and trace how 'the interpretative flexibility of the artefact "Penny-farthing" [was] materialized in quite different design lines',[6] finally achieving 'closure' on the design that we now find so familiar.

One difficulty involved in moving from the analysis of how scientific consensus arises, to apparently similar processes in technology, has been succinctly expressed by Pinch:-

> One of the problems in studying technology is that there seem to be more diverse groups involved than with science.[7]

In science, to use Collins' term, one often need only study the 'Core-Set' [8] in order to trace the social process of closure. But to investigate the evolution of the bicycle, the fluorescent light, or the electric car (to quote just three examples of recent studies),[9] is to enroll a far wider constituency. One obvious strategy, therefore, is to study the evolution of techniques of instrumentation in science itself, since this should limit the number of relevant social groups involved. This Boelie Elzen has done, in his analysis of the emergence of two different designs of 'ultracentrifuge'.[10] He shows in detail how the designs stemming from the work of Svedberg, a Swedish colloid chemist, differed from those devised by Beams, an American physicist - with the latter eventually achieving commercial success, and hence (effectively) social 'closure'. He also argues that the *meaning* attached to the ultracentrifuge (UC) by both these scientists *changed over time*: Svedberg, with more limited aims within colloid chemistry, shifted from a 'velocity-UC' to an 'equilibrium-UC'; and Beams shifted from seeing the UC as a stable platform for rotating mirrors for experiments in optics, to a much more general notion of the UC as a device for enhancing gravitational forces, with applications in a wide range of sciences (including colloid chemistry). This interpretative flexibility was associated with striking differences in the designs of the artefacts. And other social groups, outside science, became involved: Svedberg contracted the manufacture of his instruments to an industrial firm with specific 'traditions', which were imposed on the designs; and Beams found an expanding range of 'customers' for his machines, with similar effects. But, as Elzen points out, many researchers would consider the Svedberg UC to be the more 'successful' and technically sophisticated of the two designs: 'closure' by domination of the market cannot simply be equated with the emergence of any agreed 'best' version. Indeed, in specialised enclaves of the user community, we can find 'deviant closure' on quite different designs (we will see an example of this below, in the continuing use of the KPNO cubical dewar).

Elzen's focus of study certainly helps to simplify the social analysis: but he still finds himself forced to consider a complex network of relevant social groups. In particular, the UC was deployed in many quasi-independent scientific specialties, and had immediate commercial applications. The intention of the project reported on in this paper was to simplify the social context still further: by holding more variables constant, those bearing upon differences in design can be delineated more sharply. The study therefore concerns the development of one technique within a single specialty (infrared astronomy), with a relatively well-defined 'core-set', and no immediate commercial application.

3. THE ASTRONOMICAL AND TECHNICAL BACKGROUND

Astronomers who map the sky at infrared wavelengths have, until recently, used single detectors, or simple linear arrays with a handful of elements. Two-dimensional 'pictures' have then had to be built up by combining hundreds of individual observations, with great expenditure of time and labour. But, in the past decade, an alternative technology has emerged. It was pioneered by optical astronomers.[11] It involves the use of what is called a 'charge coupled device', or CCD: this allows the radiation falling on the flat surface of a silicon 'chip' to accumulate charge in discrete 'wells', from which it is periodically removed. With suitable switching circuits and sophisticated computer software, this process can lead to the display, very rapidly, on a VDU screen, of a picture reproducing the pattern of the incident radiation. Set in the focal plane of a telescope, this has proved to be a most powerful astronomical technique. It rapidly produced 'closure' in the astronomical community: the CCD is clearly accepted as a 'success'.

The question then arose : could this optical technique be extended to infrared wavelengths? The notion had not escaped those working in military R&D, and several industrial groups had, by the late 1970s, devised mosaic arrays of infrared sensors, with CCD readouts, for locating military targets. These are 'hybrid' arrays, in which a layer of sensors is 'bump-bonded' to an underlying, separate switching chip, or 'multiplexer' (for a variety of reasons, a direct 'monolithic' analogue to an optical CCD is not feasible). Although such arrays were not designed for astronomical purposes, and many of their specifications were far from ideal for such uses,[12] their existence, and the possibility of their applications in astronomy, eventually came to light. Writing in 1978, Soifer and Pipher stated that:-

> Such devices should provide a major increment in the sensitivity of infrared observations. A major challenge to infrared astronomers will be to adapt these devices to fully exploit their capabilities.[13]

The military industrial researchers pursued a range of technologies, differing in both the chemistry of the basic sensor material (silicon 'doped' with various elements; indium antimonide [InSb]; 'mercad telluride' [HgCdTe]; and germanium-based chemistries) and in the kind of multiplexer techniques (e.g. CCD, or 'charge injection device' [CID], or 'direct readout' [DRO]). These arrays can be designed to be sensitive to different parts of the infrared spectrum: they also differ in several properties of interest to astronomers - in their inherent sensitivity ('quantum efficiency' [QE]), in 'readout noise' (i.e. noise in the switching circuits), in 'dark current' (which flows at low temperatures in the absence of external radiation), in the linearity of their response to radiation, and in their switching speeds. A camera using such an array, besides incorporating appropriate lenses, mirrors and filters, associated switches and motors, and peripheral electronics, must also be *cooled* (with liquid helium or nitrogen) within (or in thermal contact with) an evacuated cryogenic ('dewar') vessel, or 'cryostat': much of the skill of infrared astronomers (and of the technicians in their teams) consists in suppressing extraneous thermal emissions by cooling and baffling.

The sequence of events from which this project stems concerns the marketing of an indium antimonide array, with 58x62 elements (or 'pixels'), by the Santa Barbara Research Center (SBRC), a subsidiary of Hughes (since June 1985, owned by General Motors). The array is sensitive to the shorter infrared wavelengths (1-5 microns), and was designed for military uses. An earlier 32x32 InSb array, using a CCD multiplexer, had been loaned to infrared astronomers at the University of Rochester by Alan Hoffman, a physicist who was working on their development at SBRC, and who had earlier been a postdoctoral fellow (in a group pursuing gravitational radiation) at Rochester. The Rochester group demonstrated the astronomical potential of this array,[14] and hence of infrared arrays in general. Meanwhile, Carol Oania, a development engineer at SBRC seeking new career possibilities, moved to the marketing department: the potential of astronomical 'consumers' for such arrays was pointed out to her, and she began actively to promote a programme to develop and sell arrays adapted to astronomical uses. A colleague at Hughes (Carlsbad) had been developing DRO multiplexers, which can be used with a range of detector materials. SBRC therefore proposed using these multiplexers, with InSb pixels (some variation in the doping was possible). Given the specifications offered, several groups of astronomers signed contracts, and laid plans for cameras incorporating what I will now refer to simply as 'SBRC arrays'. These groups and their activities were the main focus for this project. As a first approximation (but see below), the array around which each camera is designed and built can be taken to be a 'given', beyond the power of astronomers (and their technical teams) to alter. We can therefore observe the way in which the various teams harness what they take to be the potential of a common sensor technique, within what would normally be held to be a common social context of shared scientific aims, theoretical goals, established empirical data, reliable techniques, and other agreed points of reference. As we shall see, the groups developed instrumentation with their own distinctive 'signatures'.

The SBRC arrays had attractive specifications for ground-based astronomy in the near infrared, but not the least attraction of them was that they were commercially available *then* - a stage which other candidates had not yet reached. Some astronomers held back from signing SBRC contracts since, in their judgment, better options would soon be available: to opt for an SBRC array would be something of a compromise - it gets on with the job, but may be 'outflanked' relatively soon. Thus many astronomers (including some with SBRC contracts) were still actively pursuing contacts with other potential suppliers. In particular, there were many supporters of mercad telluride arrays, and several had already been tried out. The University of Arizona (Tucson), through contacts with Rockwell International (Thousand Oaks), had a camera successfully incorporating a 32x32 HgCdTe array, operating at 1-2.5 microns, with low dark current and good linearity;[15] it was reliably rumoured that a 64x64 array would soon be available. The University of Chicago had HgCdTe arrays from the same source.[16] The University of Hawaii, having achieved some successful observations with a 32x32 indium-doped silicon array supplied by Rockwell International (Seal Beach),[17] was at that time engaged with Honeywell in developing a new HgCdTe array.[18] The University of California at San Diego (UCSD) were cooperating with Aerojet ElectroSystems on bismuth-doped silicon arrays at longer wavelengths;[19] Berkeley, NASA Goddard and the Jet Propulsion Laboratory (JPL) were other groups with an active interest in infrared array imaging.[20] On the industrial side, Rockwell

International Science Center (Anaheim) were actively promoting their arsenic-doped silicon 'blocked impurity band' (BIB) arrays,[21] and Hughes (Carlsbad),[22] Cincinnati Electronics and General Electric (Syracuse) were among companies pursuing astronomical markets. There were also some European initiatives. (This list is not exhaustive.) All this activity, often associated with strident and conflicting claims, constitutes the background context to what follows.[23]

4. THE GROUPS

At that time (early October 1986), SBRC were several months behind schedule in the delivery of their 'astronomical' arrays - although they had supplied several groups with 'engineering' quality test arrays, to allow them to develop their camera circuitry. Five groups were known to have signed contracts with SBRC for the supply of astronomical arrays: it is convenient to list these in the order in which they signed contracts. This gives us, first, two national observatories, committed to the development of 'common-user' cameras for a range of applications, and, second, three smaller university groups of more modest means and more specific aims.

4.1. Royal Observatory, Edinburgh (ROE)

This national observatory, financed directly by the UK Science & Engineering Research Council (SERC), manages the British instruments on Mauna Kea (Hawaii), including the UK 3.8m Infrared Telescope (UKIRT), which is the largest 'dedicated' infrared telescope in the world. An earlier survey of British astronomers having given clear indication of the need for an 'infrared camera', the potential of infrared mosaic arrays was being monitored at ROE from the early 1980s.[24] In late 1982, Ian McLean, a staff astronomer at ROE with experience in optical CCDs, toured the USA to investigate the state of the art in infrared array technology.[25] He contacted, among others, Alan Hoffman at SBRC. His report intensified efforts to secure usable arrays - first with Rockwell, but later, following further contacts between Carol Oania and UKIRT staff in Hawaii, with SBRC. In the summer of 1984, while a contract was under negotiation, Ian McLean was appointed as 'project scientist' and leader of the ROE team to develop an infrared camera (IRCAM) using SBRC arrays.

The ROE's Technology Unit is organized into project teams which are allocated to specific instrumental developments, with one of the team designated as 'project engineer': each project consists of designing, making, installing and commissioning a common-user instrument, a task which may take two to three years. The team is then re-allocated to another project. In this case, the UKIRT organization asked the Director of ROE to assign a team to build and deliver IRCAM to Mauna Kea, where it would then be maintained by UKIRT as an instrumental facility for visiting astronomers. The staff costs come from the ROE budget, with other costs from the UKIRT budget. The basic Technology team was supplemented by other specialist staff with specific tasks. The group had 11 listed members essentially working full time on the project,[26] but as many as 20 people attended group meetings. The camera achieved 'first light' with a test array at 1.45pm on 26 March 1986, and was shipped to Hawaii at the end of September 1986.

4.2. Kitt Peak National Observatory (KPNO)

This national observatory is operated by the Association of Universities for Research in Astronomy, Inc., under contract with the US National Science Foundation (NSF). It maintains the many instruments on Kitt Peak, some 90 minutes' drive from Tucson, Arizona, where its laboratories are based. Until mid-1986, its organization had been structured more in terms of broad functions than of specific projects: 'scientific steering committees' interacted with technical teams embedded in engineering or support divisions. The team we are concerned with is the 'R&D' group, led by Al Fowler, which was the group within the Engineering and Technical Services division charged with the experimental development of new techniques, to the point at which their potential for astronomical use is established. Al Fowler is an electronic engineer, recruited from Rockwell, where he had worked on the development of multiplexers and sensor arrays. His early work at Kitt Peak was in optical CCD techniques, in which KPNO made a heavy investment. In August 1983, he began to circulate possible suppliers:-

> We are endeavoring to locate infrared arrays which could be used here for ground based astronomy. I believe that you may currently be producing suitable arrays for other customers. Since we do not have the funding to develop such arrays, I would like to determine whether copies of existing designs could be purchased by us ...[27]

The letter suggested how the problems of 'classified hardware' could be circumvented, and was accompanied by a detailed paper outlining the scientific needs.[28] In fact, this survey produced a disappointing response, but by mid-1984 arrays based on Carlsbad DRO multiplexers had been evaluated at KPNO, a contract was under negotiation with SBRC, and, by the end of the year, an infrared camera project had been agreed:[29] this was later refined in the summer of 1985.[30]

Until the summer of 1986, Fowler's group was under no obligation to 'polish up' a 'user-friendly' instrument, and had other projects in train.[31] However, under a reorganization then instituted, the group was renamed the 'IR Program (R&D)' group, and charged solely with completing two 'deliverable' versions of a common-user infrared camera - one for Kitt Peak, the other for the Cerro Tololo Inter-American Observatory, their 'sister' observatory in Chile. Previously, the group had just 5 members (including Fowler): 10 members were listed in the new regime. The 'Scientific Priorities & Steering Committee' for the project was chaired by Fred Gillett, a senior staff astronomer. The camera system achieved 'first light' with a test array early in May 1986; by midsummer, it had been tested on Kitt Peak, and was nearly complete.

4.3. University of Rochester

This group consists of two astronomers within the Department of Physics & Astronomy at the University of Rochester (Judy Pipher and Bill Forrest), with one postdoc, two or three associated students, and access to some technical support.[32] As mentioned above, this group pioneered the use of InSb CCD arrays from SBRC, and has published images of high quality derived from them. Pipher was trained in infrared astronomy at Cornell (some two hours' drive away), in the group led by Jim Houck, who is a NASA principal investigator: she retains close scientific and technical contacts with Cornell. Forrest did his graduate work in infrared astronomy at UCSD, followed by six years at Cornell, before joining Pipher at Rochester in 1981, at the time when the possibility of using an SBRC array was just emerging: he has since worked almost entirely on array systems, and is particularly interested in developing the imaging circuitry and software. A graduate student worked at SBRC in the summer of 1981, familiarizing himself with the use of mosaic arrays: the first (crude) images were presented in a poster paper at the AAS Meeting at Troy, NY, in June 1982. The group's work has been financed by NASA (Ames)[33] and the US Office of Naval Research. Once it was clear that they could afford one of the new SBRC DRO arrays, the group commissioned a relatively simple dewar/camera from George Gull, a technician based at Cornell who has several years' experience in building cryogenic devices. The process was initiated by a telephone call from Judy Pipher on 4 April 1985. The group take their camera to Kitt Peak, and to the University of Wyoming's 90" telescope near Laramie, for observations: one of the design specifications which George Gull had to meet is that the dewar must fit underneath an airline seat!

4.4. University of Texas (Austin)

This group is led by Paul Harvey, in the Department of Astronomy in the University of Texas, at Austin. Harvey gained his PhD in infrared astronomy at CalTech, then spent eight years at the University of Arizona (Tucson) before moving to Austin in 1981.[34] In 1983, the University of Texas Regents announced that they had available a fund for capital items, and asked for bids. Harvey had heard of the availability of infrared arrays, and had ideas for their possible use. He was successful in his bid for money, and this, topped up by a grant from a US Department of Defense (DoD) fund for university instrumentation, gave him enough for his hardware.[35] Originally, he had in mind purchasing the Rockwell array with which the Hawaii group were involved, but, after meeting Carol Oania at a conference in Wyoming in 1984, he signed a contract with SBRC early in 1985. He purchased an 'off the shelf' dewar from Infrared Laboratories, Inc.,[36] basic computer equipment, and standard software packages for imaging. Meanwhile, with funding from the US Air Force (secured in February 1986), he recruited two postdoctoral fellows (one from Hawaii), to join one postdoc (specializing in optical and mechanical designs) and one graduate student; the group has access to technical support within the Department. Harvey said that he aimed to "have the best group in real time infrared 'speckle' interferometry":[37] he was aiming to have the camera on the 107" telescope at the University's McDonald Observatory (five hour's drive away) by the end of 1986. But this was *not* his *only* project.[38]

4.5. California Institute of Technology (CalTech)

The CalTech infrared group, led by Gerry Neugebauer, is one of the oldest and most influential in the field. They make ample use of the 200" telescope at Mount Palomar (some three hours' drive away), and now have access to their very large (10-metre) 'Keck' telescope, in Hawaii. They have collaborated for many years with JPL over the use of arrays in spectroscopic work. They have employed Rockwell and Cincinnati arrays, and had been in touch with developments at SBRC since well before Alan Hoffman's initiative at Rochester: this experience bred a certain caution in evaluating the possibilities:-

> Infrared arrays aren't quite there yet. There's no array in hand that has really given significant results ... There's no really useful device you can go out to a telescope and really do something with ... But it's coming ...[39]

The CalTech group has always attempted to maintain as wide a range as possible of the best instrumentation: this led naturally to plans for an array camera:-

> We've realized for a long time, without thinking about other groups at all, that when there came a time that one could image well, we'd want to do it ... We have to get into the business, and we have to understand what goes on, before we build [instruments for the Keck].[40]

So a contract had been signed with SBRC. The camera was to be built by Keith Matthews, with one technician. As of early 1986, there was no hardware (or software, since this was the group's first attempt at imaging), but a graduate student had made some optical designs. No striking innovations were intended, but the "straightforward" design had not yet begun to take shape.

5. THE ROE AND KPNO GROUPS AND THEIR DESIGNS COMPARED

From the perspective laid out in the first section of this paper, a comparison of the two national observatories and their designs is the most immediately interesting. I will therefore attempt this in some detail, before going on to some comments on more general issues.

Figures 1 and 2 show what the ROE and KPNO cameras actually look like. To the uninitiated eye, the contrast is striking: disregarding the peripheral electronics, the ROE camera consists of a conventional, cylindrical dewar set on a rectangular, flat camera box, while the KPNO instrument (camera plus dewar) is totally enveloped in a cubical container. But those involved dismiss these differences as 'superficial' or 'insignificant'. The ROE team leader, for instance, put it like this:-

> Your first impression when you look at [the KPNO] cryostat compared to ours is that it must be very, very different - there must be a *fundamental* difference - because it's a different *shape* - it *looks* different. But in fact that isn't the case. We could easily have adopted their kind of cryostat, and they could easily have adopted our kind of cryostat - in fact that's a very minor, surface effect.[41]

And the KPNO team leader agrees: but he sees some advantage in the KPNO cubical design, in terms of ease of access to the camera interior. On this, the ROE leader commented:-

Figure 1

The ROE array camera
(dewar above)

Figure 1a

The ROE array camera mounted on the
UKIRT flexure rig

Figure 2

The KPNO array camera

Various views of the cryostat
(dewar + camera)

Figure 2a

The KPNO array camera (cont.)

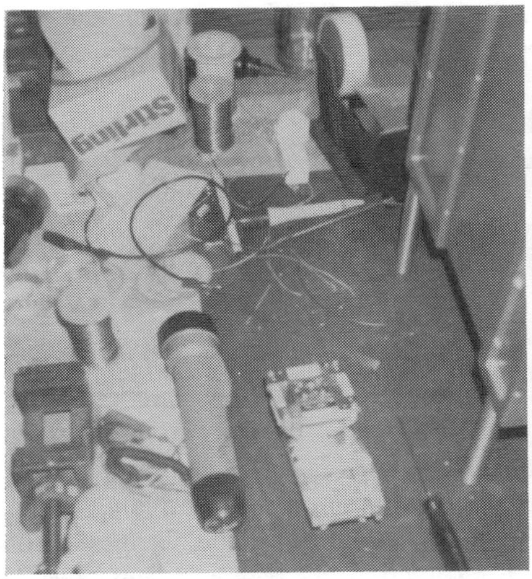

The 'lunar lander' ...

.... and two views of the box +
electronics in which the camera
is mounted on the telescope

> That's true. There is ease of access in a square box like that. And it's for that *same* reason that we made the bottom chamber of our cryostat a box ... just for that reason. Because [previous ROE] cryostats ... are all cylindrical extensions to [Frank Low-type] cylindrical dewars, and they're all devilishly hard to get access to. So that's why we also adopted a square box. The [dewar] then just becomes a reservoir of cold, nothing more.[42]

However, when one presses further into other detailed differences between the two designs, more significant features emerge. Since these reflect differences in the social and physical circumstances of the groups, it would be helpful to present these latter differences, in outline, first, before returning to the designs.

5.1. 'Social' and 'geographical' differences

Overwhelmingly the most important difference between ROE and KPNO concerns the **relative accessibility of the telescopes**. The KPNO labs are only a short drive, on a good road, from the Kitt Peak telescopes: Mauna Kea and Edinburgh are on opposite sides of the world. More than that: Mauna Kea is on the relatively remote Hawaii Island, 30 minutes by plane from Honolulu; and, worse, the telescopes are at 14,000 feet, up a rough drive. Even to take instruments from the peak of Mauna Kea to the UKIRT lab in Hilo for maintenance is a major operation. Liquid helium, for instance, has to be manufactured on the island: UKIRT runs a liquefaction plant for all the Mauna Kea telescopes. All this, as we shall see, has an effect on the teams and their camera designs.

> At 14,000 feet you don't want to have to think. Things must be easy, both from the maintenance point of view, and from the users' point of view.[43]

But one effect of this relative inaccessibility is 'positive': it forces ROE to place a much greater emphasis on the **laboratory testing of instruments**, before they are shipped to Mauna Kea. In Edinburgh, for instance, there is a dummy test rig which simulates the UKIRT mounting, and on which any instrument can be subjected to flexure tests (see Figure 1a): there is no equivalent at KPNO. Facilities in Hilo are relatively limited: instruments *must* leave ROE in a fully tested state. This emphasis on lab testing extends to the electronics, and the arrays themselves: after all, Santa Barbara is not near Edinburgh, either. ROE are forced to do in the lab what KPNO may, other things being equal, be able to do at a telescope, or to sort out by a quick trip to Santa Barbara: and this, too, makes a difference. It also results in the ability of ROE to test arrays in ways that may not be possible at SBRC: since one of the conditions for the release of arrays from the industrial suppliers is that astronomers make test results available to the companies, this can be a strong bargaining counter in negotiations for contracts.

Another striking difference between the two groups is the **different background and training of the group leaders**. At KPNO, the leader is an *electronic engineer*, to whom the camera might be thought to have a primarily technical 'meaning': at ROE, the leader is an *astronomer*, to whom the camera has a primarily 'scientific' meaning.

But it is very difficult, in practice, to trace many specific design features to this apparently clearcut difference - which is another way of saying that it is difficult to distinguish, either 'intellectually' or 'institutionally', between 'scientific' and 'technical' (and, for that matter, 'political') aims in the design of instrumentation. There are also differences in the relationships between astronomers and engineers in the two teams. I will return to this issue again later.

The two groups also differ in their **size**, in the **formality** of their organization, and in the style of **leadership** exercized in them. The ROE group has always been relatively large, its activities clearly differentiated, and its coordination characterized by a certain formality.[44] From the start, there had been a monthly team meeting, with agendas and minutes: the first, on 2 July 1984, was convened in these terms:-

> ... it is appropriate to initiate a formal series of monthly meetings associated with the Infrared Array project. Having prepared an initial specification and plan we must now re-assess these and begin to set up the Project on a more secure footing.[45]

On the other hand, no such formality or regular 'meetings' were required by the smaller, more intimate KPNO team, which was led in an involved, 'hands on' style.[46] However, with the later reorganization at KPNO, from October 1985, a 'group meeting', with minutes, was instituted, taking place every 2 to 3 weeks.[47] This change went hand in hand with Al Fowler's enlarged role and remit (see above), and the greater formal involvement of the KPNO scientists in the management of the project.

With these differences in mind, let us now return to the designs of the cameras.

5.2. Some specific design differences

As we have seen, the KPNO camera has a distinctive, cubical dewar. This dewar design is something of a KPNO 'tradition':[48] it was devised by Richard Capps, who has since moved to Hawaii, taking the design with him - it is now also in standard use at the NASA Infrared Telescope Facility (IRTF) on Mauna Kea. But an astronomer in Frank Low's Department at the University of Arizona (which is only 50 yards, over the road, from the KPNO labs) told me that he would *never* use KPNO dewars - they contain too many sources of stray reflection, and hence raise problems with baffling; and even the smaller Low dewars contain twice as much liquid helium. I put this point about the **amount of liquid helium** to the ROE team leader:-

> That's not fundamental. Again that's a matter of logistics on their mountain [KP]. Our feeling was that we chose the much larger reservoir of both liquid helium and liquid nitrogen to make filling just less frequent on Mauna Kea. We will fill only once a day with liquid nitrogen, and only every *two* days with liquid helium.[49]

Yet this consideration does not weigh heavily against the use of KPNO dewars at the IRTF on Mauna Kea: why this is so has yet to be investigated. Clearly, the perception of this difference in dewar design as 'insignificant' is problematic. The 'coexistence' of two dewar designs - of which one (the Low dewar) is the 'commercial' success, on which closure might *seem* to have been achieved, and the other represents a 'deviant tradition', sustained by its designer - suggests another case study for Pinch and Bijker's programme.

But one difference which both team leaders agree is of some significance is that the ROE design uses a **'gas switch'** to **control the temperature of the detector** (Figure 3).[50] The more conventional way of doing this job (adopted at KPNO) would simply involve a fixed copper braid connection to the liquid helium container, trimmed to an appropriate width. But the gas switch offers a more sophisticated method of controlling the temperature continuously during operation - which allows an array's performance to be evaluated in ways which the industrial suppliers cannot replicate. As the ROE team leader put it:-

> That is a difference. But again, the gas switch is just a way of establishing an intermediate temperature stage, which [can be] trivially done by having a little copper block and isolating it with fibreglass pillars from the liquid nitrogen, and connecting it by a copper braid to the liquid helium, and literally *trimming* that copper braid until you get it right. Presumably that's what [KPNO] have done here. We decided to take the gas switch route because it gave us much more flexibility in control of the temperature. We decided early on that we wanted to have that because ... one of the things we 'sold' to the company - Santa Barbara, and other companies - was our ability to stabilize at *any* temperature between liquid helium and liquid nitrogen, and evaluate the performance as a function of temperature.[51]

In other words, this design feature increases ROE's bargaining power in negotiations with SBRC. He went on:-

> And the second reason why we moved away from the 'fixed link' type of approach, which we did investigate - it *was* studied - is that it *is* a fixed link, and if you want to change it, then you have to open up and go in there and do something about it. Now what we were particularly keen - what *I* was particularly keen to see - was that we never get into a situation where we could not use the camera on Mauna Kea because we'd run out of liquid helium ... Because it is always possible to put in nitrogen and pump ... evacuate ... it and get it solid. And that gets you down to just below 50°K, which is quite healthy ... But with a *fixed* link connector, you might find that, if you had optimized that fixed link for a *lower* temperature - 20° or 30°K - that it would hold you off of 50°K - it just would not let the detector get down to that - it would keep it warm by 10° or so. But a gas switch *doesn't work* [as a variable conductance] until [the cryostat] has liquid helium in it - it's a dead *short* - and that will short out and give you solid nitrogen temperature operation.[52]

This constitutes a genuine 'user reason' in the conditions pertaining on Mauna Kea:-

Figure 3

The ROE array camera, showing the gas switch

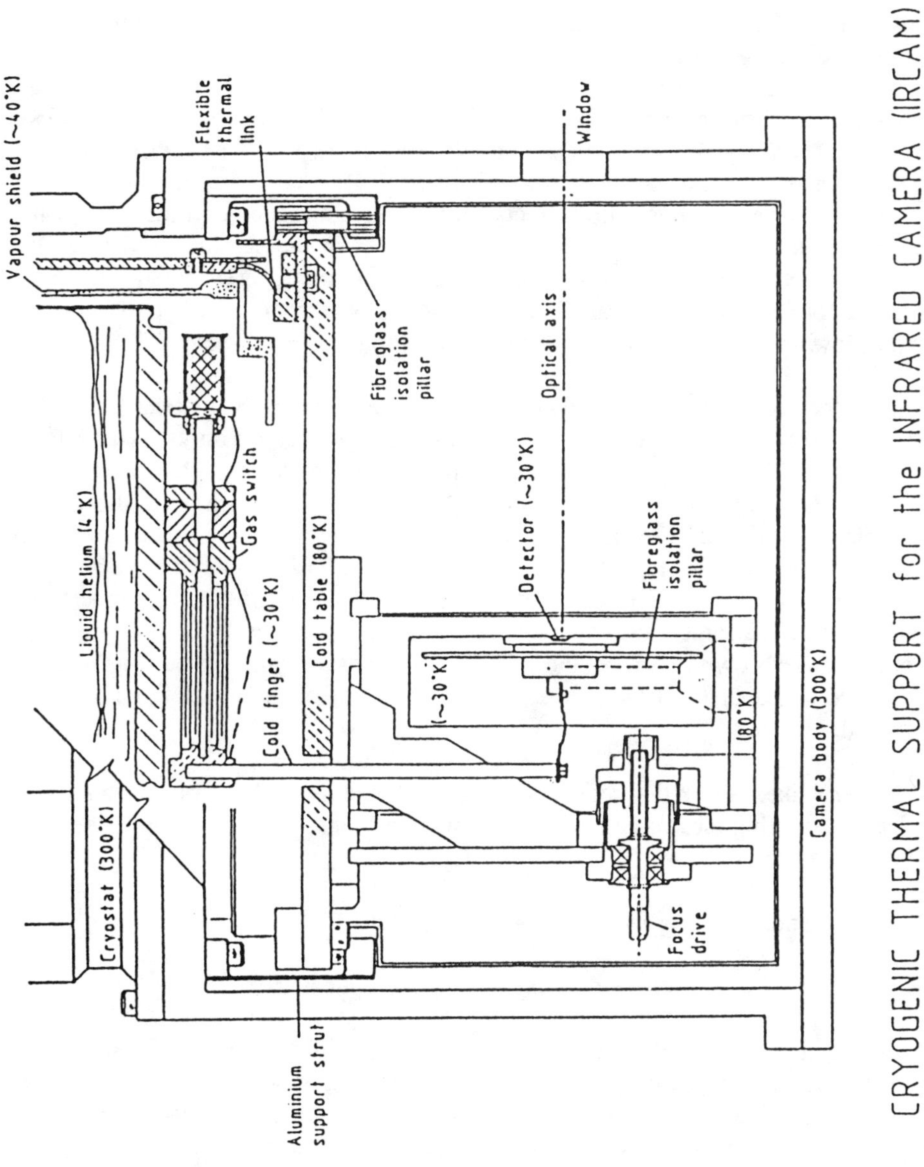

> ... you need to work in this intermediate temperature region, so you want to be able to do that efficiently. You don't want to waste helium ... and you want to do it in a flexible manner, so that if you *are* out of helium, you *can* use nitrogen. Now that's a logistical problem, again, for Mauna Kea - but it probably isn't a problem on Kitt Peak.[53]

So this design difference is partly shaped by a need to satisfy industrial suppliers, and partly by user needs in particular circumstances.

Another aspect of 'user friendliness' concerns the relative effort devoted at the two centres to **software** - and, in particular, to software related to the **user/instrument interface**. Of the eleven main members of the ROE IRCAM team, five are used as software specialists, of whom three are concerned with interfacing: on the other hand, the KPNO team only recruited one software engineer in the summer of 1985. This obviously reflects the original location of the KPNO group in 'R&D': the development of software for user friendliness at KPNO was the job of a separate, 'Support Services' division. How this will affect the use of the cameras remains to be seen.

Another difference which it is conceded may be of significance is the fact that the KPNO camera has **completely redesigned and repackaged electronics**. The KPNO camera uses no SBRC ancillary electronics: after some years of experience testing the Carlsbad DRO multiplexers, the KPNO team were in a strong position to design and build their own drive and readout units. The ROE camera, on the other hand, uses SBRC electronics for driving and reading the arrays. The KPNO camera also has a specially redesigned mounting for the array itself (nicknamed the 'lunar lander': see Figure 2a): the ROE camera uses commercial mounts supplied by SBRC. These aspects reflect Al Fowler's own expertise: the KPNO electronics are much more compact, and are designed to reduce electronic noise levels. Expressing his admiration, Ian McLean commented:-

> That I *do* like. I think that reflects greater and longer experience of low noise, dense packaging electronics, which is something that Al Fowler's quite well known for.[54]

On the other hand, the 'lunar lander' drew the comment:-

> I think it's a little complicated - *unnecessarily* complicated.[55]

And this raises a point of potential tension between development engineers and astronomers in such projects: the astronomers are often happy when things 'work', even if an ideal design has not been attained, and are anxious to 'get the camera on to the telescope and do science' - they can taunt the technical staff about 'wasting time' trying to 'Rolls-Royce' the instrument. Such rhetoric has been used at ROE. But a common-user instrument *has*, at least in some respects, to be 'Rolls-Royced', and the team know of this requirement. There is also widespread admiration for a 'well engineered' piece of equipment, and observatory staff often criticize others' hardware for its 'poor workmanship', even if they concede that it 'works satisfactorily'. So there is potential conflict, and what was described to me at ROE as "a kind of schizophrenia - astronomers run projects, but engineers build". This is presumably an institutionalized reflection of different 'meanings' attached to the artefacts, and the conflict may be evident at different times and in different places at ROE and at KPNO, given the differences in their work organization. I will return to this theme later.

Another substantial difference between the two instruments is that the KPNO camera is designed for **'single instrument' mounting** - that is, when in use it will be the *only* instrument on the telescope - whereas the ROE camera is designed to be **mounted as one of an 'instrument cluster'**. The immediate effect of this (apart from the requirements imposed by space restrictions) is that the KPNO camera is 'up looking', while the ROE camera is 'side looking'.

> Again, it's a demand of science ... We [at ROE] think that it's a very powerful feature to be able to switch wavelengths on the same telescope very rapidly ... So, for example, you could take an image with the camera and, in a matter of two minutes or so, you could be taking a *spectrum* of the same object ... Or you could be monitoring the wavelength dependence of flux, polarization, whatever, in an object from 1 micron to 1 millimetre.[56]

This 'scientific' rationalization marries well with the 'geographical' tendency to leave instruments mounted on UKIRT atop Mauna Kea, where they are designed for maintenance and operation *in situ* - whereas at Kitt Peak, installing and demounting instruments are more routine operations. And this is reflected in another design difference: in the ROE camera, filter wheels can be changed *in situ*, without taking things out of the cryostat - it has to be opened, but not *dismantled*, for the ROE design emphasizes *modularity*.

One final difference is worth mentioning. The KPNO camera, for what its designers present as essentially 'scientific' reasons, is designed for use at wavelengths from 1 to 20 microns: the SBRC array covers 1 to 5 microns, and it was anticipated that a different array, on a different 'lunar lander', will cover the rest. The optics of the KPNO camera are therefore entirely 'reflective' - i.e., they consist of mirrors only - whereas the ROE camera, designed only for 1 to 5 microns, uses both mirrors and lenses. What effect (if any) this has in practice remains to be seen. But it raises a matter which is common to both observatory groups, and on which they may be 'vulnerable' to initiatives from smaller, more adaptable university groups. Both ROE and KPNO have to satisfy a *range* of potential astronomical users. There is "good science to be done" from 1 micron to 20 - and beyond. Compromises have to be made. Merely to opt to start out with an SBRC array is, in some respects, a limitation. Astronomy at 1 to 2.5 microns has a somewhat different character from astronomy at 2.5 to 5 microns. In particular, the ambient background is lower at the shorter wavelengths, and this limits the usefulness at these wavelengths of a detector with a relatively high dark current: in 1986, the SBRC arrays seemed likely to be 'dark current limited' in this band. In addition, the design of the SBRC multiplexer is notably non-linear : this raises problems for astronomers, who require their detectors to be carefully calibrated. A non-linear array will produce perfectly acceptable *images*, but if calibration (by means of corrective software - it may not be necessary to modify the detector *hardware* itself) is not possible, such images will be of limited astronomical use. This was perceived as a potentially crucial problem at ROE, and was being addressed in 1986. (Whether it was seen as problematic at KPNO is not known.) However, as the use of the Rockwell HgCdTe array at Arizona (see above) was demonstrating, arrays *are* becoming available with at least comparable performance in the very near infrared, and less prone to these particular problems (although they may well be limited by *readout noise*). There is some agreement that the 1-2.5 micron band is "becoming more competitive" (although it does not necessarily follow from this that the main use of IRCAM will be at 2.5-5 microns - as an ROE team member put it, 'good science' is often to be found where the competition is fiercest!). But such developments were always foreseen. As already mentioned, SBRC arrays were often ordered merely because they were available. Both national observatories continue actively to monitor alternatives (ROE with Mullards, for instance), and both the ROE and KPNO cameras are so designed that their SBRC arrays can be replaced, as required, by a range of other options - as (and if) other arrays come on to the market and demonstrate their astronomical virtues. The cameras are designed to outlive any current generation of arrays, and they will not be rendered useless by the obsolescence of the particular arrays around which they were originally designed.

6. RELATIONSHIPS BETWEEN ASTRONOMERS AND SUPPLIERS

Not surprisingly, the smaller groups approach the design of their cameras in a somewhat more opportunistic and *ad hoc* manner. Since they are not providing common-user instruments, they need only consider their own needs. Since they have more limited resources, they tend to make more use of 'off-the-shelf' items (dewars, computers and electronics, software packages) in putting their systems together, and are less inclined to take technical risks, or to plan major innovations. Since they are smaller, they can work faster. And since they have little bargaining 'muscle' with industrial suppliers, they might be thought to be more inclined to 'take what they are given', without demur.

In fact, on this latter point, the Austin group is the only clear exemplar. Paul Harvey, when asked why he had signed a contract for an SBRC array, answered:-

> I think that almost every area of infrared astronomy is going to be revolutionized by this ... I would feel very bad if we didn't have an array here ... I mean, that's why I want to get this array to keep us in the game ... There are just too many things to do.[57]

Asked whether he had argued with SBRC over the specifications of the array, he replied:-

> It was clear that we could do *tons* of fantastic astronomy with the specs, and I just didn't think it was worth haggling ... The understanding was that they were going to *beat* the specs, and they were hoping they would, but all they would guarantee was the specs ... And we looked at them and said we can do lots of astronomy with that, so ...[58]

In fact, the 'specs' that SBRC offered were the result of earlier negotiations between SBRC, ROE and KPNO: this now constitutes 'the package'. And although the SBRC 58x62 InSb DRO programme is specifically aimed at an astronomical market, with some optimization of the detector properties accordingly, these specs are still severely constrained by the suppliers' *military* customers, who lay down the design principles, and remain the basic consumers, of the *underlying technology*. As Al Fowler's letter rightly said, astronomers "do not have the funds to develop such arrays".[59] To the suppliers, astronomers are a tiny market: profit (if any) is only icing on the cake, and, although there is some public relations benefit in the business, it alone cannot justify the effort. So the modifications which suppliers are willing to make to the arrays that they are already manufacturing for their military customers are strictly limited.[60] This was a constantly recurring interview theme. Yet it does not follow that there is no point in astronomers developing close relations with suppliers: quite the reverse. On the one hand, suppliers value the 'feedback' astronomers can provide on the performance of their devices in special test conditions (and the demands of the Strategic Defense Initiative tended to make this rationale more compelling): on the other, astronomers need constant advice from the suppliers' experts on how to extract the best performance from the arrays. But, more subtly, several astronomers argued that it was *essential* to develop such relations, for they allowed the transmission of an *understanding of the physics of the detector process* without which 'good science' was very difficult to do. It is hard to evaluate this claim, but it was made by many interviewees. Several groups of astronomers cultivate such relations: Hawaii, CalTech, JPL, UCSD, Rochester, Arizona, Berkeley, Chicago, and many others, have such 'inside tracks'. What it implies can best be illustrated by an example.

Since the formation of the IRCAM team, ROE were in conversation with the SBRC development engineers. At first, contact concerned the performance of the SBRC electronics, and possible modifications to them: later, it extended to the test arrays and their behaviour. SBRC engineers visited ROE to discuss problems. But all this, open and amicable as it was, could still be thought of as slightly 'arm's length', supplier/customer relations. Two unexpected problems forced the interaction on to a 'deeper' level.

The first problem came to light early in 1986. SBRC engineers, testing an array at liquid helium temperatures, found a striking fall in the 'quantum efficiency' (QE), or basic sensitivity, of the detectors at the lower temperatures. This effect had not been observed with any other arrays, and was entirely surprising. It was large enough (from about 75% to below 10%) to make the arrays of limited use for astronomical purposes. Measurements on the test array which had previously been delivered to ROE failed to reproduce the effect: similar tests at KPNO were inconclusive, but suggested that the effect was real. The situation was a classic 'replication' dilemma.[61] SBRC sent an array which had demonstrated the effect to ROE: ROE's tests on it replicated the effect. Why, then, had the ROE test array not demonstrated it? ROE argued that their test array had been treated with an anti-reflection coating: perhaps that was the cause of the problem? But this seemed physically implausible. Meanwhile, SBRC discovered that the newer arrays had a *thicker* detector layer than the specifications laid down, while the ROE test array had the 'correct' thickness. The SBRC team therefore conducted tests in which the performance of each element in an array was compared with its thickness, as measured by a laser technique: the results suggested a clear relationship between thickness and fall in QE with temperature. Perhaps the effect was a problem of diffusion of charge within the body of the detector? But further tests at ROE seemed to rule out this possibility as the major factor: emphasis shifted back to the array surface, and the ROE team proposed a physical mechanism by which 'trapped' charges in the surface layer produced the effect. Their evidence seemed (in mid-1986) to be producing 'closure': SBRC were planning to coat their next array with a new formula, only used to date in military applications, which promised to solve the problem.

The second problem arose when the ROE team were carrying out routine tests of the noise levels in the circuitry associated with the arrays. They discovered that the noise level increased dramatically when the apparatus was left for a period, only 'recovering' when it had been running again for some minutes. At first, they thought that the effect must be in the SBRC electronics: but they eventually decided that it was located in the array itself, and arose when the array was left 'unaddressed' - that is, with its elements 'unread'. The answer was to redesign the addressing mode, by rewriting the software. It transpired that ROE had 'stumbled upon' a new way of operating the array that had not arisen in the SBRC labs. A telephone conference on 16 July 1986, involving two of the ROE team and three at SBRC, produced 'closure' on this solution.

In both these cases, the interaction between ROE and SBRC reached what one might call the 'basic physics' level. From the point of view of SBRC, this involved invaluable 'feedback', and helped to justify their investment. But, from the point of view of ROE, the interaction forced the astronomers into a basic understanding of the physical processes from which their observations derive: normal 'customers' are not in this position. It remains to be seen what benefit (if any) this leads to in practice, on Mauna Kea. Meanwhile, the examples serve to illustrate the 'inside track' relationship that my respondents see as essential to effective astronomical research. But, in the experience of struggling to produce a 'working' instrument around a new detector, astronomers *learn* at a fundamental level, whether or not they are in close touch with the suppliers. To quote Paul Harvey again, this was his response when I asked him whether he was designing his camera so that, after three years or so, he could unplug his SBRC array, and plug in a new one - or whether he would have to 'go back to square one', and design a new camera for the new array from scratch:-

> My experience with every other instrument I've ever built is that by the time it's two or three years old I know so many things I wish I'd done differently that I just wouldn't want to do anything but start again from scratch![62]

And when asked whether he had other users in mind in his design, he replied:-

> I know when it starts working, lots of people are going to want to start using it ... We had the same problem with [our Cincinnati linear array spectrometer] ... The spectrometer is just starting to produce good data, and we're already getting lots of requests from people to use it ... In principle I'd like to turn it into a user instrument eventually ... but my experience is that it's hard enough to do that ... Technology advances fast enough that by the time you've got

round to finding the energy to turn it into a user instrument, you might as well go out and buy a different array.[63]

7. COMPETITION AND FUNDING [64]

It might be thought that one implication of Paul Harvey's belief that arrays offer "lots of astronomy" for everyone is that this relieves the field of **competitive pressure**. But Paul Harvey himself implicitly denied this. He spoke of a "natural tendency":-

> I want to see this on the telescope as soon as possible. I don't like months and months testing things in the lab without taking it out to the telescope to see what I can do with it. So we will take it out to the telescope sooner than some other people might. And we'll tend to do our testing simultaneously with our science.[65]

This sense of scientific *drive* is usually associated with the idea that science is 'competitive'. But it would be misleading to think of this as a simple 'competition', or 'race', between five groups, all aiming for one 'goal'. Obviously, everyone wants to be a 'front runner': there is talk of being the first to get an infrared mosaic image on the front cover of *Nature* as a symbol of 'winning'.[66] But when groups are clearly in competition for one 'prize', their relations are characterized by a degree of *secrecy*: and, when competition is intense in a field, groups will try to learn what they can of the others' strategies so that they can modify their own plans to gain some tactical advantage.[67] One striking feature of this case study is the almost total absence of such behaviour. Apart from some complaints that the British, early on, were somewhat tight-lipped (see below), there seems to be no sign of conscious control of the flow of information: all the evidence suggests that the groups are entirely 'open' with each other. Paul Harvey, for instance, received informative letters "every two months or so" from Bill Forrest in Rochester. Nor is there any evidence that, apart from a natural curiosity, anyone exhibits any *need* to know what other groups are doing: there is no case, to my knowledge, where any specifiable detail of a camera design has been altered *because it became known what another group was intending.* A kind of 'scientific indifference' seems to pervade the field.[68] When Paul Harvey told me that he thought that his emphasis on real-time speckle interferometry would give him "an advantage", he prefaced this remark by saying that he "didn't know too many details of the other groups" (apart from Rochester). On the other hand, when told that both ROE and KPNO had achieved 'first light' on their cameras, he said that this meant that he would "have to get moving".[69] Judy Pipher, at Rochester, expressed her particular 'small group' perception:-

> I don't really consider it competition because I think they haven't had a chance to even know what the problem areas are. And I think we have. I don't get perturbed. I wanted arrays to be generally available to people. I didn't like being a 'preferred category'. It seems to me that the sky is a big place. There's interesting astronomy to be done by a lot of people ... I don't worry about that ... I'm not the slightest bit worried about [being scooped] because when I look at these places with the big technical resources they're *awfully slow* at getting

> anything done. The observatories, for example - they spend *ages* testing, they work 9 to 5, they have to idiot-proof their stuff for visiting astronomers ... I'm not worried.[70]

The perception that the arrays offer "lots of astronomy" for all may help to explain this lack of secretive behaviour. But there is more to be said. With ROE and KPNO, national pride is at stake. The UK SERC have made a large investment in the IRCAM project, and the team are well aware that its success is important for the future of astronomy in Britain - indeed, the future of ROE, and of their own jobs. The early British reluctance to talk freely about their detailed plans may stem, in part, from this (but see below). Certainly, the rhetoric of competition with other groups (particularly KPNO) who might "get there first" has been used at ROE as a team motivator. (The ROE team leader admits that he has used it consciously as a management tactic, and denies that it expresses his "real views".) It pervaded a 'crisis meeting' of the group held on 18 March 1985, at which the "political importance" of the project was stressed, and a target (later relaxed) of having IRCAM on the telescope "by the end of 1985" was set.[71] And at the celebratory champagne party on 10 April 1986 to mark IRCAM's 'first light', Ian McLean, having noted that ROE had been first to sign an SBRC contract, and were now first to 'first light', added:-

> But being 'first' or 'best', by itself, is not enough : I want us to be first *and* best![72]

However, some in the ROE team exhibit less enthusiasm for 'being first'. As one of them expressed his belief to me, early in 1985, ROE have the resources to "do the job properly", and it is clear that they *will*, eventually, do just that - so they can "let some other group take the front cover of *Nature*" while they concentrate on doing the best longterm *scientific* work. He did *not* see the situation as particularly competitive. This strongly suggests that discourse involving the notion of 'competition' is a deployable resource, and its use controlled by social circumstances.

Of course, since the arrays are developed in labs covered by the restrictions of national security, and derive from military contracts, which are themselves the focus of intense competition among aerospace firms, a pervasive air of secrecy surrounds the detector technology itself. Groups with an 'inside track' access to military arrays may be bound by restrictions on how much detail they can present to less fortunate astronomical colleagues. This seems to have been the main reason for the early British secretiveness, noted above. Since SBRC were in negotiation with a *non-American* group, they imposed very tight conditions of secrecy on ROE staff: although these conditions were later removed, the discipline they engendered set the tone for some early contacts, and these memories remain. Sometimes, a group in such a 'preferred category' can be seen as having an unfair monopoly of a new technique - hence Judy Pipher's unease. One distinguished astrophysicist described such restrictive agreements to me as "illegal" - although, on closer questioning, he modified the word, through "immoral", to "improper". It was put to me that the application of optical CCD techniques was hindered by the relatively 'restrictive' policy of Texas Instruments: in the infrared sensor field, Cincinnati and SBRC were complimented on their relatively 'open' approaches, with devices on commercial sale. The matter arouses strong emotions, especially among those excluded: the issue requires further investigation. However, what is clear is that the basic rationale for the secrecy here is *industrial*, not military: what worries suppliers is the possible leakage of crucial *proprietary* information to their commercial competitors - not (except in the ROE case) any threat to 'national security' (although the latter determines the US Government's reluctance to grant export licenses).[73]

Despite the pervasive sense of financial crisis in the Western democracies, and the predations of the US Gramm-Rudman Act, the finances of the two national observatory teams are (or at least were, in 1986) *relatively* secure - indeed, relatively lavish. Neither was, at that time, in intense competition *for cash*. And, despite complaints that unless your group has 'status' it cannot command funds, the smaller US groups can put together the resources for an array camera project without too much difficulty, as the experience at Austin demonstrates. However, there is one aspect of the US scene of overriding significance, which dominates those involved in this field - and, indeed, American astronomers in general. I am referring to the planning process for the major NASA projects - notably, the Space Infrared Telescope Facility (SIRTF), and, more distantly, the second generation of instruments for the Hubble Space Telescope (HST). SIRTF's planned payload included, *inter alia*, a camera working in the 2 to 8 micron band. Through a number of routes (of which NASA Ames is one), NASA funds research intended for the evaluation, design and development of instrumentation for SIRTF. (Work of this kind is carried out at KPNO.) Many astronomers believe that, to remain in the scientific 'big league', one must secure a *space* instrument contract. There is no doubt that bidding for this goal is *fiercely* competitive. The Rochester group is involved in this competition. Much of the money to support their work with the SBRC array has come from NASA Ames : and, in 1986, they were engaged in a bid to secure a place in the NASA contract for the 2 to 8 micron instrument on SIRTF. Their 'competition' was therefore with more traditional infrared astronomers who prefer high quality single detector technology, and are suspicious of arrays - and, more recently, with advocates of HgCdTe arrays. The Rochester group's 1985 paper reporting their results with the 32x32 InSb CCD array culminates in a sentence that some see as having especial point:-

> Thus each of the approximately 900 working pixels on our array performs as well or better than the best single element InSb detector currently in use for astronomy.[74]

This NASA/space goal dominates their strategy : it is the 'meaning' they give to their array camera. While, of course, they plan to 'do science' with their arrays, even astrophysical goals can be overridden. As one of the Rochester team put it:-

> I'm more concerned about getting information for the SIRTF studies and being able to argue with [the NASA panel] with *real data*.[75]

And their team leader spelt out what this implies:-

> With the DRO, we're getting into it for a different reason. We're going after getting these things to operate in space. We're going to spend a lot of time in the lab. People may have them on the telescope before us. So what? That's not

> what we're after. We're very [soon] going to get optimized InSb arrays that Santa Barbara's building *just for us*. And that's what *we're* after. We're after a different output.[76]

And as to the race to 'get the camera on to the telescope', this is the meaning given to that goal at Rochester:-

> We know that there's a lot of things that have to be done, and we have to, by the [NASA] advocacy review, be ready for it. There are a lot of things we don't know the answers to. We know from experience that things will come up we won't understand. And you've just got to test it at a telescope ... In fact our NASA Ames proposal this time is emphasizing how important it is to take a device to a telescope. Because that's where you uncover problems. We found some real problems with ours that had to be adjusted after we had it in the field, that we didn't see in the lab.[77]

The broader scientific, political and financial context is here giving a distinctive 'meaning' to the camera this group is building - and hence influencing its design. At Rochester, a 'satisfactorily working' camera is, ultimately, one that provides "real data" to convince a NASA panel.

8. SCIENTIFIC AIMS AND TECHNICAL GOALS

Asked whether his desire to build an infrared array camera had a scientific or a technical source, Paul Harvey replied:-

> Well, to be perfectly honest, I find it easier to do astrophysics ... *interesting* astrophysics, by building instruments which make new kinds of observations possible. I enjoy that. Certainly, the science drives everything. My own particular interests are in star formation ... and we just have a lot of fun building instruments that can detect fainter stars, can see more detail in the clouds around younger stars, things like that ...[78]

Later, he ventured this comment:-

> I've found in the reviews we get of our NASA-sponsored research in the [Kuiper airborne observatory] that we always get good reviews when we do something interesting technologically. If we decide to just do science for a couple of years without developing new instrumentation, our reviews start getting worse. And I think they just like to see new technology development.[79]

Whatever the truth or justice of this last remark, it raises the question of the relative 'weights' accorded to technical innovation and to 'scientific steering' in the design of new instrumentation.[80] Does a purely technical 'momentum' carry all before it? Or is the process significantly 'shaped' by scientific aims? Is it useful to conceive of the process in such 'either/or' terms at all?

It is obvious that astrophysical aims inform every astronomer's work: without the belief that a new instrument was to be, in some sense, *scientifically valuable*, the work could not be continued. The project proposal documents at both ROE and KPNO list, with impressive specificity, the scientific rationale for devising infrared array cameras.[81] Such aims and 'user needs' are kept continually before the ROE and KPNO teams by, at ROE, the representative of the national UKIRT Users' Committee,[82] and, from mid-October 1985, by small 'Science' Sessions, which were designed by Ian McLean "to discuss the kinds of observations which might be possible with the 1-5 micron camera (IRCAM)":-

> Basically the purpose of these meetings is to:-
>
> (i) concentrate our minds on projects which really need a camera;
>
> (ii) enlighten the camera team to the expectations of users, and vice versa.
>
> This process will help me to formulate good plans for testing, modelling and data processing.[83]

At KPNO, this function is carried out by the Scientific Priorities and Steering Committee - and, of course, at both centres also by the incorporation of active astronomers in the teams. But it is equally obvious that, once such projects are under way, the day-to-day tasks often consist of *technical problems* which pose tactical hurdles - and these technical goals can come to dominate the thinking of those involved. If, for instance, you are concerned to lower the noise in some circuit, how low is 'low enough'? If no clear *scientifically informed* compromise is agreed, 'as low as you can get it' is likely to be the answer. In this way, a scientific aim can be translated into a technical goal. Hence the tendency to want to 'Rolls-Royce' the instrument: hence the potential tension between team members. The judgment of 'how far to go' is a matter of social negotiation, within constraints. The characteristics of the context can shape the outcome.

We saw that, at ROE, the team leader is an astronomer, who mediates his sense of 'user needs' to the team, and hence to their design judgments, from his own personal experience: while at KPNO, the team leader is an electronic engineer whose 'user' experience is more limited. In both locations, mechanisms for 'scientific steering' are in place. However, I raised the question as to whether the differences between the team leaders may have specifiable effects on the designs. I suggested that examples were not easy to find - but the repackaged electronics at KPNO, and the extensive 'interface' software at ROE, might be two cases. Do the two team leaders attach different 'meanings' to the apparatus? This exchange with Al Fowler suggests that they do. I mentioned imaging:-

> Oh, imaging is important for me. I get an awful lot of information out of imaging ... I might not be interested in the same *form* of data (how will I put this?) ... The kind of things *I* want to do with the data may or may not be the same things as an astronomer might want to do. I want to do imaging. But what I do with the images may be somewhat different than an astronomer's ...

So the *kind* of imaging I might want to do and the information I get *out* of the images is possibly different than what an astronomer does. But, in the imaging area, most of it's pretty much the same. You look at the image and you want to get something out of the image, and an astronomer does the same thing.

What do you want of an image? Basically, you're an engineer who wants to get a machine to work with the best parameters - noise levels, QEs, and everything. You're not specifically interested in astronomical results - I mean, you're not an observer ...

I'm not an observer, that's right ...

So that means that you look at an image and it tells you something about the characteristics of the system that you've built. And so the imaging to you is important because it highlights deficiencies in the system which you can then go back and try and correct ...

There are certain things you can see in it ... An image can give you a tremendous amount of reduced information which you can assimilate. I can get a computer to print out an inch and a half of paper and never get much out of it, but I can take that same thing and maybe put it on a display and I can immediately get ... I can get a tremendous amount of useful information or see problems that I might not *see* any other way. I mean, there are times ... for example, if you look at the thing and you'd understood there's a high correlation of pixels together ... Well, OK, that shouldn't be ... Noise should be *noise* ... That should *look noisy* ... Now how noise is supposed to look I suppose is in the eye of every beholder, but noise is supposed to be *random*. If I display a picture of noise up there it had better look random ... And ... the eye is a wonderful tool for seeing things like that. If there's *any* coherence in there, between the eye and the brain, you can really *see* ... You can get an awful lot of information that golly knows how the hell you'd process it any other way. I mean, image processing's not my thing ...

But you're interested in image processing that highlights that sort of technical thing ... [Yeah!] *... You say : 'That's noise'. Whereas, at, say, the radio Very Large Array in New Mexico, the image processing is to get the fine structure of the jet of M87 ... And you and an astronomer looking at the same image - the astronomer would say : 'Look at the filamentary structure, it looks helical, there's a magnetic field there', and all that kind of thing - whereas you'd look at it and say : 'Fantastically low noise level in this instrument' ...But you'd need an image to do that. What you're saying is that an image is an important shorthand.*

It's a way you display information. For example [*summons up an image on the VDU screen*] ... That is a plot of a row and a column. You can see immediately from here that there's obviously something very wrong here. OK. But I can only see *one* row and one column. I can put that up on a 2D image and I can see *immediately* that this particular square section of the image looks

> OK, and this whole area over here is a mess. I can see that ... by looking at the *bands*, that probably the *reason* ... that, first off, half of the array is definitely no good and the other half is better ... I know the structure of the array, so I say: OK ... It obviously must be ... And because it's *bands* it's addressing shorted or defective MOSFETs in the address structure on one half of the array ... I can get all that information *instantly* out of one picture. Or I can get these *ideas* out of one picture. I can see that the array has a crack in it. Why? Because it's got a very nice thing [*indicates feature*] ... Now there could be a *lot* of reasons for that. But the electronics ... I know the design. And there's no reason the electronics will see anything like that. The only reason I'd see something like that is a crack ... It has to be the thing that most immediately pops to your mind. These kinds of things you can instantly get out of an image. Where ... Yeah, you can wander thro' this, and maybe you would find the crack. Yes, you'd notice a bad pixel in a row. But would you *correlate* all this in your mind? Probably not ... [The astronomer] is actually, I suppose, in his own way trying to get the same thing, he's trying to compress information down to a level where ... He's trying to reduce the enormous amount of numbers he gets down to something his brain can handle.
>
> *But he wants it for a different purpose from you.* [That's right] *You want to look at it to infer the technical characteristics of the machine ...*
>
> That's right ... I want to get the instrument working as well as possible, and he wants to do astronomy.[84]

Who decides what 'working as well as possible' means is the key question! Obviously, an engineer and an astronomer may differ in their judgments on specific cases. Responses to the non-linearity of the SBRC arrays might reveal such underlying differences (although my impression is that that 'problem' was quickly disposed of).

These two possible 'meanings' of an image - these two ways of 'reading' it - are not mutually exclusive, of course. A technically trained astronomer can adopt either, depending on the context and circumstances. But whether or not this is a possibility, and which 'meaning' is predominant in any situation, will be socially shaped - by the past history and socialization of the actors, and by the particular contingencies which evoke the 'appropriate' mode. This ambiguous, dialectical social process, Janus-like, can adopt either a 'scientific' or a 'technical' face.

9. CONCLUSIONS

This paper has given some preliminary details of unfinished research on an episode, within a scientific specialty, of an instrumental development that had itself not yet reached 'closure'. At the time of writing, four different camera systems were in the later stages of their construction: but they had yet to be 'loaded' with their 'astronomical' mosaic arrays, and to start serious observing.[85] Only then can the question of 'how well they are working' be raised: and it will be some time before any 'closure' will emerge - any agreed judgment on the 'success' of the various group's designs. Whether any of them will provide a model which is later extensively copied - embodying the implicit definition of 'working successfully' - remains to be seen. Many of my respondents doubted that it would: citing the fact that all the major groups using optical CCDs use them in different ways, and "none of their cameras look alike", they suggest that 'closure', if any, will be confined to individual institutions, or small networks. (The example of the two coexisting dewar designs may be suggestive here.) I am assured (in 1991) that this instrumentation has now reached "the third generation", with any basic design lasting, at most, five years. If so, the pace of development of such techniques within science is too fast for Pinch and Bijker's programme to 'catch'.

Some detail has also been given of subsidiary episodes within this development in which 'closure' was reached on solutions to problems which, as one ROE team member put it to me, "just went away" - that is, the actors *perceived the problem to be solved.* Whether these (and others within this data base) can be fruitfully pursued and documented is not yet clear. At one time, it seemed possible that the non-linearity of SBRC InSb DRO arrays might prove to be a major 'reverse salient'[86] for this development - a problem that 'stalls' progress, demands central attention, and may eventually prove insurmountable. If so, a close study of how this problem was perceived and tackled might be worthwhile. But if, on the other hand, HgCdTe (or other) newer arrays can be shown to have taken over peacefully, with InSb expiring intestate, without a whimper, that, too, will have been a process of great interest! The ability of the ROE and KPNO cameras to adapt to such a change (which, as mentioned above, they are designed to do) will also be of interest.

In the meantime, this paper has presented some details of the kind of 'interpretative flexibility' demonstrated by those involved in the design of array cameras, and of the different 'meanings' that these artefacts can have for the actors. It has suggested how these aspects can be seen to be 'embedded' in a social context, with a range of 'significant social groups'. The dominant meaning of the arrays themselves is that of the military, and is only slightly modified by the industrial suppliers: the astronomers see arrays that are 'optimized' for the military as less than ideal for their purposes, because arrays have a different 'meaning' to them. Again, the arrays and cameras can have somewhat different 'meanings' to scientists and engineers. And different groups of astronomers assign different 'meanings' to the cameras they are building, since they are addressing somewhat different 'audiences', and serving different 'customers': the observatories are preparing instruments for a wide range of users; the Rochester group is aiming to convince NASA; and Austin just want to use their camera to 'do good science' within their own network. Each group thus builds a camera with its own distinctive 'signature'. Perhaps this case study is as complex as Elzen's two ultracentrifuges, after all.

10. ACKNOWLEDGMENTS

This research was supported by the UK ESRC, Grant No A 33 25 0036, in the 'Science Studies and Science Policy' initiative. I am most grateful to all the astronomers and industrial scientists who have given of their time so generously, and whose words I have quoted. I would especially like to thank Malcolm Longair, for suggesting this research topic, and giving me his support; and Ian McLean, Al Fowler, Alan Hoffman, Craig McCreight, Judy Pipher and Mark McCaughrean for particularly helpful advice and information. I would also like to thank Susan Cozzens for rescuing this old working draft from oblivion!

11. NOTES AND REFERENCES

1 This paper was originally drafted in 1986 as a preliminary report on research which was then still in progress. The initial interviews had not been systematically annotated, and some archival information had still to be retrieved. The scientific work under investigation was itself well behind schedule. This paper was therefore intended to give no more than an indication of the *kind* of results that this research might yield. However, since then, for reasons largely beyond the author's control, further work on the project has had to be suspended indefinitely. The paper should be read with this in mind.

2 Trevor J. Pinch and Wiebe E. Bijker, "The Social Construction of Facts and Artefacts : or How the Sociology of Science and the Sociology of Technology might Benefit Each Other", *Social Studies of Science*, vol.14, no.3, pp.399-441, August 1984. This paper was the first to advocate the use of methods and concepts developed in the sociology of scientific knowledge in the social analysis of technology. For a later commentary on this move, see Steve Woolgar, "The Turn to Technology in Social Studies of Science", *Science, Technology, & Human Values*, vol.16, no.1, pp.20-50, Winter 1991.

3 Pinch & Bijker, "Social Construction", p.414.

4 Pinch & Bijker, "Social Construction", p.427.

5 Pinch & Bijker, "Social Construction", pp.422-423.

6 Pinch & Bijker, "Social Construction", p.424.

7 Trevor Pinch, "Some Possible Implications of the Sociology of Scientific Knowledge for our Understanding of New Technology: The Case for a New Sociology of New Technology', paper presented to the Industry & Employment Committee of the UK ESRC, unpublished mimeo, p.29, November 1985.

8 H.M. Collins, "The Place of the Core-Set in Modern Science: Social Contingency with Methodological Propriety in Science", *History of Science*, vol.19, pp.6-19, 1981.

9 W.E. Bijker, J. Bonig and E.C.J. van Oost, "The Social Construction of Technological Artefacts", paper presented to the EASST Conference, Deutschlandsberg, Austria, unpublished mimeo, 24-26 September, 1982; M.Callon, "The State and Technical Innovation: A Case Study of the Electrical Vehicle in France", *Research*

Policy, vol.9, pp.358-376, 1980; for more on the bicycle, see Pinch & Bijker, "Social Construction".

10 Boelie Elzen, "Two Ultracentrifuges : A Comparative Study of the Social Construction of Artefacts", *Social Studies of Science*, vol.16, no.4, pp.621-662, November 1986.

11 For an account of the development of the CCD, see Robert W. Smith and Joseph N. Tatarewicz, "Replacing a Technology: The Large Space Telescope and CCDs', *Proceedings of the IEEE*, vol.73, no.7, pp.1221-1235, July 1985.

12 The military arrays are designed (for obvious reasons) to work in conditions of high background radiation, and at very fast switching speeds. The military are also, as one respondent put it, "not interested in energy": they only need 'quick and nasty' analogue pictures, with maximum contrast, sufficient for *recognition* - whereas astronomers use digital methods, and are anxious to measure flux levels in absolute terms. Astronomers therefore worry more about 'fill factors' (i.e. the proportion of the array surface occupied by active and reliable pixel material - in other words, they want the number of 'dud' pixels to be kept to a minimum, and the space between pixels to be as small as possible): but the latter raises problems of 'crosstalk' between pixels - which the military simply solve by increasing the gap between them!

13 B.T. Soifer and Judith L. Pipher, "Instrumentation for Infrared Astronomy', *Annual Review of Astronomy and Astrophysics*, vol.16, pp.335-369, at 365-366, 1978. By this date, some details of these arrays were being made public: see, e.g., K.J. Ando, "Assessment Study of Infrared Detector Arrays for Low Background Astronomical Research", *NASA-CR-152*, 1978; E.M. Hoendervoost *et al.*, "Hybrid InSb Focal Plane Array Fabrication", and W.J. Parrish *et al.*, "Characterization of a 32x32 InSb Hybrid IR Focal Plane", papers delivered at the IEEE International Electron Devices Meeting, Washington, DC, 1978. Their existence and likely usefulness were already well known to infrared astronomers, who thought them to be unavailable, for reasons of national security. (Soifer & Pipher's optimism that the "devices" themselves could be "adapted" to astronomical purposes has hardly been borne out by events [see below].) Once NASA began planning for the Space Infrared Telescope Facility, Craig McCreight, at the NASA Ames Center, was given a contract to investigate the possibilities of infrared sensor arrays, and to encourage their use: he has a budget out of which he commissions research (he has financed the work at Rochester). As it happens, he has mainly helped to develop already existing contacts between astronomers and industrial suppliers, many of which came about serendipitously: one *ab initio* contact he made was to put Arizona in touch with Rockwell (see below). There is an active group at NASA Ames evaluating a range of infrared arrays. Craig McCreight has also organized two 'Infrared Detector Technology Workshops' at NASA Ames that have been important for the field: the first, on 12-13 July 1983 (proceedings published August 1983, see note 14, below), had 93 participants, of whom 36 (39%) were from university and observatory groups, 30 (32%) from various space and military labs and NSF, and 26 (28%) from industry; the second, on 13-14 August 1985 (proceedings published February 1986, see note 15, below), was attended by 122 people in all, of whom 55 (45%) were from univs/observs,

34 (28%) from space/mil labs, and 31 (25%) from industry. As one respondent remarked, "the basic cast doesn't change much".

14 W.J. Forrest and J.L. Pipher, "Evaluation of a 32x32 InSb CCD for Use in Astronomy", in Craig R. McCreight (ed.), *Infrared Detector Technology Workshop*, NASA Ames Research Center, Moffett Field, CA, August 1983 [hereafter referred to as *Ames 1983*], pp.16-1 to 16-69; William J. Forrest, Andrea Moneti, Charles E. Woodward, Judith L. Pipher and Alan Hoffman, "The New Near-Infrared Array Camera at the University of Rochester", *Publications of the Astronomical Society of the Pacific*, vol.97, pp.183-198, February 1985.

15 M.J. Lebofsky, E.F. Montgomery and W.F. Kailey, "Evaluation of Rockwell HgCdTe Arrays for Astronomical Use", in Craig R. McCreight (ed.), *Proceedings of the Second Infrared Detector Technology Workshop*, NASA Ames Research Center, Moffett Field, CA, February 1986 [hereafter referred to as *Ames 1986*], pp.17-1 to 17-14.

16 Mark Hereld and D.A. Harper, "Tests of HgCdTe-on-Sapphire Focal Plane Arrays', *Ames 1986*, pp.19-1 to 19-9.

17 D. Pommerrenig, D. Enders, L. Trousil, R. Capps, E. Irwin, E. Tollestrup and E. Dereniak, "Hybrid Silicon Focal Plane Arrays", *Ames 1983*, pp.15-1 to 15-9; Eric Tollestrup and Richard W. Capps, "A Study of Si:In Hybrid Surface Channel IRCCD Devices", *Ames 1986*, pp.21-1 to 21-16.

18 A. Sood, N. Hartle, J. Marciniec and P. Zimmermann, "Shortwave Photovoltaic HgCdTe Detector Arrays for Space Astronomy", *Ames 1986*, pp.15-1 to 15-17; D.L. DePoy and D.N.B. Hall, "Characterization of 2.5 micron HgCdTe Detectors for Low-Background Applications", *Ames 1986*, pp.16-1 to 16-9.

19 C.M. Parry, "Conceptual Design of a Multiplexed Ge:Ga Detector Array", *Ames 1983*, pp.10-1 to 10-12; B. Jones, R. Puetter, R. Heir, A. Schmidt, J. Casler, S. Nelson and J. Hujsak, "UCSD IR Detector Development Programs", *Ames 1983*, pp.12-1 to 12-6.

20 John F. Arens and Michael C. Peck, "Evaluation and Characterization of Hybrid Arrays", *Ames 1986*, pp.26-1 to 26-21; Gerald M. Lamb, Peter S. Shu, Donald C. Lokerson, Daniel Y. Gezari and Jeffrey Bowser, "Evaluation of a Si:Ga Direct Readout Array", *Ames 1986*, pp.25-1 to 25-16; Gary C. Bailey, "Integrating Indium Antimonide Linear Arrays at 65°K and Below", *Ames 1983*, pp.18-1 to 18-12.

21 R. Bharat, M.D. Petroff, J.J. Speer and M.G. Stapelbroek, "Si:As BIB Detector Arrays", *Ames 1986*, pp.20-1 to 20-43.

22 Stuart Worley and Steve Gaalema, "Interim Report for Development of 58x62 Si:Sb Detector Arrays", *Ames 1986*, pp.23-1 to 23-6; A.M. Fowler, J.P. Britt, R.R. Joyce, R.G. Probst and J.L. Gates, "Preliminary Test Data Using the MOS DRO with Si:In Detector Material", *Ames 1986*, pp.24-1 to 24-12.

23 For further details, see David Edge, "Competition in Modern Science", in Tore Frängsmyr (ed.), *Solomon's House Revisited: The Organization and Institutionalization of Science*, Science History Publications, USA, Canton, MA, 1990, pp.208-232.

24 The survey was carried out by the UKIRT Users' Committee. At the UUC meeting on 17 July 1981, a range of options for possible UKIRT common-user instruments was reviewed, and three items related to array cameras (a 8-13 micron array photometer, a monolithic array, and detector development) were placed among the six items with 'top priority'. Ironically, a proposal for a 1-5 micron array photometer was placed among four with 'second priority' [Document UUC(81)17, Minutes of UUC meeting of 17 July 1981, p.14]. UUC papers at that time show, in 1-5 micron array work, UKIRT and Hawaii in active contact with SBRC (2x32 InSb), ROE and JPL with Cincinnati (InSb linear arrays), Leicester with General Electric (1x32 InSb CID), and 'the French' at Kitt Peak with a GE 32x32 InSb CID array: at longer wavelengths, Rockwell and Aerojet are noted as active, with doped silicon arrays [Document UUC(81)23, "New Instruments for UKIRT", 4 November 1981]. It was noted that "the SBRC array has not been built yet, but they had quoted on the understanding that it could be built, and are keen to work with UKIRT" [Document UUC(82)1, Minutes of UUC meeting of 6 November 1981, p.13]; the quote was $550,000 for one 2x32 array [Document UUC(81)17, p.12]. There is no doubt from the records that the options for a camera were being actively scanned at that time. Somewhat earlier, in an ROE Internal Report "On the Application of a Cooled, Slow-Scan CCD Camera to Imaging and Low Dispersion Spectropolarimetry", March 1980, Ian McLean noted:-

> Large format [optical] CCD's are being made under contract for NASA for the Space Telescope and other astronomical applications. If theoretical expectations of performance can be realised in practice then these devices will find wide application in optical astronomy. It should also be noted that charge coupled action has been produced in materials other than silicon, and "true" infrared CCD's may yet be possible.

And he later commented:-

> This writer is primarily interested in astrophysics rather than solid state detectors per se and thus will be a staunch supporter of any technique that works and improves the astronomer's lot! It is easy to fall in love with the CCD concept, but clearly, other alternatives may be possible.

25 Ian McLean, "Infrared Detector Arrays : A Preliminary Report of the Current Status in the USA", ROE, mimeo, November 1982. By early 1984, the commitment had hardened: see I.S. McLean, "A Specification for an Infrared Imaging System for UKIRT Using Focal Plane Arrays", ROE, mimeo, January 1984. The Appendices to the latter paper give details of arrays from Cincinnati Electronics, Rockwell, Mullard and SBRC, *in that order*. The paper closes by stressing the *flexibility* necessary in the camera design:-

> Although it seems unlikely that CIDs will be used, we should take care to do nothing that precludes them. Since the proposed UKIRT program splits

naturally into 1-5 microns and 8-14 microns it is acceptable to treat the detectors for these 2 wavebands separately and to develop two similar but independent drive/control systems. Work on a 10 micron imaging system will certainly be needed soon. Information on the principal types of array which we must be able to drive is appended. In many instances the manufacturer will want to provide us with drive electronics. This poses a problem if (a) we cannot interface to that electronic package and (b) the electronics are unacceptably poor for our application. Clearly however we should try to make use of the manufacturer's electronics where appropriate even if only as a guide to what we must design ourselves.

26 Infrared Array Project Team, *IRCAM : An Infrared Camera and Infrared Array Control System for UKIRT : A Technical Design Summary*, ROE, mimeo, August 1985, p.19.

27 Text of letter supplied by Al Fowler from his computer file. The letter was sent to Hughes (Carlsbad), SBRC, Ford Aerospace, Electro-Optek, Honeywell, Cincinnati, General Electric, Rockwell International, RCA and Aerojet: apart from SBRC, only Aerojet replied with any offer.

28 R&D Department, Engineering Technical Services, KPNO, *Request for Information on Infrared Detector Arrays for Use in Ground Based Astronomy*, Tucson, AZ, final draft 26 September 1983.

29 KPNO Memorandum from R. Probst and R. Joyce to SOC, *1-14 micron Infrared Array Camera System*, 20 September 1984.

30 KPNO Memorandum from F. Gillett, R. Joyce and R. Probst to Sidney Wolff, *Instrument Proposal for Use of SBRC Arrays*, 30 May 1985, forwarded to J. Jefferies 3 June 1985. The proposal was to build "a straightforward duplicate of a dewar now being fabricated by R&D [Group] for array testing", despite the fact that the additional features incorporated in that dewar for test purposes made it less than ideal for a common-user instrument. "Although not the ultimate in versatility, the proposed camera ... [is] capable of realizing the full sensitivity of these arrays for important scientific problems".

31 See Al Fowler, Ron Probst, Jay Gallagher and Dick Joyce, "Diodes in the Dome!", *NOAO Newsletter*, no.2, p.18, 1 June 1985, which reports on tests of an experimental 62x58 PtSi Schottky Diode Array on the 1.3-m telescope. The team was also evaluating arrays from (e.g.) Hughes (Carlsbad): see Fowler *et al.*, "Preliminary Test Data'" (note 22). They were also, before the latest KPNO reorganization, developing a spectrometer, using a Cincinnati linear array, for the Chile observatory, but this project has now been abandoned.

32 A second postdoc was later appointed. The group also included a lab dog - but no-one believes me! The group have access to a 12" telescope on the campus, and a 24" telescope at the C.E.K. Mees Observatory, nearby.

33 See note 13, above.

34 His post is now tenured, w.e.f. 1 September 1986.

35 The University grant was $130,000; the DoD grant $80,000. An SBRC array, in 1986, cost around $100,000.

36 This is a small firm in Tucson run by Frank Low, the senior Professor in the infrared astronomy group at the University of Arizona. Its dewars have a reputation for reliability, and they sell widely: one might call them the 'safety bicycles' of cryostats.

37 Paul Harvey, interview, Austin, TX, 26 May 1986. The 'speckle' technique requires very fast Fourier transforms of the array output: the DoD grant was specifically to develop this capacity.

38 Paul Harvey was also working on the development of a spectrometer using a Cincinnati linear array, observing on the Kuiper airborne observatory, which is based at Moffett Field, CA (NASA Ames).

39 Keith Matthews, interview, Pasadena, 2 June 1986. It is perhaps worth emphasizing here that these astronomical arrays will involve *some* essential innovation: it is not the case that the military arrays have removed all the uncertainty over their performance. The SBRC arrays were on offer for commercial contract, but were not yet actually available for use, although 'engineering' arrays had been tested: *one* of the later Rockwell HgCdTe arrays was in a camera, and had been on a telescope, but this array was (in 1986) not yet on commercial offer, and was still under development. Keith Matthews' caution had good grounds.

40 Matthews, interview.

41 Ian McLean, interview, ROE, 25 July 1986.

42 McLean, interview.

43 McLean, interview.

44 Ian McLean had been on an SERC Management Course, and taken its lessons to heart.

45 Agenda, Infrared Array Project, Group Meeting No.1 (Monday, 2 July 1984), document ISMCL3 Q/2.

46 One is reminded here of the contrasting leadership styles of Martin Ryle and Bernard Lovell in the radio astronomy groups at Cambridge and Jodrell Bank: Fowler is reminiscent of Ryle, McLean of Lovell. See D.O. Edge and M.J. Mulkay, *Astronomy Transformed : The Emergence of Radio Astronomy in Britain*, Wiley-Interscience, New York, 1976, Chapter 9; and D.O. Edge, "Styles of Leadership in Three Early Radio Astronomy Groups", in W.T. Sullivan, III (ed.), *The Early Years of Radio Astronomy:*

Reflections Fifty Years after Jansky's Discovery, Cambridge University Press, Cambridge, 1984, pp.351-364.

47 Fred Gillett, interview, KPNO, 29 May 1986.

48 The evolution of this KPNO 'tradition' is outlined in illuminating detail in A.A. Abraham and W.G. Robinson, "Integration of Visible CCDs into Cryogenically Cooled Dewars for Astronomical Applications", paper delivered at the SPIE Conference on "Instrumentation in Astronomy", Tucson, AZ, mimeo, 3-8 March 1986.

49 Ian McLean, interview, ROE, 25 July 1986.

50 In the 'gas switch' (see Figure 3), an evacuated copper vessel is inserted into the 'cold finger' which stretches from the liquid helium container to the detector mounting. Some helium gas is injected into this vessel, and this is then partly liquefied when the dewar is filled with liquid helium: a small heater can then control the amount of gas in the switch, by evaporating some of the liquified gas, as required. The thermal conductance of the vessel (and hence of the cold finger) varies with the amount of helium gas in the vessel: hence the equilibrium temperature of the detector can be continuously varied. When the liquid helium supply in the dewar is exhausted, this regulatory function ceases, since the dewar temperature rises, and none of the helium gas in the switch vessel is then liquefied: the vessel is filled with gas, and offers very little resistance to the passage of heat. One of those involved at ROE, while admitting the force of the argument for this device, also admitted that the switch "added a complication" - it was "something else that could go wrong" - and thought that a case could be made out that "we would be better off without it". This is just one of a multitude of cases of judgment in design detail.

51 Ian McLean, interview, ROE, 25 July 1986.

52 McLean, interview.

53 McLean, interview.

54 McLean, interview.

55 McLean, interview. In the light of the comment about the "added complication" of the ROE gas switch (see note 50), perhaps the KPNO 'lunar lander' is not the *only* "unnecessarily complicated" device on these cameras!

56 McLean, interview.

57 Paul Harvey, interview, Austin, TX, 26 May 1986.

58 Harvey, interview.

59 See note 27. The larger space astronomy contracts can include sums large enough to persuade suppliers to undertake major adaptations of array design and

fabrication (particularly of the multiplexers) to optimize their use for astronomical purposes. Sums of $2m or more are required to commission such a 'customized' array development.

60 Some alteration of the doping of the detector material, anti-reflection coatings, and the like, *are* possible - but basic changes in array and multiplexer design are *not*.

61 For the classic paper on the problems of replication in science, see H.M. Collins, "The Seven Sexes: A Study in the Sociology of a Phenomenon, Or the Replication of Experiments in Physics', *Sociology*, vol.9, pp.205-224, 1975.

62 Paul Harvey, interview, Austin, TX, 26 May 1986.

63 Harvey, interview.

64 For further discussion of these aspects, see Edge, "Competition" (note 23).

65 Harvey, interview.

66 "Getting the first image on to the cover of *Nature*" was often quoted to me as a goal of scientific competition in this field, an indication of "winning". To my knowledge, this particular goal has not yet been reached. But the KPNO team was first on to the cover of *Science*, vol.242, 2 December 1988 - ironically, with an infrared image taken by an array using a later technology than the earlier InSb arrays! (It uses a 256x256 platinum silicide Schottky barrier diode array, manufactured by a separate division of the Hughes Corporation, and not then commercially available.)

67 See Edge & Mulkay, *Astronomy Transformed*, Chapter 7.

68 Recall Keith Matthews' remark that CalTech would want an imaging array "without thinking about other groups at all".

69 Paul Harvey, interview, Austin, TX, 26 May 1986.

70 Judy Pipher, interview, Rochester, NY, 20 September 1985.

71 Notes on meeting at ROE, 18 March 1985.

72 Notes on meeting at ROE, 10 April 1986.

73 When ROE first negotiated with SBRC, it was thought that the US Government's restrictions on the export of militarily sensitive technology would mean that no export license would be granted for an infrared array. In these circumstances, the fact that UKIRT is in Hawaii, on American soil, would be a distinct advantage! But this restriction has since been relaxed, and arrays *can* be exported.

74 Forrest *et al.*, "New Near-Infrared Array Camera" (note 14), 192.

75 Bill Forrest, interview, Rochester, NY, 20 September 1985.

76 Judy Pipher, interview, Rochester, NY, 20 September 1985.

77 Pipher, interview.

78 Paul Harvey, interview, Austin, TX, 26 May 1986.

79 Harvey, interview.

80 For some other comments on this theme, see Edge & Mulkay, *Astronomy Transformed*, Chapter 4, and pp.437-438, 451-452.

81 See notes 25, 26, 29 and 30.

82 The UKIRT Users' Committee first met on 29 January 1981, and thereafter every four months or so. One ROE member was Richard Wade, who, although he had no specific tasks on the IRCAM project, attended most IRCAM Group meetings, which he occasionally chaired. His advice and opinions were assigned great weight by the group.

83 Ian McLean, *Memorandum: Infrared Camera Science Sessions*, ROE, 15 October 1985. Some minor details of design, and of plans for operational schedules, arose from these meetings.

84 Al Fowler, interview, KPNO, AZ, 29 May 1986. For a very interesting discussion of imaging in optical astronomy, see Michael Lynch and Samuel Y. Edgerton Jr., "Aesthetics and Digital Image Processing: Representational Craft in Contemporary Astronomy", in Gordon Fyfe and John Law (eds), *Picturing Power: Visual Depiction and Social Relations, Sociological Review Monograph* no.35, Routledge & Kegan Paul, London, 1988, pp.184-220.

85 Astronomical results from these array cameras are now being published widely. For early surveys, see Ian Gatley, D.L. DePoy and A.M. Fowler, "Astronomical Imaging with Infrared Array Detectors", *Science*, vol.242, pp.1264-1270, 2 December 1988; C.G. Wynne-Williams and E.E. Becklin (eds), *Infrared Astronomy with Arrays: Proceedings of the Workshop on Ground-Based Astronomical Observations with Infrared Detectors, University of Hawaii at Hilo, 24-26 March 1987*, Honolulu, 1987; M. Mitchell Wardrop, "A Window Looking Out on Creation", *Science*, vol.250, p.32, 5 October 1990.

86 This concept is developed by Thomas P. Hughes, notably in his book *Networks of Power: Electrification in Western Society, 1880-1930*, Johns Hopkins University Press, Baltimore, MD & London, 1983.

Chapter 8

Institutional problems surrounding the acquisition of detectors in high-energy physics at CERN in the early 1970s

John Krige
European University Institute (Italy)

ABSTRACT

The paper describes the conflict between CERN staff and visitors over who should build heavy detectors to be used at the new 300 GeV accelerator, due to be commissioned in 1976. The arguments used by the protagonists are described, as well as some of the deeper forces at work in the debate. It is suggested that the conflict could only be resolved when detectors became so large that CERN and its outside users had no option but to collaborate in building them.

1. INTRODUCTION

It is a commonplace that the post-war development of high-energy physics has demanded the construction of accelerators on a national or, in the case of Europe, international scale.[1] Before, and even immediately after the war, it was possible for some universities to consider acquiring their own particle accelerators. In fact in Britain in 1945, no less than five universities embarked on accelerator construction programmes. By the end of the 1940s the difficulties faced by such programmes confirmed that most, if not all, universities did not have the engineering and technical staff, the interdisciplinary teams, the workshops, the managerial experience or the funds to build an accelerator of more than a few hundred MeV.[2] The new generation of machines, the machines in the GeV range, had to be built in national institutions like the Brookhaven National Laboratory in the USA (set up in the late 1940s) or international organizations, like CERN in Europe (set up in the early 1950s). Correlatively, and in contrast to a university accelerator, these machines became **facilities** expected to serve a community of physicists far wider and more numerous than that located inside the institution itself.

This is common knowledge, as I have said. What is perhaps less well appreciated is that the development of the detectors used to exploit an accelerator has followed a similar path. In the early 1950s a group of physicists wanting to do an experiment at a laboratory like CERN could build their detector in their own workshops, typically a university, take it with them to the laboratory, do their runs, and return home with their instrument and their data. The growing popularity of large hydrogen bubble chambers (i.e., of active length of about 1m or more) changed all that. On the one hand, a large bubble chamber was a costly item of heavy engineering and it could take anything from 3-5 years to build. On the other hand, its operation called for major infrastructural support from the host laboratory. A beam of particles had to be transported from the accelerator to the detector, and purified en route. Provision had to be made for developing and analyzing the film of the processes occurring in the chamber. A major security system had to be set up to prevent an explosion occurring due to the unexpected release of hydrogen into the air. A large bubble chamber, once installed, occupied a considerable amount of space on the experimental floor (at the expense of other experiments), was not dismantled or moved without good reason, and was expected to have a useful life of some 5 years.[3] Like an accelerator, it became something of a fixture, intended to supply data to physicists from many institutions. Like an accelerator the detector became a facility.

What was true for bubble chambers from the late 1950s onwards was true for electronic detectors (counters, spark chambers and later wire and drift chambers, etc.) a decade later. In the early 1960s at CERN an experiment with an electronic device was conducted by a team of five or six people who conceived and built their apparatus themselves, tested their idea, and then dismantled their equipment. The spark chamber, which was to revolutionize the field, was adequately suited to these 'artisanal' techniques. As one enthusiast explained in 1961, it was easy to construct -- "anyone can build one in the basement and make it work" -- and cheap and versatile -- "[...] you can build a device for a particular experiment and give it away, store it in the basement, or forget it and build another one for the next experiment."[4] By the late 1960s, as the technology developed, and experiments became more sophisticated and complex, so did the equipment become correspondingly large, costly and time-consuming to build. In fact what we see at the end of the decade are the antecedents of the huge electronic detectors in use at an accelerator like CERN's Large Electron-Positron Collider (LEP), which take some six years to build, weigh thousands of tons, occupy an entire underground experimental hall, cost (at least) $100 million -- and are used by 400-500 physicists from tens of institutions all over the world. Oversimplifying rather, in the late 1960s we have the beginnings of the transformation of electronic detectors from university-scale instruments used by a small group into international (or national) research facilities supporting multi-institutional collaborations.[5]

In a previous study Dominique Pestre has discussed the debates surrounding the arrival of big bubble chambers at CERN in the late 1950s, and the struggles to set up a committee structure to manage their use by the physics community.[6] In this paper I want to make a similar, though more modest study of the conflicts surrounding the acquisition of heavy equipment for doing electronic experiments which took place about a decade later. My documentation is somewhat meagre, and its scope circumscribed to an enquiry made in the early 1970s about the relationship between CERN and its user community. This limits me to presenting the arguments invoked by two groups of physicists -- the 'in-house' CERN staff, and the CERN visitors in national laboratories and universities -- over who should build big equipment; the position of the CERN management is not discussed in any detail.[7]

Visitors, of course, have been a perennial feature of scientific cooperation since the 17th century. The motives for making visits have been many -- to serve an apprenticeship, to exchange information and ideas, to check the reliability of the host's knowledge claims, to use a piece of equipment one does not have at home.[8] The interest of the situation which we will describe below is that the visitors to CERN were not simply **users** of instruments owned and managed by the host laboratory. They were also, or also wanted to be, instrument **builders**.[9] In particular they wanted to build big electronic detectors. These devices would consume considerable human and material resources, and would be operational for a lengthy period of time at the main laboratory. Above all, and related to these considerations as well as to the fact that CERN was an international laboratory, physicists other than those who built them would have the right to use them. To construct a detector in a competitive situation like this was to ensure that one had certain privileges. Builders had the (unwritten) right to do the first runs -- and so to skim the cream in the domain opened up by the detector. Through their understanding of its functioning and its caprices they could expect to participate in every important experiment done by other groups with it. Through their experience they could play a key role in shaping the experimental programme which gradually unfolded around it. In short, to build a facility was to gain some measure of power over its use, to have 'objective' reasons for domination is a context where the emphasis was on participation.

The organization of the paper is simple. After briefly setting the scene, I will expose successively the arguments used by the CERN staff and by the physicists in institutions in the member states in defence of their claim that they should build big electronic equipment for use at CERN. I shall also probe deeper, exploring some of the broader issues at stake in the debate, and some of the factors which enflamed the passions of the visitors in particular. In conclusion, I shall show how this particular dispute was 'resolved' -- that resolution, I hasten to add, being due, not to the 'desire' for peace or to the 'spirit' of internationalism, but to the progressive removal of the material conditions for conflict.

2. SETTING THE SCENE

Early in 1971, after a decade of deliberations, the CERN Council agreed to build a 300 GeV Super Proton Synchrotron in Geneva. The European high-energy physics community was delighted. At last they would have access to a machine which could compete with that already under construction at Fermilab near Chicago. At the same time they were worried. For the siting of yet another major facility at CERN could only increase their dependence on the Geneva laboratory. In 1966 CERN had accounted for 30% of the total European high-energy physics budget. By 1972 this had climbed to 46%. The funds granted to national laboratories and universities had decreased proportionately, the latter being particularly badly hit.[10] As two senior officers of the European Committee for Future Accelerators (EFCA) put it, there was a danger of CERN becoming "a huge splendid enterprise for the enjoyment of a selected class of persons, with thousands of technicians working for them."[11]

At its fifteenth plenary meeting on 4 May 1973 ECFA, which comprised the élite of the European high-energy physics community, established a working group to look into the relationships between CERN, the national laboratories and the universities. Its chairman was Professor J.C. Gunn of Glasgow University.[12] The working group met several times during the subsequent year, set up four subcommittees to deal with specific topics, and solicited written statements from the physics community. Its final report was published on 2 August 1974.[13]

Two related themes emerged repeatedly during these deliberations. The first was that resources and facilities for high-energy physics in Europe were being concentrated at CERN, at the expense of activities in the member states' laboratories. "I feel strongly", wrote one user, "that we [i.e. national groups] must resist forces that completely centralise all life and initiative in the subject at CERN."[14] To that was coupled a concern, a resentment even, among the outside users that this concentration of resources was going along with a concentration of privileges for the in-house physics staff, that CERN was becoming an 'oasis' cut off from its base in the national laboratories and universities. These twin concerns were translated into a number of specific complaints to the ECFA working group, complaints about CERN salary scales, its employment policy, the rules it had imposed on access to workshop facilities, the degree to which it consulted with outside users and, of particular concern to us, CERN physicists' attitude to the construction and exploitation of big equipment for use at the SPS.[15]

3. WHO SHOULD BUILD BIG EQUIPMENT?

Outside users found CERN's position on this question objectionable on two counts. Firstly, they felt that the laboratory's physicists wanted to hoard important, big-equipment projects for themselves. As one British physicist put it in a written statement to Gunn's working group, he had been struck by some "rather startling comments" made by CERN senior staff at a meeting to plan the SPS experimental programme. "The question asked by a senior CERN physicist", the correspondent wrote was "whether all **major** experimental projects at the SPS should come under the control of CERN paid project leaders." CERN Director-General W. Jentschke confirmed the tendency in a presentation to ECFA in December 1972. "Big experimental projects", he said, "were usually financed by CERN", and the laboratory planned to spend a good deal of money building "major equipment for SPS physics in the West Area [...]".[16]

Secondly, visitors objected to being excluded from the decision-making processes leading to the choice of major experimental facilities. These called for important investments of time and money and, while opening new horizons, also imposed constraints on the physics that could be done by staff and users. The normal procedure at CERN at the time was for an in-house group to define the scientific aim of a particular piece of equipment, design and cost it, and seek approval from the laboratory's Scientific Policy Committee (SPC) for the package. By this time, of course, considerable resources had already been spent on the project, and the SPC generally accepted the proposal. What outside users wanted was to participate in this process from the beginning, "to get involved in the day-to-day dialogue on what physics to back and what projects to start."[17]

Two more or less 'classical' arguments were used by CERN staff to defend their dominant position regarding the acquisition of big equipment. Firstly, they insisted on the need for **continuity**, on the desirability of the ongoing presence of a core of applied physicists and engineers not only during the construction and running-in of the equipment, but also during its subsequent exploitation and (possible) upgrading. It was simply not possible for outside users, people who were by definition not based at CERN, to ensure this. Scientists in national laboratories and universities, besides having other responsibilities, would find it difficult to leave their home institutions for the lengthy periods of time required to install, maintain and develop heavy and complex detectors. The whole process, they insisted, from conception to implementation, should be the prerogative, and the responsibility of CERN staff.

This argument was reinforced by another: that such equipment should come under 'neutral' control, by which was meant the control of CERN. Being big facilities, the detectors built around the SPS would have to serve the entire European high-energy physics community. No national group could be counted on to respect the wishes of users from other countries -- even though it profited from an infrastructure paid for by all member states. CERN, on the other hand, being a European laboratory, had an **obligation** to make its equipment available to outside users. This, it was claimed, provided something of a guarantee that big detectors built by CERN staff, though tying up resources, and imposing certain directions and a certain inertia on the experimental programme, were at least available for any competent group who wanted to use them.

Outside users, for their part, advanced three important arguments why big detectors should be built at home. Firstly, they insisted that the viability of national laboratories (in those countries that had them) depended on it. These institutions had been equipped with reasonably important accelerators in the late 1950s and 1960s. Correspondingly, they had built up impressive technical and engineering facilities. At Saclay in France, for example, "physicists [were] assisted by engineers and technicians who [had] acquired a large amount of experience both in very fast electronics, on-line computers, and the construction of all kinds of equipment involved in electronic experiments such as spark chambers, wire chambers, scintillators, Cerenkov counters, etc. [...]". At Frascati in Italy there were groups specialized in magnets, in cryogenics, in ultra-high vacuum techniques, and in electronics, as well as a workshop specialized in building detectors for electronics experiments. The Rutherford laboratory in England had "a strong applied physics support programme".[18] As far as the users were concerned, "it would be a madness to ignore this capacity [...] and to go instead to a further reliance on CERN engineering and technical facilities."[19] This was not simply because national resources might lie idle, but because the long-term survival of the institutes depended on it. The decision to go ahead and build the 300 GeV SPS effectively sounded the deathknell for national high-energy accelerator programmes in all but the most wealthy European country, the Federal Republic of Germany. By the end of 1973 Saturne (Saclay) had already ceased to do high-energy physics, and in the UK there were plans to shut down two major accelerators NINA and Nimrod, within the next five years. National laboratories had to redefine their roles, and to carve out a new niche for themselves to ensure their survival. The construction of heavy equipment for use at CERN was seen to be one important way of keeping them in business, and of keeping up the level of resources dedicated to them in the past.

Coupled to this argument the outside users had another: that electronic detector development required and enhanced strategically important skills and knowledges. These were technically advanced projects, relying on know-how in fields like fast electronics, computing, special materials, and so on. Member states had a "legitimate desire" to see those skills fostered in their technicians and engineers, be they in national laboratories and universities, or in local industry.

Finally, there were the advantages to university staff of having electronic detectors built at home. Physicists were involved in all stages of these devices (see later) from design to data analysis. If all the work was done at a geographically remote laboratory the strain on their professional and personal lives became intolerable. If they made their work at CERN a top priority, and asked for reduced teaching loads and administrative duties in their universities, they risked alienating their colleagues and damaging their career prospects in the physics department. If they put their university life first they would inevitably be relegated to a subordinate role in the development of their experiment at CERN. And it was not only jobs but also families that suffered. As one lecturer who submitted evidence to Gunn's working group explained, a counter physicist had to envisage spending "roughly equal amounts of time at the university and distant laboratory, often on a day by day basis." And, he went on, as far as he, a married lecturer at Manchester University was concerned, "I see no way of continuing to make significant contributions in this branch of research after NINA closes. The personal problems", he added, were "just too great under the present circumstances."[20]

4. SOME WIDER ISSUES AT STAKE IN THIS CONFLICT

Two sets of arguments, then, both of them reasonable and well-founded from the point of view of the group of actors who advanced them, the in-house CERN staff or the visiting physicists. Yet behind both there were deeper, more or less conscious forces at work. These forces did not necessarily impugn the validity of the cases made by our main protagonists. But they did charge them with a certain amount of emotional energy, and transform them from 'mere' arguments into 'weapons' used to defend and advance particular group needs and interests.

To begin with the CERN staff physicists. Their wish to control the equipment and the implementation of the SPS experimental programme sprung from their determination to develop 'their' laboratory, their careers, and their status in the high-energy physics community. Behind their appeals to 'neutrality', their claims to be collaborators who could be counted on to protect the interests of all European physicists, lay a strong competitive drive. For them, CERN was at once an international facility, and a laboratory which had to compete with others in Europe and elsewhere in the world. The groups of physicists based there were as determined as any other to have the best equipment and to be the first to make a major discovery. They would collaborate, yes, but preferably in the most interesting and important physics experiment that was on the agenda. They had the time and the resources to work at the leading edge of the field, to be ahead of all their rivals. They were not going to have that advantage wrenched from them to 'save' national laboratories or to 'ease' the lives of university lecturers. For them, CERN was a laboratory in its own right, like any other national laboratory. It was not, or it was not simply, a service centre, providing facilities for European high-energy physicists, and under community 'control'.[21]

This attitude may seem objectionable to some. Yet, looked at pragmatically, it has undoubtedly contributed to CERN's success. Consider, for example, the comparable situation which arose in the early 1960s when the European Space Research Organization (ESRO) was being set up. Under strong British pressure it was accepted very quickly that the organization should be a supplier of services, and not what we have called a laboratory in its own right. Correlatively, its initial programme, as spelt out in the famous 'Blue Book' produced in October 1961, specifically laid down that "it was not intended that ESRO should compete, either in personnel or funds, with universities and other research institutes in carrying out pure scientific research".[22] One consequence of this policy was that the organization never built up a strong in-house scientific staff pushing their projects over and against and, if necessary, at the expense of, national programmes. For some commentators it is partly for this reason that ESRO failed, where CERN succeeded, to forge its own identity in the 1960s as a first-rank scientific organization, and could only survive in the 1970s because of the commercial and military importance of space.

Let us now turn to the CERN user community. Beneath their arguments, their militancy, lay the feeling that they were being treated by the in-house staff as "second-class citizens", that they were dominated and looked down on by an élite who cared for little else than the protection and extension of their already considerable privileges. This sensitivity, in turn was sharpened by the fact that more and more physicists were using electronic methods of detection in the early 1970s, and so were brought into a much closer, and more frequent, relationship with the Geneva laboratory and its staff than had been the case for bubble chamber physicists. This was true for all stages in the life of an experiment.

A bubble chamber was essentially a piece of heavy engineering, and physicists had little to do with its construction and installation. By contrast, an electronic detector was conceived by physicists to do specific tasks, they participated in its construction, in its assembly, and in its testing on the experimental floor. A bubble chamber run was a more or less routine affair, and the one or two physicists who were present were there essentially to monitor performance. The physicists' relationship to their electronic detectors was interactive, and it was imperative that they be present while taking data. Finally, the results of a bubble chamber experiment were recorded on film, and could be analyzed long after the run was done at any institute which had adequate film scaming and measuring equipment and computers. By contrast, data-taking and data-analysis were often part and parcel of the same process with electronic devices; there was a constant on-line feedback between the one and the other. In short, as one staff member put it, "The symbiosis CERN-outside group(s) has to be rather close for electronic experiments, where much work occurs on the spot at the time of the exposure, and where apparatus has to be built and assembled [...]".[23]

Hence the resentment we have alluded to. Rubbing shoulders and sharing tasks with CERN physicists day in and day out, users came increasingly to object to their exclusion from the everyday life of the laboratory, to demand that they "be fully integrated into the CERN machinery". As the chairman of ECFA, Guy von Dardel, said publicly in 1975, "The physicists of the European High Energy Physics Community, which are the users of CERN, [...] look at CERN as **their** laboratory, and not as an ivory tower for a select body of physicists on site", going on to insist that the great difficulties then being experienced by the home laboratories rendered "the ideal of CERN as **the laboratory of us all** even more essential".[24]

For outside users, too, then, the identity of CERN, and their place within it was at stake in the debate over who should build big equipment. If they found themselves at odds with the CERN physicists, if in fact the two groups were talking past each other, it was because each had quite different, even conflicting concepts of what CERN was. It was not a dispute which could be settled easily, or which could be settled at all by argument and debate.

One last remark to close this section. The growth in numbers, size, and complexity of electronic equipment in the late 1960s and early 1970s did not simply herald a change in the relationship between the users and the in-house staff in a high-energy physics laboratory. It also brought about a change in the practice of physics itself and, accordingly, in a physicist's self-definition. Put crudely, the task of a bubble chamber physicist was essentially to conceive an experiment, decide what beam conditions were required to do it, and then to spend hours analyzing the film produced by the detector. He or she was, to use Pestre's nice phrase, a "white collar worker", who was seldom present on the "shop floor", where the data was produced. Physicists doing electronic experiments, by contrast, were also "blue collar workers", who conceived, assembled, tested, and ran their detectors **in situ** themselves, in the experimental halls.[25] The evolution of this technique thus called for the overcoming of the distinction between head and hand, between physicist and engineer, it called for the emergence of a new kind of person, an applied physicist, who was at once capable of having a good idea, of designing a detector to test it, of soldering thousands of wires or calibrating hundreds of photomultipliers, and of writing computer programs to analyze results. If the CERN visitors demanded the right to build big equipment for use at the host laboratory, it was also because this kind of activity was becoming central to their conception of **what it meant to be a physicist** in the early 1970s.

5. CONCLUDING REMARK

The conflict that we have described, if not fully resolved, grew less intense with the passage of time. In particular CERN staff were increasingly less inclined to 'hoard' big equipment projects for themselves, and actively sought collaboration with physicists in other laboratories. (The same is true, of course, for the outside users, who became less concerned about protecting their national laboratories and universities.) The main reason for this renewed spirit of cooperation was essentially pragmatic. The new generation of electronic detectors built from the late 1970s onwards, and particularly the 4pi 'hermetic' detectors used at colliding beam machines, were so complex and costly in terms of both money and manpower, that physicists from different institutions, including CERN, simply had to pool their resources and to collaborate with one another. This does not mean that CERN staff lost their dominant position in the evolution of the laboratory's experimental programme. The fact that they could be continuously on site ensured that they had a better appreciation of the detector's caprices, and were in a better position to capitalize rapidly on the data it produced. At the same time visitors, while aware of their disadvantaged position, apparently found it more easy to accept. After all their contributions were essential, the **sine qua non** for the construction and operation of major items of equipment. And they no longer believed that they were regarded as, or treated like, "second-class citizens".[26]

6. ACKNOWLEDGEMENTS

This paper was written within the framework of the History of CERN Project, with additional financial support for an interview programme provided by The Andrew W. Mellon Foundation.

7. REFERENCES

1. For national laboratories see, for example, L. Hoddeson, "Establishing KEK in Japan and Fermilab in the US: Internationalism, Nationalism and High Energy

Accelerators", Social Studies of Science, Vol. 13, No. 1, pp. 1-48, February 1983, and C. Westfall, "Fermilab: Founding the First US 'Truly National Laboratory'", in F.A.J.L. James, ed., The Development of the Laboratory. Essays on the Place of Experiment in Industrial Civilisation, Macmillan, London, 1989, pp. 184-217. For international collaboration in Europe see A. Hermann, J. Krige, U. Mersits and D. Pestre, History of CERN. Vol. I. Launching the European Organization for Nuclear Research, and Vol. II. Building and Running the Laboratory, North Holland, Amsterdam, 1987, 1990.

2. J. Krige, "The Installation of High-Energy Accelerators in Britain After the War: Big Equipment but not 'Big Science'", in M. De Maria, M. Grilli, and F. Sebastiani, eds, The Restructuring of Physical Sciences in Europe and the United States 1945-1960, World Scientific, Singapore, 1989, pp. 487-501. The Lawrence Berkeley Laboratory was an important exception, of course -- see J. Heilbron and R. Seidel, Lawrence and his Laboratory. A History of the Lawrence Berkeley Laboratory, Volume I, University of California Press, Berkeley, 1989.

3. For the transformations induced by bubble chambers, see P. Galison, "Bubble Chambers and the Experimental Workplace", in P. Achinstein and O. Hannaway, eds., Observation, Experiment, and Hypothesis in Modern Physical Science, MIT Press, Cambridge, 1985, pp. 309-73, and P. Galison, "Bubbles, Sparks and the Postwar Laboratory", in L.M. Brown, M. Dresden and L.M. Hoddeson, eds, Pions to Quarks. Particle Physics in the 1950s, Cambridge University Press, Cambridge, 1990. See also G.M. Swatez, "The Social Organization of a University Laboratory", Minerva, Vol. 8, pp. 36-58, 1970.

4. Remarks made by Bruce Cork at a Spark Chamber Symposium held on 7 February 1961 at the Argonne National Laboratory. The proceedings were published in Rev. of Sci. Instruments, Vol. 32, pp. 479-531, 1961.

5. For a useful description of one of these electronic detectors see P. Watkins, The Story of the W and the Z, Cambridge University Press, Cambridge, 1986.

6. D. Pestre in Hermann et al., History of CERN. Vol. II, chapter 8.

7. This paper is a reworked and refocussed version of an unpublished internal report produced within the framework of the History of CERN Project, and entitled J. Krige, "The Relationship Between CERN and its Visitors in the 1970s", Studies in CERN History CHS-31, CERN, Geneva, 1990, 38pp.

8. I am indebted to Simon Schaffer for drawing my attention to this point in his comments on an earlier draft of this paper.

9. For the importance of the difference between being an instrument **builder** and an instrument **user** see the comments by D. Pestre in Hermann et al., History of CERN. Vol. II, chapter 13, section 6. See also R.F. Hirsh, Glimpsing an

History of CERN. Vol. II, chapter 13, section 6. See also R.F. Hirsh, Glimpsing an Invisible Universe. The Emergence of X-Ray Astronomy, Cambridge University Press, Cambridge, 1983, at p. 63.

10. Data presented by ECFA Secretary Dick Harting to an ECFA information meeting on 30 November 1972. The minutes of the meeting are CERN/ECFA/73/3 (undated), and a copy can be found in file DG21220 in the CERN archives.

11. Quotations from a circular letter by Georgio Salvini and Dick Harting to ECFA members, document CERN/ECFA/72/10, 21/12/72, and attached as Appendix II to the Report of Working Group 3 of the European Committee for Future Accelerators, dated 1 August 1974 (DG21222-CERN).

12. For the composition of the Group and its terms of reference see the minutes of the ECFA meeting on 4/5/73, CERN/ECFA/73/5, and the Report of Working Group 3.

13. It is Report of Working Group 3.

14. Letter Lipman to Gunn, 4/5/73 (YGC23572-CERN). Lipman had spent the previous ten months at CERN working as the leader of an experiment at one of the big facilities built by CERN. His experience, his feeling that he had been "rather well involved in CERN", and the generally objective tone of his letter, have led us to place considerable weight on this document.

15. For more detail on the other complaints see J. Krige, CERN and its Visitors.

16. Quotations respectively from letter Lipman to Gunn and from minutes of ECFA meeting on 1/12/72, CERN/ECFA/73/4 (undated), (DG21220-CERN).

17. Letter Lipman to Gunn.

18. For information on the national labs see A. Berthelot, "Elementary Particle Physics in Saclay", paper presented at ECFA information meeting, CERN, 26/5/72, "Frascati National Laboratories. A Report to ECFA. May 15, 1972", Servizio Documentazione dei Laboratori Nazionali di Frascati del CNEN, and G.H. Stafford, "The Rutherford High Energy Laboratory", all in file (YGC23572-CERN).

19. Letter Lipman to Gunn.

20. Letter Allison to Baugh, 10/9/73 (YGC23572-CERN).

21. For more on this see D. Pestre in Hermann et al., History of CERN. Vol. II, chapter 8.

22. See Report of the Scientific and Technical Working Group to the European Preparatory Commission for Space Research, 2nd ed., December 1961, p. 12, in the European Space Agency Archives, European University Institute, Florence, Italy.

23. See Goldschmidt-Clermont's report prepared for ECFA, Structure of CERN Research Groups and Support Provided, ECFA/WG3/RLP4/73, 7/9/73 (DG21221-CERN).

24. Farewell Address to Professor W. Jentschke Pronounced at the CERN Council Session on December 17 1975 by the Chairman of ECFA, Guy von Dardel, undated, unsigned (DGR21311-CERN).

25. D. Pestre, in Hermann et al., History of CERN. Vol. II, chapter 8, section 4.1.

26. This has emerged clearly in a set of interviews made with physicists who worked on two of these large electronic detectors in the late 1970s (experiments UA1 and UA2) at CERN.

Chapter 9

Introduction of laser techniques in turbulence and combustion domains: a case study on the impact of instrumentation on science

Iskender Gökalp
Centre National de la Recherche Scientifique (France);

Lusin Bagla
Group d'Etudes sur les Sciences et la Technologie (France);

Susan E. Cozzens
Rensselaer Polytechnic Institute

1. INTRODUCTION

This paper concerns the introduction of a new class of experimental techniques into the instrumentation space of two scientific domains. It reports on the progress of a collaborative research effort between French and American teams[1] on how the introduction of laser-based optical techniques modified the organizational and cognitive structures of the domains of turbulence and turbulent combustion.

The introduction of a new instrument or technique can constitute a major change in the development of a scientific domain[2]. The diffusion process involves both the adoption of the instrument and its consequences, and therefore consists of interactions between the diffusing technology and the medium it invades[3]. The advent of a new technique may modify the complex network of interactions between the various sequences of experimental work, as well as the interaction between experiment and theoretical/numerical efforts.

From the examples of instrument change investigated in the science studies literature, one may distinguish three different paths for the diffusion process[4]. One path consists of borrowing an already-developed technique from another field. (To be sure, the borrowing is almost always accompanied by some modification of the original technique to accomodate the application constraints of the borrower field.) A second path is the evolution of an existing technique by successive modifications of its operating principle or by using it in conjunction with another instrument. A third path is the adoption of a totally new technique based on a different operating principle than the techniques previously in use.

The case studies investigated here are examples of the third path. They concern the introduction of laser-based optical techniques into two domains, turbulence and turbulent combustion[5]. These techniques are used to probe the structure of several thermodynamic and fluid mechanical quantities such as velocity, temperature, density, or species concentration, in either fluid or gas flows (in turbulence studies) or in flames (in turbulent combustion studies).

The comparison of the introduction patterns of laser techniques in these two domains allows us to contrast two essentially different situations. In pure turbulence studies, probe techniqes such as hot-wire thermo-anemometry, have been in use since the 1920's to determine fluctuating quantities such as velocity or temperature. These techniques, however, are useless in the combustion situation for two reasons. First, the material probes and essentially their sensors (see Section 2) are too fragile to be used in a hostile medium such as a flame. Second, as all the thermodynamic quantities are fluctuating simultaneously and as the operating principle of the hot-wire technique is convective heat transfer, the output signal is sensitive to velocity, temperature, and concentration all together. The isolation of one type of information is thus almost impossible because of the strong correlations between the fluctuating quantities. Consequently, in the turbulent combustion domain, laser based techniques, such as laser Doppler anemometry (or velocimetry) have found a totally virgin territory in which they are warmly welcomed by the whole community. The generalization of the use of laser based optical techniques in combustion studies (which also invaded non-turbulent or laminar combustion studies) during the 1980's, drastically modified the cognitive structure of the combustion field by allowing combustion scientists to acquire important experimental information that was lacking in the pre-laser area[6]. On the other hand, in pure turbulence studies, the reactions to these new techniques were much more complex because of the presence of a well confirmed earlier technique and of the associated skills.

In this paper, we discuss preliminary findings that describe the diffusion dynamics of laser-based techniques from two perspectives. The first focuses on adoption per se. After introducing the problems of pure turbulence and turbulent combustion, we turn in the third section of the paper to a delineation of the adoption in the two areas of one particular laser based technique, laser Doppler anemometry (LDA). LDA is a useful example because it has been applied in both turbulence and turbulent combustion studies, and because velocity, which it measures, is important in characterizing turbulent fields. We use quantitative indicators to portray the broad-scale pattern of adoption of LDA in turbulence studies, and case studies to show the factors that went into the switch from hot-wire techniques to LDA among turbulence and turbulent combustion researchers.

In the fourth section, we then explore the way LDA interacted with experimental and cognitive structures within turbulent combustion studies. We argue here that LDA helped consolidate TC as an autonomous field of inquiry, distinguishable from other turbulence and combustion studies. It did this through three effects: model-unifying, role-unifying, and place-unifying.

2. EXPERIMENTAL PROBLEMS IN TURBULENCE

Fluid mechanical turbulence is characterized by random fluctuations, in space and time, of all the thermodynamic and fluid mechanical parameters of the flow field. The experimental determination of the fluctuating part of these quantities has always been a major challenge in turbulence studies. The reason is the following. Until the expected generalization of the full numerical simulation of turbulent flows (which is currently limited to low velocity and simple geometry flows because of the huge computer power and time required), the prediction of turbulent flows continues to be based on the numerical solution of conservation equations (conservation of mass, momentum, enthalpy, and species). The number of available equations, which are derived from the present understanding of the physics of turbulence, is less than the number of unknown terms, which are generated by the correlations between fluctuating quantities. The solution adopted to "close" this open set of equations is to introduce physical models which heavily rely on experimental information.

Since approximately the 1930's, the only method for obtaining instantaneous (time varying) velocity measurements in turbulent fluid flows was the hot-wire anemometer (HWA). The sensor of this intrusive diagnostic method consists of a fine wire, only a few microns in diameter, stretched between two probe support needles. An electronic circuit passes a current through the wire, causing its temperature to rise above the temperature of the flow. The response of the hot wire to changes in the flow velocity is governed by the heat balance between the wire and the flow. The feedback system of the incorporated electronic circuit maintains the sensor at a constant wire resistance by responding instantaneously to changes in the characteristics of the flow past the wire. In incompressible and isothermal turbulent flows, the wire response is a function of velocity alone, hence it is relatively easy to obtain the mean flow and turbulent velocity information by performing a straightforward calibration and monitoring of the voltage across the electronic bridge.

As mentioned above, in pure turbulence studies, hot-wire anemometry was a mature technique when laser techniques were first introduced. Turbulence scientists had been familiar with this technique for decades, and they accumulated much experimental data using it. In parallel, the theoretical representation of turbulence was developed in conjunction with these experimental data.

Limitations of the hot-wire technique, however, become particularly severe in low speed flows, in flows close to walls, and in non-isothermal or reacting turbulent flows. For example, in heated flows, the simple relationship between the bridge voltage and velocity gives way to complex calibrations and calculations since the hot-wire output is a function of temperature, velocity, and density, which must be separated. Moreover, hot wires are extremely fragile and cannot be used conveniently in "dirty", high density, high temperature, or corrosive flows.

Laser based optical techniques were introduced into turbulence studies in the early 1960's. These techniques are based on the use of laser light and various optical principles to determine the parameters of a flow system. For example, the Doppler effect associated to Mie or Rayleigh light scattering allows the direct determination of velocity; Rayleigh light scattering enables the determination of the total density; and Raman light scattering allows the determination of the concentration of various species and of temperature. Also, as these various quantities can be determined quite independently from each other, their correlations can also be determined by combining two techniques. Finally, as the laser techniques are non-intrusive, they are not influenced by the hostility of the medium they explore and they do not perturb this medium.

Laser Doppler Anemometry (LDA) is one of these recent optical techniques that measures velocity directly. Yeh and Cummins (1964) first proposed the idea of using laser light scattered from small particles in a flowing gas to determine the flow velocity[7]. The principle of LDA relies on the measurement of the Doppler shift of laser light scattered from small particles (less than 1 micron in diameter) carried along with the fluid. The actual physical processes underlying the laser Doppler method can be described as a measurement of the phase difference of two optical waves scattered by small particles in the probe volume. The time rate of change of this phase difference can be interpreted as a measure of the suitable defined fluid velocity relative to the interference fringe planes in the probe volume. This interpretation requires that the particles follow the fluid motion with negligible slip due to fluid shear or acceleration.

The instantaneous flow velocity is inferred from the particle velocity by analyzing the collected light with sophisticated electronic and digital signal processors. By using more than one incident laser beam and additional optics, velocities along two or three directions can be measured simultaneously.

The following table presents a comparison of some technical features of HWA and LDA.

	HWA	LDA
Nature of the instrument	mechanical probe	optical probe
Intrusiveness	intrusive	non-intrusive
Compactness	high	low
Energy Needs	low	high
Other utility needs	none	cooling water
Home repairability	high	low
Safety contraints	low	high
Cost	low	high

In this comparison, we used the characteristics of a two-color LDA system which permits the determination of two components of the velocity vector, as this is also achieved by classical HWA systems. If the comparison is based on single component systems, both for HWA and LDA, the relative compactness of LDA will be greatly increased in the case of integrated and miniaturized systems. Also, for such systems the energy and other utility needs are closer to those of a conventional HWA.

The table indicates that there are both benefits and costs involved in using LDA. The cost of the system itself is high, depending essentially on the ability of the system to determine simultatneously one, two or three components of the velocity vector. More safety precautions need to be taken since lasers are involved. And there are parts of the system that remain "black boxes" to those that use them, in the sense that they can neither build nor repair them in the laboratory. This is so for the laser light source, but also for the processors of the Doppler signal. The most friendly part of the LDA system is its optical part, where with some knowledge in optics, the user may partly modify it. Producing reliable results with LDA demands a greater range of knowledge, some of it more specialized (in physics and optics) than with HWA (which requires knowledge only in fluid mechanics, heat transfer, and electronics). Some teams may have this knowledge; others have to call in outside help. In short, we may summarize this comparison by saying that the "control distance" between the user and the instrument is longer in the case of LDA. LDA users warn each other to expect to invest a year in setting up and calibrating their equipment before regular experiments can be run. Even then, some users report that they must devote a professional person on a nearly full-time basis to the equipment to keep it producing reliable information. All this creates, or has created during the earlier phases of the LDA development, obstacles to the diffusion of the technique; yet those obstacles have been overcome. How broadly is the subject of the next section.

3. ADOPTION PATTERNS

3.1. Broad-scale patterns

A first element of the diffusion dynamics of a scientific instrument or technique is the quantitative and temporal dimensions, or its broad-scale patterns. These dimensions tell us about the penetration rate of the instrument in different scientific domains, in different national settings, or in different media in a single national context, such as universities, public research institutes, or industrial/private research centers. Below, we present some global evidence on this evolution by focusing on the penetration rate of LDA in turbulence studies.

Our description is based on a search in computer abstracts and titles for the terms turbulence(t) in combination with either flow or velocity. The latter terms were used to distinguish articles on mechanical turbulence from ones on, for example, turbulent economic conditions. As a rough indicator of the use of HWA or LDA in these articles, the computer also searched for the terms hot-wire anemometry or velocimetry and laser Doppler anemometry and velocimetry in titles and abstracts[8]. We used the two data bases that have the best coverage of engineering literature, the Engineering Index Compendex Plus (1970-1989) and the Science Index INSPEC (1970-1976) followed by INSPEC (1977-1989). For the twenty years explored, the total cumulative number of papers indexed in the two data bases is more than 6 million[9].

For the 1970-1989 period, two data bases explored included 1633 items dealing with hot-wire anemometry. When coupled with the key words Turbulence(t) and Flow or Velocity this number reduces to 901, 203 of which are duplicated items between the two data bases. It may be argued that the 732 items (44.8 % of the total sample) that were left out when coupled with the previous key words concern the use of hot-wire anemometry in non-turbulent (or laminar) flows. The total number of items dealing with laser Doppler anemometry identified from the two data bases is 3738. When coupled with the key words Turbulence and Flow or Velocity, this number reduces to 1309, 263 of which are duplicated items. The number of items left out when LDA is combined with these terms is 2429 (65% of the total). This seems to indicate that LDA is more extensively used than HWA in non-turbulent situations. However, some of these items may also have a merely technical content and thus correspond to papers reporting on the development and improvement of LDA, without any specific reference to the application area of the technique.

Fig. 1 shows the proportion of turbulence papers that are known (from the keywords) to use HWA. The ratio decreases during the 1970's and remains approximately constant during the 1980's. In constrast, the LDA ratio (shown in Fig. 2) increases during the whole period, but at a higher rate during the 1970's.

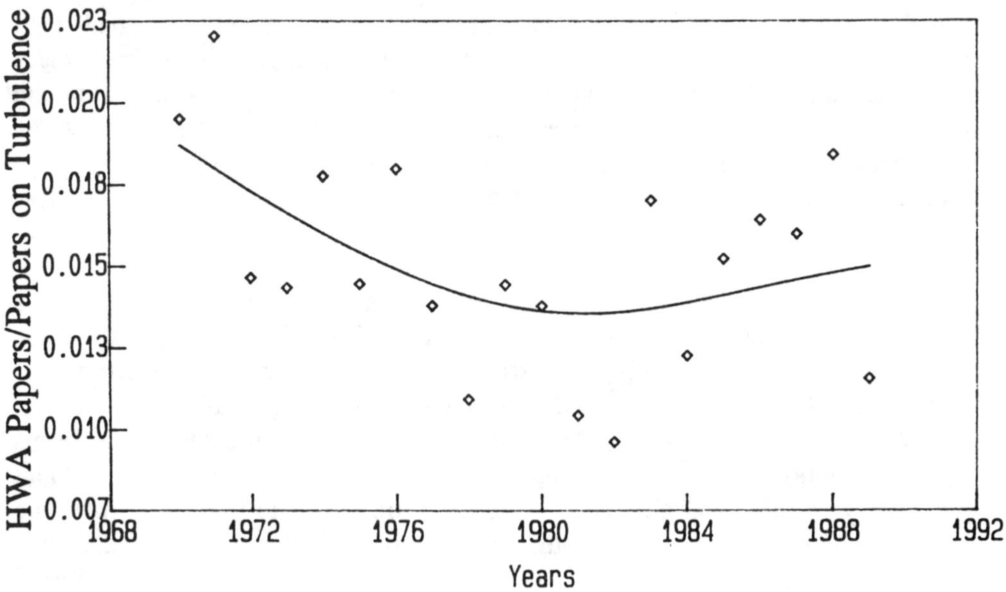

Figure 1

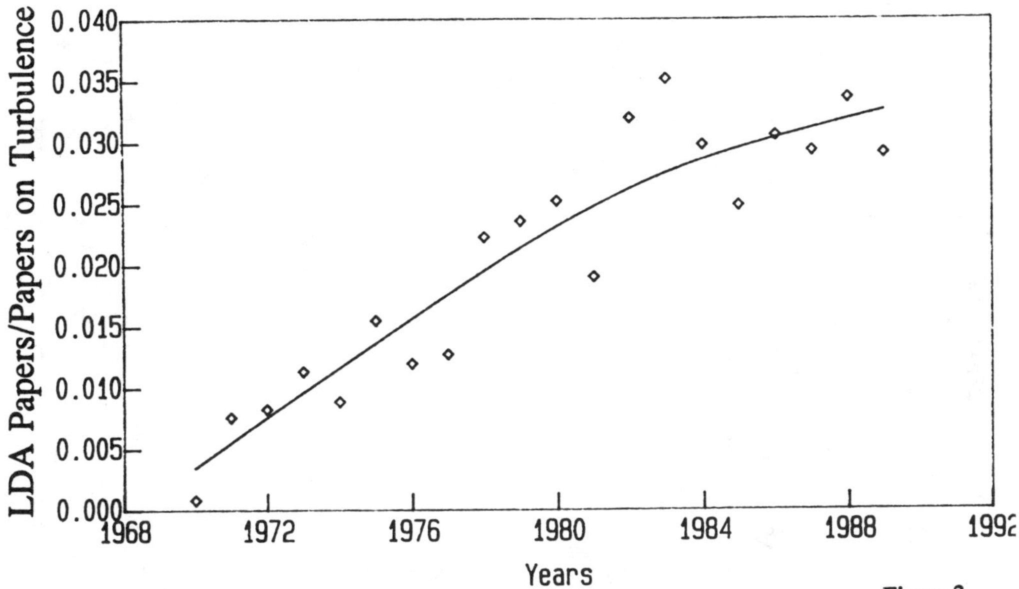

Figure 2

The proportions displayed in these two figures indicate that this indicator is far from perfect. Clearly, since these two techniques are the only ones available to measure the velocity fluctuations in a turbulent flow, one would expect them to be used in every experimental article dealing with the structure of turbulence, in non-reacting and reacting flow situations. Thus, if keywords were a perfect indicator of use, the percentage of articles in which they appeared could be taken to be the percentage of experimental literature in the field. But that figure is only 5%, an estimate which must be low. The keyword indicator, therefore, probably captures only part of the overall use of HWA and LDA. There is no reason to expect, however, that the underestimate is different between the two instruments, and the profile of adoption (presented below) is remarkably similar to data provided from another source. We thus considered it reasonable to proceed with this approach.

The balance between HWA and LDA use appears in Fig. 3, where the yearly variation of the ratio of the papers using LDA to those using HWA is plotted. LDA gains on HWA steadily during the 1970's, rising to about double the rate of use during the 1980's. The same information is put differently in Fig. 4, which shows the relative importance of HWA and LDA in the experimental characterization of the turbulence. This figure is constructed as the time evolution of two ratios: I_{HWA} = HWA/(HWA + LDA) and I_{LDA} = LDA/(HWA + LDA), so that for each year $I_{HWA} + I_{LDA} = 1$. The data show clearly that the fraction of turbulence literature known to use HWA is dropping while the LDA fraction rises. The curves cross in 1980.

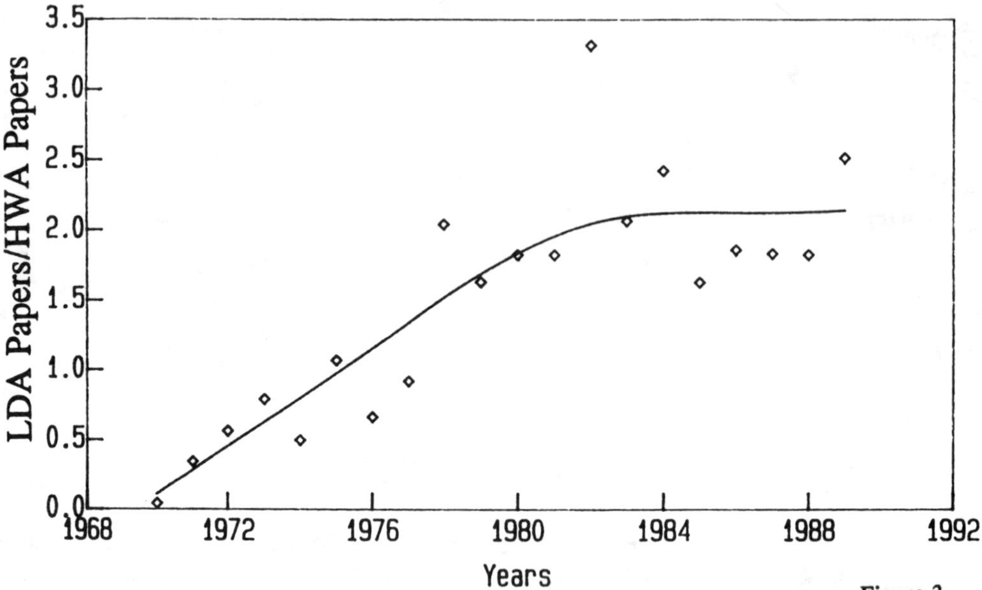

Figure 3

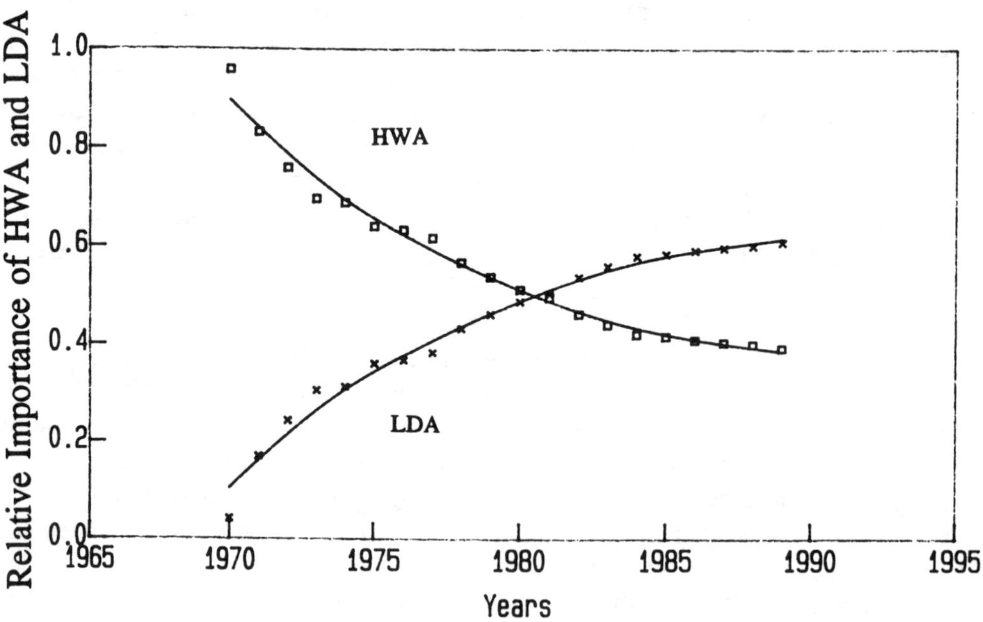

Figure 4

This picture of substitution dynamics is confirmed by data showing percentage of sales of HWA and LDA provided by DANTEC, one of the two major developers and vendors of the techniques. In 1978, HWA sales were close to 60%, but by 1984 only about 33%. Both literature-based data and sales figures thus tell a similar story: rapid change-over in the years surrounding 1980.

3.2. Case studies of adoption

The broad-scale patterns just presented are made up, of course, of decentralized decisions made in a number of laboratories, decisions influenced by the attitude toward LDA in the field, the competitive situation, and local contingencies of various kinds. Our preliminary results indicate that a major portion of the adoption of LDA, at least in the smaller area of turbulent combustion, consisted of the adoption of the instrument by new researchers in this domain at the time they set up their laboratories. This is an interesting phenomenon in its own right, which we will explore further in later reports on this project.

Some of the increase in LDA use, however, must have taken the form of switch-over within existing laboratories. In order to understand this phenomenon, the French part of our team has developed a set of case studies of individuals (senior scientists, research engineers, or technicians) who once worked with HWA but switched to LDA, or are using both instruments now. These individuals, some TC and some pure turbulence workers, were all working in one of the three major French turbulence or combustion laboratories where LDA was introduced. These public research centers, all affiliated with the French National Centre for Scientific Research (CNRS), have comparable sizes (between 70 and 100 persons). One is a pure turbulence laboratory located in Marseille. The second one has strong interests in the role of turbulence in combustion and is located in Rouen. The third one, located near Paris, has a much more diversified research programme such as pure turbulence, combustion, high speed flows, heat transfer, etc.

Preliminary results suggest three models of LDA adoption. They are the early adoption model, the progressive adoption model, and the resistance model. It is interesting to try to correlate these models with factors that characterize the individuals involved in the instrument substitution process, the research programs, and the history of the concerned research centers. Three observations on this issue are summarized below.

The early adoption model shows two sub-models. The first one corresponds to the case where the scientists adopt LDA totally and apply it to the problem that they were investigating with HWA. In TC, this behavior can be explained as a natural consequence of the earlier period, characterized by the lack of any appropriate instrument to measure turbulence in a reactive flow. Indeed, within the few years around 1975, all the major combustion laboratories of the United States, Europe, and Japan working on turbulent combustion switched to LDA. This early adoption was driven by necessity. In pure turbulence, the necessity is less general. Nonetheless, some scientists switched quickly to LDA for various and complex reasons. For some of them, the new technique appeared to be an opportunity or investment for the future; it enabled them to extend the range of their investigation either to some regions of the flow they were unable to investigate with HWA, such as the frontiers of turbulent jets mixing with the surrounding medium where the turbulence intensity is too high, or to some more drastic flow conditions, such as high temperature flows. On the other hand, some of the pure turbulence scientists became highly specialized in this new technique itself; in some cases their scientific productivity dropped significantly. Some of them continued their careers by organizing meetings and training courses on LDA.

The second sub-model is the case that corresponds to a drastic change of the scientific research program of the scientist concerned, such as a move from pure turbulence studies to turbulent combustion studies. An example is the case of a scientist who was very well known in the application of hot-wire techniques to the investigation of turbulent boundary layer structure in classical fluid mechanical problems. He adopted LDA very early and developed a new research program on turbulent combustion.

The second observation concerns the resistance model. The age of the scientist is not the major factor that explains this behavior. For example, one senior scientist, very highly specialized in the application of hot-wire techniques to high speed flows, adopted LDA some ten years before retirement and applied it to heat transfer problems, which was a new domain for him. One important factor that accompanies the resistance of some scientists to switching to LDA seems to be the presence of a technician very highly skilled in hot-wire anemometry and who works almost exclusively with the scientist. We observed this situation in two laboratories. In both cases, but also in general, we did not observe the case of the transformation of an HWA technician into a LDA technician.

A factor that seems important in explaining the resistance model is a cognitive one. Indeed, HWA delivers a continuous signal that one can easily use to determine the frequency spectrum of velocity fluctuations. LDA delivers a discrete signal which, in some cases, may be a severe problem. For example, a scientist interested in the spectral approach or in the determination of the smallest turbulence scales will be reluctant to use LDA in turbulence studies, because these approaches need a high frequency and continuous signal. Some of the best-known scientists in pure turbulence studies are still continuing to use hot-wire techniques intensively for essentially this reason.

The third observation concerns the progressive adoption model. Several factors explain this behavior, but the research programme of the scientist dominates. This model corresponds mainly to the behavior of scientists active both in pure turbulence studies and in related domains where turbulence is only a part of the investigated phenomena, such as turbulent heat transfer or turbulent combustion.

4. EFFECTS IN TURBULENT COMBUSTION

We turn now from the description of large-scale patterns of adoption to a closer look at the consequences of LDA for the domain of turbulent combustion. One of the authors has described elsewhere the emergence of turbulent combustion as a autonomous field of inquiry during the 1970's[10]. In this study of LDA, we are examining the role this instrument played in that emergence.

Some hundred laboratories or research groups world-wide devote themselves to the study of turbulent combustion. Their work is essentially to analyze and model the production of chemical reactions and thus of thermal energy in a turbulent flow; hence their findings are important to manufacturers of furnaces as well as all sort of engines (jet, automobile, rocket) where the chemical production of thermal energy is used as the power source. The major sources of government support in the U.S., for example, are the Air Force Research Office and the Department of Energy; and several government laboratories are involved in the research (such as Sandia National Laboratories in Livermore, California, and Lawrence Berkeley Laboratory in Berkeley, California), along with university and industrial scientists.

The introduction of LDA and of other laser based diagnostics into TC has had a major impact on the development of interpretative and predictive models, on the division of work between model developers and experimental information producers, and on the relations between workplaces where both type of scientific work is conducted. We address these impacts in the following.

4.1. Model-unifying effects

The study of turbulent reacting flows is not an easy task. The non-linear couplings between turbulence and chemical reactions give rise to extreme difficulties in the rigourous mathematical representation of these aerothermochemical systems. This is why the physical understanding of the structure of turbulent flames, based on their experimental investigation, and the development of models to interpret these empirical information are so important. The major task of a model in a given turbulent combustion situation is to relate the turbulence characteristics of the unburned mixture to the turbulent flame structure, and ultimately to the main applied parameter, which is the turbulent flame propagation velocity or the turbulent rate of conversion of chemical energy into heat.

The cognitive structure of the turbulent combustion models during the pre-laser area can be illustrated with the example of a simple laboratory burner configuration, such as a turbulent Bunsen burner, with which most basic studies on turbulent combustion are conducted. The turbulence structure (the velocity fluctuations) is first measured with HWA in a cold flow. The mixture is then ignited and the turbulent flame propagation velocity is measured by some photographic technique, by dividing the total flame surface area by the total flow rate. Some correlations are then proposed to link these two measurements. At this point, however, two other types of information have been missed. The influence of the flame on the oncoming turbulence stucture has not been accounted for; and the detailed thermal and chemical structure of the turbulent flame was not known. Today, LDA allows the determination of the turbulence structure just ahead of the flame front and thus permits researchers to relate and unify more efficiently the turbulent and flame propagation aspects of a turbulent flame. The use of other laser techniques to map out the instantaneous internal structure of turbulent flames further allows investigators to increase the degree of unification between the turbulent, thermal, and chemical aspects of turbulent flames. Today's models incorporate these new kinds of information and thus provide a more complete description of the interactions between turbulence and combustion.

Theory shows a parallel development. In the absence of any quantitative data about the structure of turbulent flames, since the 1940's several theoreticians have proposed models of turbulent combustion based on classifications, using as criteria the comparison of temporal, spatial, and velocity scales of both turbulence and combustion. The pre-laser era classifications were quite straightforward, with the frontiers between the regimes of turbulent combustion sharp and clearcut. New classifications, proposed in the 1980's after the accumulation of important experimental data collected by laser techniques, suggest intermediate regimes and emphasize the importance of transitional situations. Thus, new experimental information, contrary to intuition, has helped blur the boundaries between various regimes. Indeed, the new information has made it clear that the early classifications were too pure and that the dominant interaction mechanisms between turbulence and combustion were actually best represented by a unified representation of the expected models.

4.2. Role-unifying effects

The role unifying effect concerns the division of scientific work between theoreticians and experimentalists within a given domain. With the increase of the available experimental information relevant for modeling purposes, some pure theoreticians of the earlier period have shown more interest in conducting or supervising experimental work, and conversely, some well known experimentalists have also shown strong interest in theoretical developments. This does not mean that a complete role reversal is occurring, but it does indicate that sharp boundaries within the scientific division of labor may be blurred by developments in instrumentation.

One example may illustrate the role unifying effect. K.N.C. Bray from Cambridge University, England was the first theoretician to approach turbulent combustion studies in terms of conservation equations of turbulent reacting flows. At the beginning of the 1970's, he addressed the question of the modification of turbulence through a propagating flame front by trying to take into account the competition between the various factors contributing to this balance. He is regarded as one of the most prominent theoreticians of the premixed turbulent combustion domain, and he is one of the most cited authors of the turbulent combustion domain.

Recently, K.N.C. Bray and his co-workers have proposed a model in order to predict the chemical production rate in premixed turbulent flames. The new model is based on simple physical ideas and directly measurable quantities. It equates the mean chemical production rate at a given point within the turbulent flame to the product of the chemical conversion rate per flame crossing and of the number of flame crossings per second. An alternative version of the model is based on spatial information about the flame crossing statistics in a unit flame area. In this version, the number of instantaneous flame fronts in a unit flame area may be determined with the laser tomography technique by visualizing the instantaneous movements of the turbulent flame. Bray has recently developed an experimental study to determine the necessary information for his new model. The flame configuration he selected is the one that corresponds best to the major hypotheses of his model, and the experimental technique he is using (laser tomography) is the one capable of giving the needed quantitative information.

Role unification of this kind is not easy to achieve. Several strategies are necessary. One strategy is to collaborate with experimentalists working in other workplaces. Another strategy is to bring one of these experimentalists to one's own workplace for some time and share skills. These strategies together may then help to orient the experimental lines of work of the domain in the directions needed by the model and to enhance the experimental orientation of a predominantly theoretical individual. This strategy may prove particularly fruitful if the collaborating experimentalists are also willing to move toward a role unification from the other direction[11].

4.3. Place-unifying effects

Before the diffusion of laser based techniques, most turbulent combustion studies were undertaken in workplaces where the principal work lines consisted of experimental or theoretical approaches to pure turbulence studies or experimental or theoretical approaches to pure combustion studies. These TC workplaces were isolated islands within combustion or turbulence laboratories. The limited extent of any contact between the larger workplaces (specializing either in turbulence or combustion) strongly influenced the research lines developed within the TC islands by enclosing them in purely turbulent or purely chemical approaches. The TC work done in these places was concerned essentially with modeling applied combustion parameters (such as the flame propagation velocity) or with developing theoretical ideas on the influence of chemical reactions on the structure of turbulence. On the experimental side, as discussed above, researchers were hampered by lack of any suitable technique for determining the structure of turbulence in a reacting medium.

With the development and diffusion of the various laser-based techniques, the workplaces where turbulent combustion studies are conducted have become progressively autonomous with respect to general turbulence or combustion laboratories. The laser instruments used in turbulent combustion thus helped to define more sharply the frontiers of turbulent combustion laboratories or of turbulent combustion research groups within combustion or turbulence laboratories.

A preliminary survey of American and French combustion laboratories that use laser techniques intensively suggests that three models of spatial reorganization are relevant. One model is the "new instrument = new laboratory" approach. This is the case of the Combustion Research Facility of Sandia National Laboratories at Livermore, California, which was considered a testing ground for various laser diagnostics. The second model is a gradual one, in which a new building or extension, which hosts the laser techniques and the experiments to which they are applied, is constructed near the old one. This is the model followed by the C.O.R.I.A. laboratories at Rouen in France. Finally, the third model is a piecemeal approach where space is made for the new techniques in existing buildings, by taking out older instruments or experiments and by rearranging the inner space and the distribution of utilities. This is by and large the most general model.

Place unification also helped encourage theoretical lines of work in turbulent combustion by suggesting new modeling ideas. New turbulent combustion groups dealing with the theoretical and numerical prediction aspects of turbulent combustion appeared and interacted strongly with experimental groups. In fact, the period during which the experimental and theoretical lines of work were conducted in separate places lasted only few years. Recently, in the 1980's, experimental turbulent combustion work and theoretical/modeling/numerical work have increasingly been done in the same workplace. The best examples of this place unification effect are the above mentioned Combustion Research Facility of Sandia National Laboratories at Livermore, California, and the C.O.R.I.A laboratories at Rouen in France. These two centers concentrate in the same place the most advanced laser techniques and the most powerful predictive numerical codes. This interaction between experimental and theoretical approaches to turbulent combustion enhances also the role unification effect of the experimental techniques.

5. CONCLUSIONS

In drawing preliminary conclusions from this work in progress, we return to the comparison between pure turbulence studies and turbulent combustion research in adopting LDA. In turbulent combustion, the adoption was sudden and relatively complete. We have seen some of the consequences of that sudden shift for the field. The new laser-based techniques, including LDA, increased the interactions between experimental and theoretical lines of work at several levels: the work place, the individual, and the cognitive structure of the new research lines. Closer interaction between experiments and theory then encouraged the rise of a new organizational structure in the domain, characterized by increasing unification of experimental and theoretical approaches.

Thus a domain which had been frustrated for decades by the lack of experimental access to its object of attention was transformed, spatially, socially, and theoretically, by the availability of the right technique. For investigators in this field, the investment in time, personnel, and money that LDA required was well rewarded. The changes that followed the introduction of LDA reveal a strong interaction between instrument and context.

In contrast, the evidence for the pure turbulence domain shows that the adoption was gradual and partial. An alternative technology was well entrenched and effective. For individual investigators, returns on the investment in LDA were less dramatic than in the turbulent combustion case. Based on the comparison with turbulent combustion, we surmise that the consequences of LDA for the structure of turbulence research were also less pronounced, reflecting a weaker interaction with the overall instrumentation space.

To explain the difference between the cases, we cannot rely solely on either the character of the instrument or the character of the field it entered. Instead, we need to call in the process of interaction between the two. Individual decisions to adopt and adapt the instrument reflected both existing constraints and the opportunity to change the structure of the field. Past, future, and present became intertwined. Further complicating the picture is the interaction between the internal dynamics of the field and its external applications; but this part of the story remains to be told in a later report from our project.

Perhaps it is more important, though, simply to note the difference between the cases than to account for it. To return to our original observations, the introduction of new instruments in science may transform fields - or it may leave them relatively unchanged. It is, in fact, the interaction between the technology and the field that receives it which is crucial.

Notes

1. The participating teams are Dr. I. Gökalp (Centre National de la Recherche Scientifique, Laboratoire de Combustion et Systèmes Réactifs, Orléans, France); Dr. L. Bagla (Groupe d'Etudes sur la Science et la Technologie, Maison des Sciences de l'Homme, Paris); and Dr. S.E. Cozzens, Franz Foltz and Bo Shen (Department of Science and Technology Studies, Rensselaer Polytechnic Institute, Troy, New York. The research is supported by the Studies in Science, Technology, and Society Program of the National Science Foundation, under the program monitoring of Dr. Ron Overmann.
Parts of this paper were presented at the Conference on Rediscovering Skills in Science, Medicine, and Technology, Bath, UK, September 14-17, 1990 and at the Tenth Meeting of the International Scientific Instruments Commission, Smithsonian International Center, Washington, DC, September 21-25, 1990.

2. See, for example, P. Galison, "Bubble chambers and the experimental workplace," in P. Achinstein and O. Hannaway, *Observation, Experiment, and Hypothesis in Modern Physical Science*. Cambridge, MA: The MIT Press, pp. 309-373, 1985; and T. Lenoir, "Practice, Reason, Context: The dialogue between theory and experiment." *Science in Context*, 2(1): 3-22, 1988.

3. The notion of "instrumentation space" calls for some explanation. The experimental scientific work is composed of several sequences. Here we shall propose a model comprising four sequences. Sequence 1 is the definition of the experimental configuration, of the parameters to be varied, of the quantities to be measured and of the parameters to be deduced from the measured quantities. Sequence 2 is the conception and realization of the experimental set-up which will "produce" the fact or phenomenon to be investigated. Sequence 3 is the selection and operationalization of the measurement techniques. Finally,

Sequence 4 is the acquisition, reduction, and storing of the data. In a given experiment, these sequences interact with each other so that modifications in one sequence may strongly affect the others. Furthermore, each of these sequences have different type of relations with the theoretical/numerical levels of scientific work in a given domain. Finally, the above model of experimental scientific work implies that the main material elements of the instrumentation space are i) the set-up, ii) the measuring instrument proper (or the detector) and iii) the data acquisition, reduction and storage system (the DARS system). (see I. Gökalp, "A four-stage model of scientific experimentation." In preparation).

4. Some examples of such studies are discussed in I. Gökalp, "Turbulent reactions: Impact of new instrumentation on a borderland scientific domain." *Science, Technology & Human Values*, 15(3): 284-304, 1990.

5. These domains belong to engineering sciences. But they are basic sciences too. This is noteworthy since the distinction between basic and applied sciences are often misunderstood. The formers are easily associated with domains such as those dealing with the structure of matter or other similar areas; the latters are generally associated with engineering sciences. However, the cognitive structure of the domains listed above is similar to that, for example, high energy physics. One distinction might be that in engineering sciences the problems are posed by some technological development and are not the direct outcome of the investigations on some natural phenomena or fact or laws. For example, in turbulent combustion the closest one can get to a natural phenomena corresponding to the type of situation investigated in this domain is a forest fire in a windy weather. But the domain of turbulent combustion is developed in relation to combustion chambers and not to forest fires. However, this does not mean that the knowledge gained in investigating turbulent combustion in relation to a technological area could not be or is not used in the control or the prediction of the propagation rate of a forest fire. This emphasizes that we need a much more elaborated understanding of the complex relations between natural phenomena and phenomena created by technology.

6. For a discussion of these modifications see I. Gökalp, "The interrelating of scientific fields: the case of turbulence and combustion." *Studies in the History and Philosophy of Science*, 21(3): 413-429, 1990.

7. Yeh and Cummins were working at the Columbia University Radiation Laboratory. Their paper appeared in Applied Physics Letters and was entitled "Localized fluid flow measurements with an Helium-Neon Laser Spectrometer." Their work was supported by the National Science Foundation, the Army Reseach Office and the Joint Services of the U.S. Army, Office of Naval Research, and Air Force Office of Scientific Research. The introduction of their short paper (2 full pages) is the following:

"Flow patterns in fluids are commonly studied by injecting dyes into the flow streams and observing the streamlines produced by the dyes as they are carried along by the current. For velocities 0.05 cm/s or less, such techniques become inaccurate due to the diffusion of dye molecules through the medium. We utilized a laser spectrometer to examine the Doppler shifts in the Rayleigh scattered light at velocities as low as 0.007 cm/s. Monodispersed polystyrene spheres of diameter 0.557 microns were placed in water to form a dilute colloidal suspension" (p. 176).

Yeh and Cummins measured the velocity profiles of a water flow in a flow tube of 10 cm diameter. They compared their results with theoretical profiles from the laminar flow theory. They concluded that "this method should be applicable to the study of many fluid flow problems. Slower rates (velocities) than those measured here may be studied by appropriate choices of scattering particle size and angle of observation (...) (S)imultaneous determination of average velocities and of velocity gradients should be possible."

It is interesting to note that the motivation of Yeh and Cummins in developing a laser based velocity measurement technique was to achieve velocity measurements in very slow flows. In their paper there is no reference to hot-wire anemometry, turbulence, or turbulent combustion. This shows clearly that the invention of the laser spectrometer for velocity measurements did not initially challenge hot-wire anemometry, as this latter technique was not suitable for low velocities (approximately less then 50 cm/s). The initial motivation of the inventors was neither to measure turbulence nor to provide a non-intrusive technique for turbulent reacting flows. Their motivation was to measure low velocity flows. This shows that, as in the case of other technological devices, the development/diffusion or "socialization" patterns of a scientific instrument/technique are largely independent of their initial conditions of invention.

8. This computer based search lumped together the uses of HWA and LDA in all areas where fluid mechanical turbulence is involved, including pure turbulence studies and turbulent combustion studies, but also other areas such as turbulence in meteorological studies or in turbulent heat transfer. Consequently, the picture of HWA and LDA use that arises from this search is a global one. A domain specific approach would be much more satisfactory, but it requires the examination of the corresponding literature directly. This approach is in progress in our group.

9. The analysis of the online search is presented in I. Gökalp and L. Bagla, Unpublished Report, 74 p., February 1991.

10. See Gökalp (1990) in note 6.

11. See Gökalp (1990) in note 4.

Chapter 10

Development of high resolution NMR spectroscopy as a structural tool

James Feeney
National Institute for Medical Research (United Kingdom)

1. INTRODUCTION

The discovery of the nuclear magnetic resonance (NMR) phenomenon and its development and exploitation as a scientific tool provide an excellent basis for a case-study for examining the factors which control the evolution of scientific techniques. Since the detection of the NMR phenomenon and the subsequent rapid discovery of all the important NMR spectral parameters in the late 1940s, the method has emerged as one of the most powerful techniques for determining structures of molecules in solution and for analysis of complex mixtures. The method has made a dramatic impact on the development of structural chemistry over the last 30 years and is now one of the key techniques in this area. Support for NMR instrumentation attracts a dominant slice of public funding in most scientifically developed countries. The technique is an excellent example of how instrumentation and technology have revolutionised structural chemistry and it is worth exploring how it has been developed so successfully. Clearly its wide range of application and the relatively direct connection between the NMR data and molecular structure has created a major market for the instrumentation. This has provided several competing manufacturers with the incentive to develop better and better instruments. Understanding the complexity of the basics of NMR spectroscopy has been an ongoing challenge attracting the attention of physicists. The well-organised specialist NMR literature and regular scientific meetings have ensured rapid exploitation of any theoretical advances that have a practical relevance. In parallel, the commercial development of the technology has allowed the fruits of such theoretical advances to be enjoyed by the wider scientific community.

2. THE NMR EXPERIMENT AND SPECTRAL FEATURES

The effects of nuclear moments were first detected in the hyperfine structure observed in the optical spectra of particular atoms [1]. On the basis of these observations Pauli[2], in 1924, proposed that certain nuclei possess magnetic moments which interact with atomic orbital electrons. Stern and Gerlach[3] used beam methods to characterise the behaviour of atomic magnetic moments in magnetic fields and later these methods were improved by Rabi and coworkers[4] who subjected the molecules in the beam to electromagnetic irradiation at the appropriate frequency to give a resonance effect. These pioneering experiments led to the description of the nuclear magnetic resonance phenomenon[3-5]. Initial attempts to observe the nuclear magnetic resonances directly in bulk materials were unsuccessful largely because of the unfavourable materials used[6,7]. However, in 1945, scientists from two separate laboratories (Bloch[8], at Stanford University and Purcell[9], at Harvard University) independently observed hydrogen resonance signals for the first time using samples of water[8] and paraffin wax[9] respectively. In 1952 these workers were jointly awarded the Nobel Prize for their discovery. In these early experiments the relaxation behaviour of the nuclear spins was also noted and subsequently well-characterised by several workers[10-18]. A survey of their contributions is given in Emsley, Feeney and Sutcliffe[19].

The basic requirement for an isotope to show an NMR effect is that its nucleus must possess a magnetic moment designated μ. There are more than 100 known isotopes with magnetic moments and some important examples are listed in Table 1. The relative sensitivity of nuclei to NMR detection at constant magnetic field strength is related to μ^3 with the proton (^1H) being the most NMR sensitive naturally occurring isotope. Each nucleus is also characterised by a spin quantum number I which takes 1/2 integral values as indicated in Table 1.

TABLE 1

Nuclear properties of some isotopes

Isotope	Spin Number I	Natural Abundance %	Larmor Frequency MHz (2.349T)	Sensitivity At constant B_o
^1H	1/2	99.985	100	1.00
^2H	1	1.56×10^{-2}	15.35	9.65×10^{-3}
^{13}C	1/2	1.108	25.15	1.59×10^{-2}
^{15}N	1/2	0.365	10.14	1.04×10^{-3}
^{19}F	1/2	100	94.08	0.833
^{23}Na	3/2	100	26.45	9.25×10^{-2}
^{31}P	1/2	100	40.48	6.63×10^{-2}

On introducing a sample containing magnetic nuclei of spin I into a magnetic field (B_o), the nuclei occupy (2I + 1) different nuclear energy levels: thus for ^1H which has a value of I = 1/2 there are two such energy levels. Each level is characterised by a magnetic quantum number m which has values (I, I-1, ... -I) so that for ^1H the two states have m values of +1/2 and -1/2. The energies of the levels are $\pm\mu_z B_o$ and thus the energy separation between the two levels is given by $2\mu_z B_o$ (where the components of the magnetic moment μ along the direction of B_o are $\pm\mu_z$). As in other spectroscopic methods transitions can be induced between the different energy levels (selection rule $\Delta m = \pm 1$) by irradiating with energy of the correct frequency to satisfy the Bohr condition

$$\mu_z B_o / I = h\nu_o \qquad (1)$$

where h is Planks constant. The frequency, ν_o, referred to as the Larmor frequency, is seen to depend on the field strength B_o: for ^1H nuclei at 2.349 Tesla the frequency is 100 MHz. At this field strength other magnetic nuclei absorb radio frequency energy at completely different Larmor frequencies (e.g. ^{13}C, 25.15 MHz; ^{31}P, 40.48 MHz). There are now commercially available spectrometers with superconducting magnets operating at 14.09 Tesla (600 MHz for ^1H). The Larmor equation can also be described by $\omega_o = \gamma B_o$ where ω_o is the Larmor frequency expressed as an angular frequency ($\omega_o = 2\pi\nu_o$) and γ is the magnetogyric ratio of the nucleus.

The adjacent energy levels are populated according to a Boltzman distribution and because the energy separation is small there is only a small excess of population in the lower energy level. This has an important bearing on the sensitivity of the

NMR experiment because only the excess population in the lower level effectively absorb energy at the Larmor frequency. The nuclei in the lower level absorb energy while those in the upper level give out energy by stimulated emission. The inherent low sensitivity of NMR detection has been overcome to some extent by using fourier transform methods[20,21] in combination with signal accumulation techniques[22].

Relaxation times

An understanding of relaxation times can best be reached by considering the classical description of the NMR experiment where we have a collection of nuclear magnetic moments precessing about the direction of B_o (z axis) with an angular frequency ω_o where $\omega_o = \gamma B_o$ (Eqn. 2). In such a collection of nuclear moments we have a mixture of those aligned along the field direction and those aligned anti-parallel to the field. Because of the excess of nuclei aligned along the field direction there is a resultant component of magnetisation along the z axis: it is changes in this component which are monitored in relaxation studies. By applying an rf field B_1 along the x axis, the value of M_z is perturbed from its equilibrium value, and after the B_1 field is removed, M_z will then return to its equilibrium value by an exponential relaxation process characterised by a time T_1 known as the spin lattice relaxation time.

When the B_1 rf field is applied along the x axis to the system at equilibrium the nuclear magnetic moments begin to precess in phase creating a bulk component of magnetisation in the xy plane. When the B_1 field is removed this bulk magnetisation disappears as the magnetic moments lose their phase coherence and the rate at which this process takes place is characterised by T_2 - the spin-spin relaxation time.

Chemical shifts

One of the first applications of the NMR technique was to attempt to measure accurate magnetic moments of different nuclei by comparing their resonance frequencies in the same applied magnetic field. During the course of these measurements it became clear that the measured values depended on the chemical environment of the nucleus under investigation. The phenomenon - known as the chemical shift effect - was first seen by Knight[23] in studies of metals and metal salts and later by Proctor and Yu[24] for ^{14}N and by Dickinson[25] for ^{19}F. Proctor and Yu, working in Bloch's laboratory, were attempting to measure the accurate magnetic moment of the ^{14}N nucleus. They chose to examine NH_4NO_3 in the hope that this sample would give an NMR signal with twice the intensity of that of a molecule containing one nitrogen. They were astonished when they detected not one, but two separate ^{14}N signals. Their initial excitement based on the possibility that the differences might have some nuclear origin soon evaporated when they realised, together with Bloch, that they were probably dealing with some 'nasty chemical effects.' Little did they know that these effects were destined to become the basis of one of the most powerful structural techniques ever developed. Such was the birth of high resolution NMR spectroscopy and the discovery of the chemical shift effect. Thus, if one examines the NMR absorption spectrum of a sample containing molecules with nuclei in different chemical environments then each different type of nucleus is observed to absorb energy at a separate radiofrequency for a fixed value of the applied magnetic field, B_o. These effects are particularly easy to detect in liquids or in solutions which give spectra with narrow absorption bands.

In Fig. 1 we see that the ^1H NMR spectrum of acetic acid shows separate absorption frequencies for the two different types of protons (i.e. in the CH_3 and COOH groups). The ^1H nuclei in the two different groups have slightly different electronic environments which effectively shield the nucleus to different extents from the applied magnetic field B_o. Because the shielded nuclei experience slightly different magnetic fields they will absorb radiofrequency energy at different resonance frequencies according to Eqn. (1). The frequencies at which various types of protons absorb energy are characteristic of particular types of proton. It is usual to measure the frequencies as different frequencies from that of a reference material (tetramethylsilane, TMS) and to quote the chemical shifts in parts per million (ppm) by using the relationship

$$\delta = \frac{(\nu_{\text{Sample}} - \nu_{\text{Reference}}) \times 10^6}{\nu_{\text{Reference}}}$$

Fig. 1. The ^1H NMR spectrum of acetic acid at 60 MHz.

These dimensionless values are referred to as chemical shifts and when expressed in ppm are independent of the applied field strength. Extensive collections of chemical shift data for protons and other magnetic nuclei in all types of molecule have been compiled and these data can be used to assist in the structural analysis of molecules. Structures of complex natural products such as antibiotics, peptide hormones, steroids, lipids and carbohydrates have been determined. Detailed conformational and stereochemical problems have been solved and molecular dynamic processes over a wide frequency range investigated.

Signal intensities

Under ideal conditions the intensity of a nuclear magnetic resonance absorption (i.e. the integrated area under the signal) is directly proportional to the number of nuclei contributing to the signal. This makes the NMR technique extremely useful for the quantitative analysis of mixtures of compounds. Relative concentrations can be determined directly from measurement of the intensity ratios of the various signals and absolute concentrations can be obtained if a suitable calibration procedure is adopted.

Spin-spin coupling constants

In 1951, Proctor and Yu[26] showed that molecules containing different magnetic nuclei give rise to NMR multiplets which are attributable to scalar interactions between the different spins (see also Gutowsky and McCall[27]). The energy of a particular nuclear spin depends on the orientation of the nuclear spin of a nearby nucleus, the effect being transmitted via indirect interactions involving the valency electrons. The spin-spin coupling effects can operate between either heteronuclear or homonuclear species. These interactions are clearly seen in the ^1H spectrum of acetaldehyde shown in Fig. 2. Here the signal for the methyl protons is a doublet, each component corresponding to the two possible orientations of the spin (+1/2 and -1/2) of the interacting formyl proton. The signal for the formyl proton is a quartet, each component of the quartet corresponding to a different value of total spin of the three interacting methyl protons (+3/2, +1/2, -1/2, -3/2). The interaction between two spins is a scalar interaction the strength of which is known as the spin-spin coupling constant, usually denoted by J. When measured in Hz the coupling constants are independent of magnetic field strength (unlike chemical shifts).

The multiplicities on the absorption bands assist in the assignment of signals in complex molecules and the measured values of the coupling constants often prove useful in structural and conformation studies[28,29].

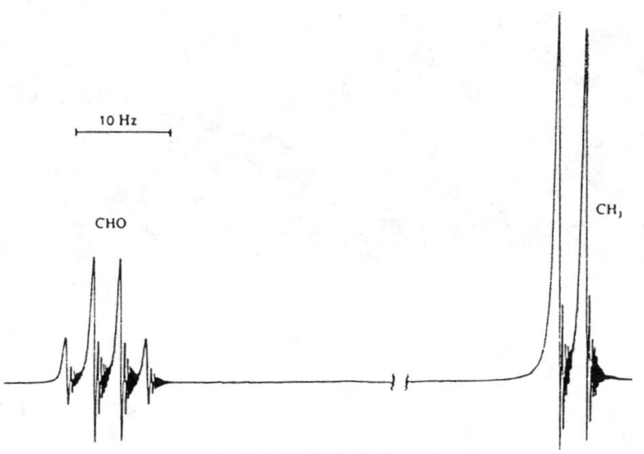

Fig. 2. ^1H NMR spectrum of acetaldehyde at 60 MHz.

Initially, NMR spectra were obtained by sweeping sequentially through all the spectral frequencies (so called continuous-wave or CW spectra) but later high powered rf pulses of short duration were used to excite all the different nuclei simultaneously and fourier transform (FT) techniques [20] were used to analyse the response to the excitation. This effectively reduces the time for acquisition of data and leads to improved sensitivity (see later).

Large molecules containing several different types of hydrogen usually give rise to complex spectra consisting of overlapping multiplets: even for a modest size molecule the spectrum can be so complex that it is impossible to assign all the signals in a simple one-dimensional experiment and it is necessary to resort to two-dimensional NMR techniques[31]. In these experiments, the nuclear spins are manipulated such that their signals can be subjected to a double fourier transform leading to a 2-dimensional (2D) array of spectra. Some of these 2D spectra contain a set of diagonal peaks with frequencies $w_1 = w_2$ corresponding to signals in the normal one-dimensional (1D) spectrum. Additionally there are several off-diagonal cross-peaks ($w_1 \neq w_2$). These cross-peaks indicate a correlation between pairs of diagonal peaks. In one type of corrleation spectrum (COSY) the cross-peaks correlate diagonal signals from nuclei which are spin-spin coupled (for example from neighbouring protons which are separated by 2 or 3 bonds in the structure). The COSY spectrum of the small protein, bovine pancreatic trypsin inhibitor, is shown in Fig. 3.[79]

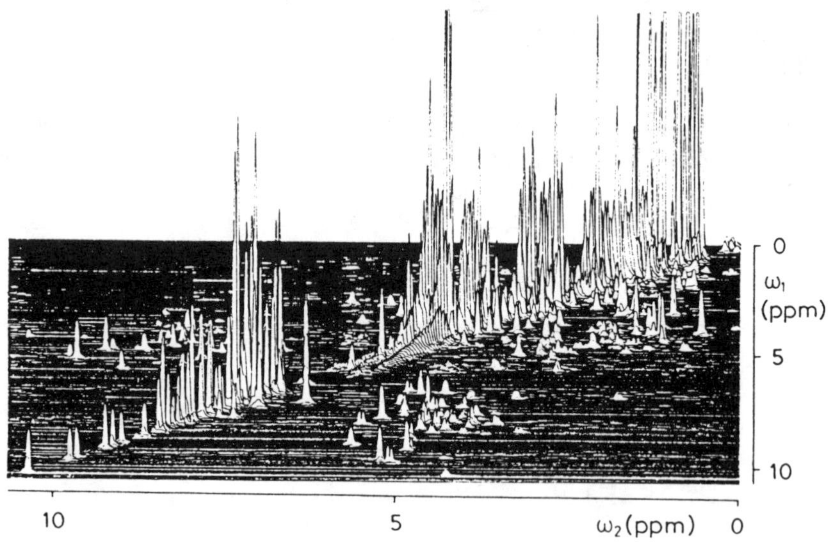

Fig. 3. The ^{1}H COSY 2-dimensional spectrum of bovine pancreatic trypsin inhibitor (from Wagner and Wüthrich [79])

Another 2D experiment detects nuclear Overhauser enhancements (NOESY) and here the cross-peaks correlate diagonal frequencies from nuclei which are close in space (< 5 Å). These 2D experiments considerably simplify the spectra by removing overlap of signals and the correlations provide vital information for signal assignment[32] and for conformational analysis[33]. Experiments which give 3D and 4D displays of the data from molecules labelled with ^{13}C and/or ^{15}N have also been devised[34,35]. The information from such experiments allows the almost complete signal assignment of proteins with MWt up to 20 KDa. Furthermore, the observed NOESY data can be used to determine three-dimensional structures for small proteins in solution by applying distance geometry[33] or restrained molecular dynamics calculations[36]. Several structures of small proteins have now been characterised by this approach and often the structures can be defined as precisely as those in a 2Å refined X-ray structure (Wüthrich)[37,38].

Some of the important milestones in the development of the NMR technique as a structural tool are listed in Table 2.

TABLE 2

Milestones in the development of nuclear magnetic resonance as a structural tool

1924-1939	Early work characterising nuclear magnetic moments and using beam methods (2,3,4).
1945	Detection of NMR signals in bulk materials (8,9).
1949-1950	Discovery of chemical shift (23,24,25).
	Discovery of spin-spin coupling (26).
1951	Discovery of ^1H chemical shifts (48).
1952	First commercial NMR spectrometer (Varian 30 MHz).
1952	Bloch (8) and Purcell (9) receive Nobel Prize for Physics.
1953	Bloch equations for NMR relaxation (8,43).
1953	Theory for exchange effects on NMR spectra (49,50).
1955	Solomon equations for NMR relaxation (14).
1953-1958	Sample spinning used for resolution improvement (43).
	Field gradient shimming with electric currents (42,80).
	Magnetic flux stabilisation (Varian).
	Spin-decoupling (51).
	Variable temperature operation ((74) and Varian).
1957	Field/frequency locking to improve stability (76).
1957	Analysis of second-order spectra (39,52,75).
1959	Vicinal coupling constant dependence on dihedral angle (53).
1961	First commercial 60 MHz field/frequency locked NMR spectrometer (Varian A60).
1961	First superconducting magnet NMR spectrometer (Varian 220 MHz).
1962	Indirect detection of nuclei using heteronuclear double resonance (INDOR) (54).
1964	Spectrum accumulation for signal averaging (22).
1964	First use of computers in NMR spectroscopy (22).
1965	Nuclear Overhauser enhancements (NOE) used in conformational studies (55).
1965	Development of fast fourier transform techniques (77).
1966	Fourier transform (FT) techniques introduced into NMR (20).
1967	Liquid crystal solvents used (59).
1968	Magic angle spinning for high resolution NMR studies of solids (61,62).
1969	First commercial FT NMR spectrometer (Bruker 90 MHz).
1969	Computer controlled pulse programmers.

1969	Lanthanide paramagnetic shift reagents (56).
1970-1975	^{13}C studies at natural abundance become routine.
1970	First commercial FT spectrometer with superconducting magnet (Bruker 270 MHz).
1971	Pulse sequences for solvent signal suppression (57,58).
1971	T_1 relaxation measurements in FT mode (60).
1971	2-dimensional (2D) NMR concept suggested (63).
1973	(First NMR analysis of body fluids (64) and tissues (65).)
1973	(Spin-imaging methods proposed (66,67).)
1973	360 MHz superconducting NMR spectrometer (Bruker).
1975	2D-NMR techniques developed (31).
1977-1980	(Spin-imaging of human limbs and organs (see 68).)
1979	Detection of insensitive nuclei enhanced by polarisation transfer (INEPT) (69,70).
1979	Detection of heteronuclear multiple quantum coherence (HMQC) (44).
1980	500 MHz superconducting spectrometer (Bruker).
1982	Full assignments for small protein (32).
1983	First 3D structure of protein from NMR data (33).
1987	600 MHz superconducting spectrometer (Bruker; Varian).
1988	2D-NMR combined with isotopically labelled proteins for full assignments (71).
1989	3D NMR on isotopically labelled proteins (34).
1990	4D NMR on isotopically labelled proteins - assignment and conformation (35).

3. DEVELOPMENT OF NMR AS AN ANALYTICAL TOOL

Following the discovery of chemical shifts and spin coupling constants, the potential of NMR as a method for structural analysis was immediately recognised. High resolution NMR spectrometers were constructed in various laboratories (for example those of H.S. Gutowsky[39], R.E. Richards and Varian Associates) and the pioneering work of scientists such as H.S. Gutowsky, J.N. Shoolery and J.D. Roberts and their collaborators soon demonstrated the scope of applications of the technique. These scientists clearly recognised the potential of the NMR method and developed it with vigour. The success of the method derived from the well-defined correlations between molecular structure and the NMR data measured under high resolution conditions. The method had wide appeal attracting scientists ranging from theoreticians to analytical chemists. The basic simplicity of the method in correlating chemical shifts, spin-coupling constants and signal intensities with molecular structure without the need for a detailed understanding of the theoretical background ensured that the method was widely used.

The early progress of NMR spectroscopy was determined mainly by novel technical developments relating directly to NMR and based on a detailed understanding of the physics of the spin systems. However, the full development of the method also relied on borrowing technology already being used successfully in other forms of spectroscopy and measurement.

The basic requirements for a good spectrometer could be defined from the earliest days in terms of providing high sensitivity and high resolution, spectral simplicity and spectral stability. High sensitivity can be achieved by increasing the magnetic field strength (signal/noise is proportional to $B_o^{3/2}$) and improving the rf electronics to reduce noise. High resolution operation requires stable, homogeneous magnetic fields to give line widths < 0.2 Hz and the need for spectral simplicity requires high fields to provide a large chemical shift dispersion. Magnetic field and rf frequency stability is required for ease of operation and for all spectral accumulation and difference spectroscopy experiments. With such well-defined goals for an emerging technique of obvious analytical potential it was not surprising that commercial companies immediately took on the dominant role in the development of the technique.

Sensitivity

The basic needs for higher field strengths and improved electronics in order to increase sensitivity have provided a continuing challenge and progress has always been limited by the available enabling technology. Thus a steady increase in the field strength has been achieved over the years (see Fig. 4A) first by using electromagnets and permanent magnets for fields up to 2.3 Tesla and then by proceeding to superconducting magnets up to 14.0 Tesla as the technology developed. The increased field strength together with parallel improvements in electronics have contributed to the steady increase in sensitivity over the years (see Fig. 4B). However these methods when used in isolation did not provide adequate

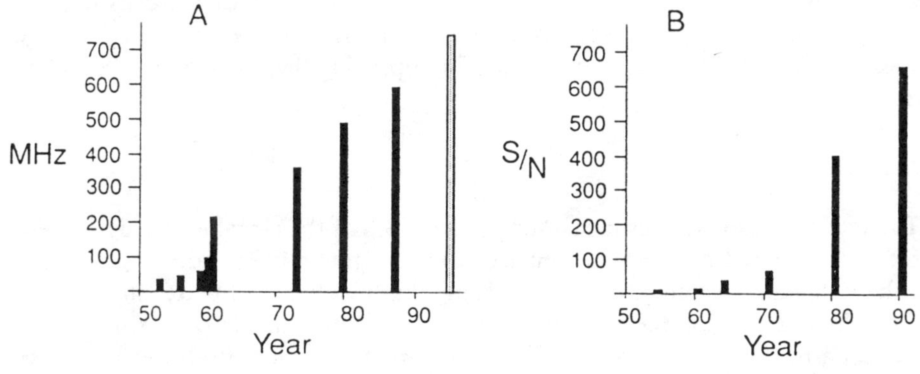

Fig. 4. The increase in NMR field strength (A) and sensitivity (B) versus time.

sensitivity for studying insensitive nuclei of low natural abundance such as ^{13}C or ^{15}N or indeed ^{1}H nuclei in dilute solutions. Initially this problem was tackled by using spectrum accumulation techniques with a computer of average transients: this gave an increase in the signal/noise ratio proportional to $\sqrt{N}$ where N is the number of accumulated scans[22]. The achieved improvement in S/N depends on the line widths and the spectral width examined but ~20 fold improvements over a single frequency-swept spectrum could be achieved routinely for complex ^{1}H or ^{13}C spectra containing several lines by accumulating data for 24 hours. The spectrum accumulation technique had previously been used in several other areas: its successful integration into the NMR method was possible only after the spectral stability problems had been overcome by using field/frequency locking (see later). Its application to NMR was an important milestone in the development of the technique because it was the first use of digital computers in NMR and it was at this stage that the problems of converting analog signals to digital signals began to be addressed.

Even the improvement provided by the spectrum accumulation technique did not meet the sensitivity requirements and a breakthrough came in 1967 when the fourier transformation (FT) technique (again a technology that had already been established with other methods) was introduced[20]. As mentioned earlier NMR spectra had been previously obtained by sweeping through the spectral frequencies with a typical overall sweep time of 500-1000 s: in the FT method the examination time is much reduced (typical acquisition times 1-5 s) because all the nuclei are excited simultaneously to provide a time domain spectrum of the resonance responses which is then fourier transformed to give the normal frequency domain spectrum. The combination of the time-saving of the FT method with spectrum accumulation techniques opened up the application of NMR not only to studies of less sensitive nuclei but also to ^{1}H studies of complex biological molecules examined at low concentrations. It is worth noting that both of these techniques were initially developed for application in other fields. Clearly it is important to have mechanisms for transferring such technology between different areas in an efficient manner. Other examples of this can be found in the field of NMR signal processing: for example, maximum entropy methods, originally devised for use in geophyscis and radioastronomy have been adapted for use in improving the analysis of incomplete NMR data sets[40,41].

Resolution

The intrinsic resolution as measured by the width of the signal at the signal half-height is determined by the homogeneity of the magnetic field and the stability of both the field and frequency. It is remarkable that resolution of better than 0.2 Hz can be achieved even on the very high magnet field instruments (at 600 MHz this represents a resolution of 1 in 3 x 10^{9}). Obviously the magnet itself needs to have a high intrinsic homogeneity but this alone would be insufficient to provide the required resolution. Such resolution can only be achieved by using shim coils and field-frequency locking [42,75,80]. Shim coils are a set of electrical coils which are designed to produce 1st, 2nd and higher order magnetic field gradients along the x, y and z axes such that a magnetic field profile can be generated which effectively counteracts the natural field gradients of the magnet to produce a 'flat' region of magnetic field over the sample volume. Proper adjustment of the currents to the shim coils ensures that the signals have narrow line widths and good line shapes (Lorentzian): the latter is essential in some applications where suppression of large signals is required.

Spinning the sample[43] at modest speeds (10-20 Hz) about one axis improves the resolution further by averaging the gradients along the other two axes. It is very difficult to stabilise the magnetic field and the rf frequency independently to the high degree required: this problem was solved in 1961 by incorporating field/frequency locking circuits (Varian A60 instrument). Instead of independently stabilising the field and frequency the field/frequency ratio is maintained at a constant value. This is achieved by carrying out a separate NMR locking experiment in addition to the analytical NMR experiment. Any fluctuations in the NMR 'locking' signal are immediately sensed and used to provide error signals which correct the field thus maintaining a constant field/frequency ratio. In the early days, the locking experiment was carried out on an external sample close to the experimental sample: nowadays, one usually locks internally onto the deuterium resonance (^{2}H) signal of the deuterated solvent used in the analytical sample.

Spectral simplicity

Higher magnetic fields give greater chemical shift dispersion and this leads to simpler spectra since the spin-spin splittings remain constant with field strength and the line-widths are not greatly affected. This is seen in Fig. 5 which compares the ^{1}H spectra of a steroid at 300 and 600 MHz. Very often the spin-spin multiplet structure becomes first order at the higher field strength which leads to further simplification in spectral analysis.

From 1955 to 1980 developments in magnet technology provided field strengths which doubled every 5 years progressing from 0.7 Tesla (30 MHz) to 11.7 Tesla (500 MHz) over this period. The last 10 years has seen an increase in field strength to 14 Tesla (600 MHz) and it seem unlikely that it will exceed 23.3 Tesla (1000 MHz) before the year 2000 unless new superconducting material technology becomes available. Clearly further large gains in chemical shift dispersion will not be achieved easily by simply increasing the magnetic field.

Fortunately, other methods of obtaining spectral simplification have been developed based on the multi-dimensional (2D, 3D, 4D) NMR techniques mentioned earlier[34,35]. By spreading the spectrum into a 2-dimensional (or more) array of cross-peaks there is less chance of signal overlap resulting from chemical shift degeneracy. Using such methods Wüthrich[32] was able to fully assign the ^{1}H spectrum of a small protein (bovine pancreatic trypsin inhibitor) and show how the NOE data could lead to the 3-D structure of the protein in solution[33]. Further improvements were obtained by examining molecules generally labelled with ^{15}N or ^{13}C (usually proteins prepared biosynthetically) in combination with multi-dimensional NMR techniques. Heteronuclear multiple quantum coherence[44] (HMQC) experiments allow proton signals from ^{15}NH groups to be detected and characterised by both the ^{1}H and ^{15}N resonance frequencies[45,46]. The ^{15}N chemical shifts are usually larger than and different from the ^{1}H chemical shifts and in a 2D-HMQC experiment the cross-

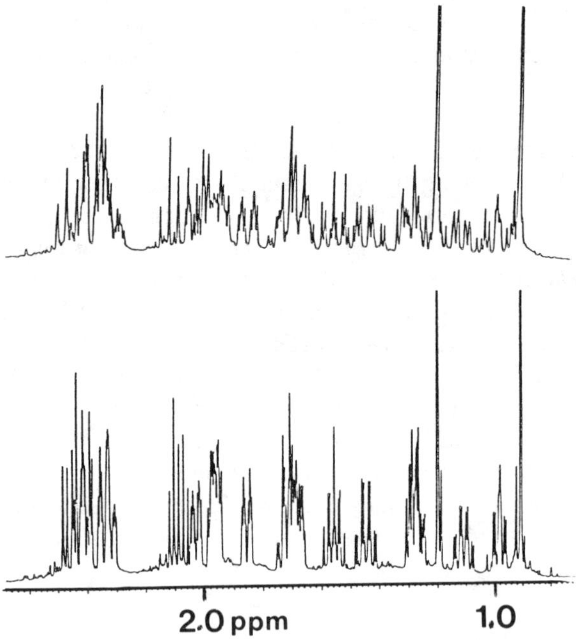

Fig. 5. The ^1H spectra of 4,5-androstene-3,17-dione at 300 MHz (top) and 600 MHz (bottom) (Varian Associates).

peaks are characterised by both frequencies, thus reducing the chance of signal overlap (which then only occurs when two NH groups have the same ^{15}N and ^1H chemical shifts). The HMQC experiment can be incorporated into 3D experiments such as NOESY.HMQC: this gives a set of ^{15}N edited NOESY spectra where the NOESY cross-peaks are spread out in a third dimension according to the ^{15}N chemical shift of the directly coupled amide nitrogen. These methods can be further extended to 4D studies in which ^1H, ^{15}N and ^{13}C frequencies are used to define the cross-peaks. Using such methods, signal assignments and NOE information can be obtained for almost all the protons in proteins as large as 20,000 MWt and the data used to determine their 3-D structures in solution[47].

The latter developments have taken place since 1988 largely in the laboratory of Bax, Clore and Gronenborn at NIH, Washington. These elegant experiments evolved from the pioneering 2D-NMR studies which formed the basis for the experiments and from the development of the necessary hardware to carry out the multipulse experiments. This major breakthrough in simplifying and assigning complex NMR spectra was thus affected by successfully bringing together several existing NMR methods.

Ease of operation

The introduction of commercial spectrometers with field/frequency locking in 1961 gave sufficient spectral stability to allow the use of spectrum accumulation techniques and the recording of precalibrated spectra. The arrival in 1970 of dedicated computers for spectrometer control and data processing simplified the operation of the spectrometers still further. Versatile pulse programmers and the associated hardware for generating rf pulses with defined properties have developed considerably since their introduction in 1970: a modern commercial NMR spectrometer can now allow the user to carry out virtually all the published multi-pulse experiments with relative ease. The need for these technical developments was demanded by the new multipulse experiments devised for multidimensional and inverse-detection studies.

4. FACTORS DETERMINING THE DEVELOPMENT OF NMR

Within seven years of the successful detection of NMR resonance signals all the important spectral parameters had been discovered and theories put forward to explain their origins. Furthermore, the first commercial high resolution instrument had already been delivered and the scope of applications was being vigorously explored. During the initial curiosity driven phase (see Fig. 6), the development of NMR was determined by the need to understand the basic principles of the technique and to construct instruments which would allow all the details about chemical shifts, spin-spin coupling constants, relaxation times and exchange effects to be uncovered and properly defined. During this period, the enormous potential of NMR as a structural tool was recognised even though the low sensitivity available at the time meant that many of the projected applications were little more than pipe dreams. By 1957 NMR was seen as a non-destructive analytical technique which in principle could report structural information about the environment of more than 100 magnetic isotopes. Already chemical shifts and coupling constants (mainly involving ^{1}H and ^{19}F at this stage) had been measured for a wide range of known structures and it had been established that characteristic values of the parameters could be used for structural and stereochemical determinations of molecules of unknown structure. It was realised that if only the sensitivity and spectral complexity problems could be overcome then NMR would provide a Pandora's box for the chemist, biochemist and material scientist. Many major structural problems were identified for NMR investigation - even though the technique was grossly inadequate for tackling them. For example, polymer scientists, acquired instruments in the late 1950's hoping to determine stereo tacticities and cross-linking in synthetic polymers; in fact it was several years later that improved instrumentation allowed such problems to be tackled successfully. Other laboratories struggled with problems in organometallic chemistry: again it was many years before the full potential of NMR could be realised in this area. Even biochemical applications were attempted in these early days: the almost featureless ^{1}H spectrum of a protein was reported in 1957 and attempts to simplify protein spectra using selective deuteration were made[73]. The technique was, however, enjoying considerable success in helping to solve structural problems involving moderately sized molecules ($<$400 MWt): it was particularly successful in natural product chemistry where it could be used for deciding between several possible structures that satisfy compositional data. It was also proving a powerful technique for defining the stereochemical details of various structures, for example, alkaloids. More challenging applications on larger molecules had to await improvements in sensitivity and spectral simplification. However the steady progress in these areas - largely maintained by the instrument manufacturers who gradually improved the electronics and increased the strength of the magnetic field B_0 - was not in itself sufficient to open up studies of large complex molecules. Thus the technique entered a phase (1956-1965) where it was driven by the problems waiting to be solved but limited by the advances in technology (see Fig. 6). By 1965 most universities were well equipped with high resolution NMR spectrometers. In 1965 the sensitivity limitation was tackled by the combined application of fourier transform and spectrum accumulation methods (both techniques having been developed previously for use with other spectroscopic techniques) and this opened up many new areas of application including routine studies of nuclei other than hydrogen (for example ^{13}C and ^{15}N). The efficient fast fourier transform algorithm developed by Cooley and Tuckey (77) made it possible to carry out the fourier transformation using small dedicated computers: without this, the introduction on FT techniques and later 2D and 3D NMR methods would have been seriously impeded. During the period 1965-1975 the main advances were achieved by applying improved technologies, in electronics and magnets and in signal collection, as they became available. This is indicated in Figure 6.

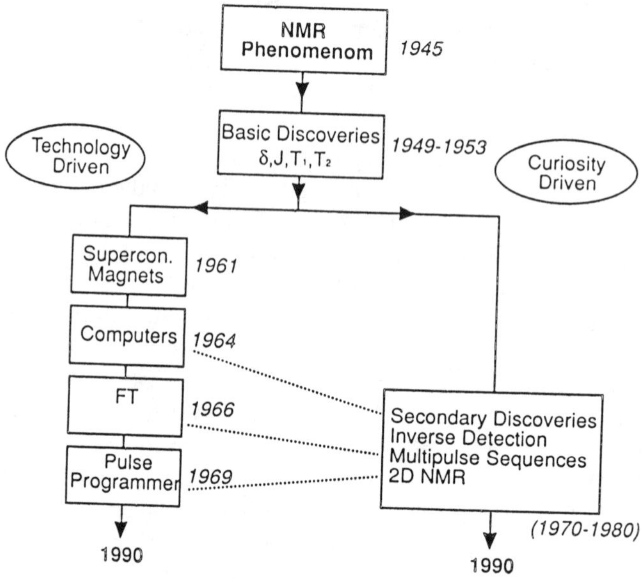

Fig. 6. Flow chart indicating development of NMR

While the period 1965-1975 did not see any major novel breakthroughs in the technique, the field strength continued to increase steadily and dedicated computers for data acquisition and processing became an integral part of the spectrometers. During this period many elegant applications in organic chemistry, organometallic chemistry, polymer science and biochemistry were published. ^{13}C studies on the natural abundant isotope became routine, not only for structural work but also for investigating rapid molecular motions (correlation times) via ^{13}C relaxation measurements. Structural determinations on fairly large molecules (~3500 MWt) became commonplace and measurements of nuclear Overhauser effect were frequently used to assist in structural determinations by identifying protons in close proximity to each other.

Fortunately, while the expansion into applications work was underway, a few research groups concentrated on understanding the NMR spin physics in more detail: such work resulted in several novel multipulse experiments which had far reaching value. For example, the INEPT[69,70] and HMQC[44] experiments (see Table 2) which provide ways of tackling the sensitivity problem for insensitive non-proton nuclei and other developments allowed the technique to be extended to high resolution studies of solids. In 1975, the most important breakthrough in NMR since the initial discoveries was provided by the development of multi-dimensional NMR techniques by Ernst and coworkers[31] (based on an idea earlier suggested by Jeener[63] but not developed by him in a practical sense). These experiments have great potential for simplifying complex spectra and for establishing correlations between signals from scalar spin-spin coupled nuclei (separated by covalent bonds) or from dipole-dipole relaxed nuclei (close in space). Extensions of these multidimensional NMR techniques when applied to $^{13}C/^{15}N$ labelled molecules provided new methods for achieving spectral simplifications and correlations which led to the complete assignment of signals and structural determinations of molecules as large as 20 KDa[34,35]. Exploitation of these secondary discoveries (see Fig. 6) was only possible because of the important technological advances that had already taken place in computer hardware and data collection and processing methods[38].

To summarise, after the initial discovery of all the basic spectral parameters, NMR development was driven by identified problems which needed higher sensitivity and spectral resolution but were limited by the available technology. Steady development of the technology, largely by the instrument manufacturers, followed during the period 1957-1975. After 1975 several important novel discoveries emerged from University laboratories working on basic NMR research and these provided the basis for a wide range of multi-pulse and multi-dimensional NMR experiments which have transformed the capability of the NMR technique.

REFERENCES

1. S. Tolansky, High Resolution Spectroscopy, Methuen, London, 1947.
2. W. Pauli, Naturwisschaften, 12, 741, 1924.
3. W. Gerlach and O. Stern, Ann. Phys. Leipzig, 74, 673, 1924.
4. I.I. Rabi, S. Millman, P. Kusch and J.R. Zacharias, Phys. Rev., 55, 526, 1939.
5. N.F. Ramsey, Molecular Beams, Oxford University Press, 1956.
6. C.J. Gorter, Physica, 3, 995, 1936.
7. C.J. Gorter and L.F.J. Broer, Physica, 9, 591, 1942.
8. F. Bloch, W.W. Hansen and M.E. Packard, Phys. Rev., 69, 127, 1946.
9. E.M. Purcell, H.C. Torrey and R.V. Pound, Phys. Rev., 69, 37, 1946.
10. F. Bloch, Phys. Rev., 70, 460, 1946.
11. F. Bloch, Phys. Rev., 105, 1206, 1957.
12. N. Bloembergen, Nuclear Magnetic Relaxation, W.A. Benjamin, New York, 1961.
13. R. Kubo and K. Tomita, J. Phys. Soc. Japan, 9, 888, 1954.
14. I. Solomon, Phys. Rev., 99, 559, 1955.
15. N. Bloembergen, E.M. Purcell and R.V. Pound, Phys. Rev., 73, 679, 1948.
16. N. Bloembergen, J. Chem. Phys., 27, 572, 1957.
17. H.S. Gutowsky and D.E. Woessner, Phys. Rev., 104, 843, 1956.
18. H. Shimizu and S. Fujiwara, J. Chem. Phys., 34, 1501, 1961.
19. J.W.Emsley, J. Feeney and L.H. Sutcliffe, High Resolution NMR Spectroscopy Vol. 1, Pergamon Press, London, 1966.
20. R.R. Ernst and W.A. Anderson, Rev. Sci. Instrum. 37, 93, 1966.
21. D. Shaw, Fourier Transform NMR Spectroscopy, Elsevier, Amsterdam, 1976.
22. R.R. Ernst, Rev. Sci. Instrum., 36, 1689, 1965.
23. W.D. Knight, Phys. Rev., 76, 1259, 1949.
24. W.G. Proctor and F.C. Yu, Phys. Rev., 77, 717, 1950.
25. W.C. Dickinson, Phys. Rev., 77, 736, 1950.
26. W.G. Proctor and F.C. Yu, Phys. Rev., 81, 20, 1951.
27. H.S. Gutowsky and D.W. McCall, Phys. Rev., 82, 748, 1951.
28. E.D. Becker, High Resolution NMR, Theory and Chemical Applications, Second Edition, Academic Press, 1980.
29. R.J. Abraham and P. Loftus, Proton and Carbon-13 NMR Spectroscopy, Heyden, 1978.
30. R.R. Ernst, G. Bodenhausen and A. Wokaun, Principles of Nuclear Magnetic Resonance in One and Two Dimensions, Clarendon Press, Oxford, 1987
31. W.P. Aue, E. Bartholdi and R.R. Ernst, J. Chem. Phys., 64, 229, 1976.
32. G. Wagner and K. Wüthrich, J. Mol. Biol., 155, 347, 1982.
33. W. Baun, G. Wider, K.H. Lee and K. Wüthrich, J. Mol. Biol., 169, 921, 1983.
34. D. Marion, P.C. Driscoll, L.E. Kay, P.T. Wingfield, A. Bax, A.M. Gronenborn and G.M. Clore, Biochemistry, 28, 6150, 1989.
35. L.E. Kay, G.M. Clore, A. Bax and A.M. Gronenborn, Science, 249, 411, 1990.
36. R. Kaptein, E.R.P. Zuiderweg, R.M. Scheck, R. Boelens and W.F. van Gunsteren, J. Mol. Biol., 182, 174, 1985: A.T. Brunger, G.M. Clore, A.M. Gronenborn and M. Karplus, Proc. Natl. Acad. Sci. USA, 83, 3801, 1986.
37. K. Wüthrich, NMR of Proteins and Nucleic Acids, Wiley, New York, 1986.
38. K. Wüthrich, Science, 243, 45 (1989).

39. H.S. Gutowsky, C.H. Holm, A. Saika and G.A. Williams, J. Amer. Chem. Soc., 79, 4596 (1957).
40. S.F. Gull and J. Skilling, IEE Proceedings, 131, 646, 1984.
41. S. Sibisi, Nature, 301, 134, 1983.
42. M.J.E. Golay, Rev. Sci. Instr., 29, 313, 1958.
43. F. Bloch, Phys. Rev., 94, 496, 1954.
44. L. Müeller, J. Amer. Chem. Soc., 101, 4481, 1979.
45. D. Marion, L.E. Kay, S.W. Sparks, D.A. Torcia and A. Bax, J. Amer. Chem. Soc., 111, 1515, 1989.
46. E.R.P. Zuiderweg, L.P. McIntosh, F.W. Dahlquist and S.W. Fesik, J. Mag. Reson., 86, 210, 1990: E.R.P. Zuiderweg and S.W. Fesik, Biochemistry, 28, 2387, 1989.
47. G.M. Clore and A.M. Gronenborn, Prog. NMR Spectr., 23, 43 , 1991.
48. J.T. Arnold, S.S. Dharmati and M.E. Packard, J. Chem. Phys., 19, 507, 1951.
49. H.S. Gutowsky, D.W. McCall and C.P. Slichter, J. Chem. Phys., 21, 279, 1953.
50. H.M. McConnell, J. Chem. Phys., 28, 430, 1958.
51. A.L. Bloom and J.N. Shoolery, Phys. Rev., 97, 1261, 1955.
52 H.J. Bernstein, J.A. Pople and W.G. Schneider, Can. J. Chem., 35, 65, 1957; 35, 1060, 1957.
53. M. Karplus, J. Chem. Phys., 30, 11, 1959; J. Phys. Chem., 64, 1793, 1960.
54. E.B. Baker, J. Chem. Phys., 37, 911, 1962.
55. F.A.L. Anet and A.J.R. Bourn, J. Amer. Chem. Soc., 87, 5250, 1965.
56. R.E. Sievers (Editor), NMR Shift Reagents, Academic Press, New York, 1973.
57. S.L. Platt and B.D. Sykes, J. Chem. Phys., 54, 1148, 1971.
58. F.W. Benz, J. Feeney and G.C.K. Roberts, J. Magn. Reson., 8, 114, 1972.
59. J.W. Emsley and J.C. Lindon, NMR Using Liquid Crystal Solvents, Pergamon, 1975.
60. R.L. Vold, J.S. Waugh, M.P. Klein and D.E. Phelps, J. Chem. Phys., 48, 3831, 1968.
61 E.R. Andrew, A. Bradbury and R.G. Eades, Nature, 182, 1659, 1958.
62. E.R. Andrew, Prog. NMR Spectr., 8, 1, 1971.
63. J. Jeener, Ampere International Summer School, Basko Polje, Yugoslavia, 1971 (unpublished).
64. R.B. Moon and J.H. Richards, J. Biol. Chem., 248, 7276, 1973.
65. D.I. Hoult, S.J.W. Busby, D.G. Gadian, G.K. Radda, R.E. Richards and P.J. Seeley, Nature, 252, 285, 1974.
66. P.C. Lauterbur, Nature, 242, 190, 1973.
67. P. Mansfield and P.K. Grannell, J. Phys. Chem., 6:L422, 1973.
68. P.G. Morris, NMR Imaging in Medicine and Biology, Oxford University Press, 1986.
69. G.A. Morris and R. Freeman, J. Amer. Chem. Soc., 101, 760, 1979.
70. D.P. Burum and R.R. Ernst, J. Magn. Reson., 39, 163, 1980
71. D.H. Torcia, S.W. Sparks and A. Bax, Biochemistry, 27, 5135, 1988.
72. M. Saunders, A. Wishnia and J.G. Kirkwood, J. Amer. Chem. Soc., 79, 3289, 1957.
73. J.L. Markely, I. Putter and O. Jardetzky, Science, 161, 1249, 1968.
74. H.S. Gutowsky, L.H. Meyer and R.E. McClure, Rev. Sci. Instr., 24, 644, 1953.
75. J.T. Arnold, Phys. Rev., 102, 136, 1956.
76. E.B. Baker and L.W. Burd, Rev. Sci. Instr., 28, 313, 1957.

77. J.W. Cooley and J.W. Tukey, Math. Comput., 19, 297, 1965.
78. R.E. Hoffman and G.C. Levy, Prog. NMR Spectr., 23, 000, 1991.
79. G. Wagner and K. Wüthrich, J. Mol. Biol., 155, 347, 1982.
80. W.A. Anderson, Rev. Sci. Instr., 32, 241, 1961.

Chapter 11

Present and future applications of NMR to medicine and materials science

Peter Morris
Nottingham University (United Kingdom)

1.INTRODUCTION

The phenomenon of nuclear magnetic resonance (NMR) was first observed in the immediate post second-world-war period by two American physicists, working independently: Bloch at Stanford and Purcell at Harvard. Their observations were reported in 1946 in the same volume of Physical Review and led to the joint award of the 1952 Nobel Prize for Physics. Once the details of the interaction had been worked out, and the chemical specificity had been appreciated, a period of instrumentational refinement followed before NMR took its place as arguably the most powerful analytical technique available to the organic chemist. The historical development of NMR and the basis of its analytical power are described in the companion article by Dr. J. Feeney.

NMR spectroscopy serves many needs, including routine chemical analysis, determination of protein structures, *in situ* diagnosis of in-born errors of metabolism and medical imaging. Although the same NMR phenomenon is observed by scientists working in these disparate disciplines, their instrumentational requirements differ widely, and there is often rather little interaction between the different NMR communities. Many specialist international societies have evolved and their annual meetings are extremely well attended - in the case of magnetic resonance imaging by typically several thousand participants. The growing diversity of applications, coupled with a unique ability to address certain key contemporary questions (for example the solution structure of biological macromolecules, especially proteins), has made NMR spectrometers the most requested of large scientific instruments. Their high unit cost means that they also account for a substantial fraction of the research budgets of national scientific funding agencies.

Of necessity, the scope of this article must be limited. It focuses primarily on magnetic resonance imaging (MRI) and discusses basic principles, present applications and possible development pathways, together with some of the implications these may have for future healthcare policy. The driving force behind the rapid and highly successful development of MRI instrumentation has been, and will remain clinical in origin. However, some non-medical applications are also considered, particularly in connection with the newly emergent technique of NMR microscopy.

The sister technique to MRI, magnetic resonance spectroscopy (MRS) is less well developed clinically. It is a very powerful research tool which is advancing our understanding of human biochemistry but has yet to make its mark in the diagnosis of major diseases. This is an essential goal in the eyes of the technologists and industrialists, who have to sell instruments in substantial numbers, if they are to recoup their enormous development investment. Their voice is often heard loudest, and there is therefore a tendency to measure the effectiveness of MRS in clinical instruments sold, rather than in terms of the scientific advances made through its use.

The application of NMR to materials science, though older in origin than MRI, is considerably less well developed. In part, this reflects the fact that NMR cannot be used to study metal structures because radiofrequencies do not penetrate into the body of the metal beyond a few skin depths. (However, it is still possible to observe surface phenomena, such as catalysis, in metal powders.) It is also a consequence of the relative difficulty of the NMR techniques involved. Accordingly only brief coverage is given to this potentially important area of application.

2. MAGNETIC RESONANCE IMAGING

2.1 Historical development of MRI

The first experiments on MRI were conducted in 1972 by Paul Lauterbur at the State University of New York at Stonybrook and, independently, by Peter Mansfield and colleagues at Nottingham.[1,2] These demonstrative experiments were met with considerable scepticism both from radiologists, for whom the introduction of the EMI (now X-ray CT) scanner[3] had recently advanced their subject more than any invention since the discovery of X-rays in 1895, and also from the NMR community. After all, how could radiowaves, with wavelengths of several 10s of metres, generate high resolution images of human anatomy? Images of the human finger[4] and wrist[5] paved the way for widespread acceptance of the technique. From the outset, the primary goal was clinical imaging, and this was hotly pursued, initially in academic institutions, notably at Aberdeen, Nottingham and Stonybrook, and later in industry too. The problem was one of scale - NMR had hitherto been used principally for the analysis of chemical structures, and so NMR spectrometers were designed for samples of test-tube proportions. The two orders of magnitude increase in linear dimension necessary to accommodate a human torso represented a considerable act of faith on the part both of the development teams and the magnet manufacturers. The first crude images of the human chest[6] and abdomen[7] were published in 1977 and 1978, respectively and commercial systems became available in the early 1980s.

Although these first clinical images were of high quality compared to those obtained by nuclear imaging or crude X-ray tomograms, they did not compare favourably with the images obtained by X-ray computerized tomography. However, the pioneering clinical studies at the Hammersmith Hospital and other centres soon established that MRI was capable of exquisite soft tissue contrast, which could be manipulated, and the images were free from the 'streak artefacts', characteristic of the X-ray images, at the level of the posterior fossa, for example. The contrast in MRI head images between white and grey matter was particularly exciting and rapidly lead to important applications in a range of demyelinating diseases, such as multiple sclerosis.[8] In a national survey conducted in 1986/7, the UK Medical Research Council identified 10 key areas in which MRI had already become established as the investigative method of choice, and a further 4 in which it was comparable or complementary to X-ray CT. These are listed in Table 1.

Areas in which MRI has been established as diagnostic method of choice

- cerebral palsy (as research tool)
- posterior fossa lesions
- foramen magnum lesions
- multiple sclerosis
- syringomyelia
- acoustic neuromas
- lumbar spine and disc disease
- great vessels (in context of cardiovascular disease)
- parotid gland tumours
- bone tumours

Areas in which MRI is comparable or complementary to X-ray CT

- tumours of the rectum, prostate, cervix, uterus, bladder, breast, cerebrum, meninges, pituitary gland
- stroke
- congenital heart disease
- dementia

Table 1. Clinical applications of MRI in the UK as of 1988.

The majority of these applications are associated with common focal diseases, particularly in the head. Recent developments in MRI, which permit real time imaging, measurement of flow velocity, diffusion and perfusion (see Section 3), and increased opportunity to study a wider spectrum of patients, promise to greatly extend the range of method-of-choice applications. This is reflected in the establishment by charitable research foundations and others of many MRI centres dedicated to the study of specific diseases.

2.2 Principles of MRI

2.2.1 Spatial localization

When nuclei with spin I are placed in a magnetic field they can adopt one of 2I+1 possible orientations with respect to the field direction. In the case of the proton, I=1/2, and there are two possibilities - parallel or antiparallel to the field. The parallel state has a lower energy than the antiparallel state, the difference (Zeeman splitting) being proportional to the field strength B. Transitions can be induced by electromagnetic radiation at a frequency which exactly matches the energy splitting. The resonance frequency ω is given by the Larmor equation:

$$\omega = \gamma B$$

where γ is the magnetogyric ratio, a constant for a particular nucleus. For normal 'high resolution' NMR where the (small) chemical shifts are observed between nuclei in different chemical environments, B must be very homogeneous (up to a few parts in 10^{10}). In contrast, for MRI, B is made to vary with position, so that nuclei in different locations have different resonance frequencies. Usually, a linear field variation is employed, and the field gradients are generated by separate gradient coils, distinct from the main magnet windings. For an 'x-gradient':

$$B = B_0 + G_x x$$

where $G_x = \frac{dB_z}{dx}$ and the main magnetic field B_0 is applied along the z-direction. The corresponding dependence of resonance frequency on position is:

$$\omega = f(x) = \gamma(B_0 + G_x x).$$

The effect of such a gradient on the NMR spectrum recorded from a water-filled test tube is illustrated in Fig. 1. In the absence of a gradient, all spins are located in the same field (B_0) and resonate at the same frequency ($\omega = \gamma B_0$). With the gradient applied, spins located on the left hand edge of the test tube lie in a relatively lower magnetic field and so resonate at a proportionally lower frequency than those on the right hand edge. Spins located along a vertical strip lie in the same magnetic field and will all contribute to the signal at the corresponding frequency.

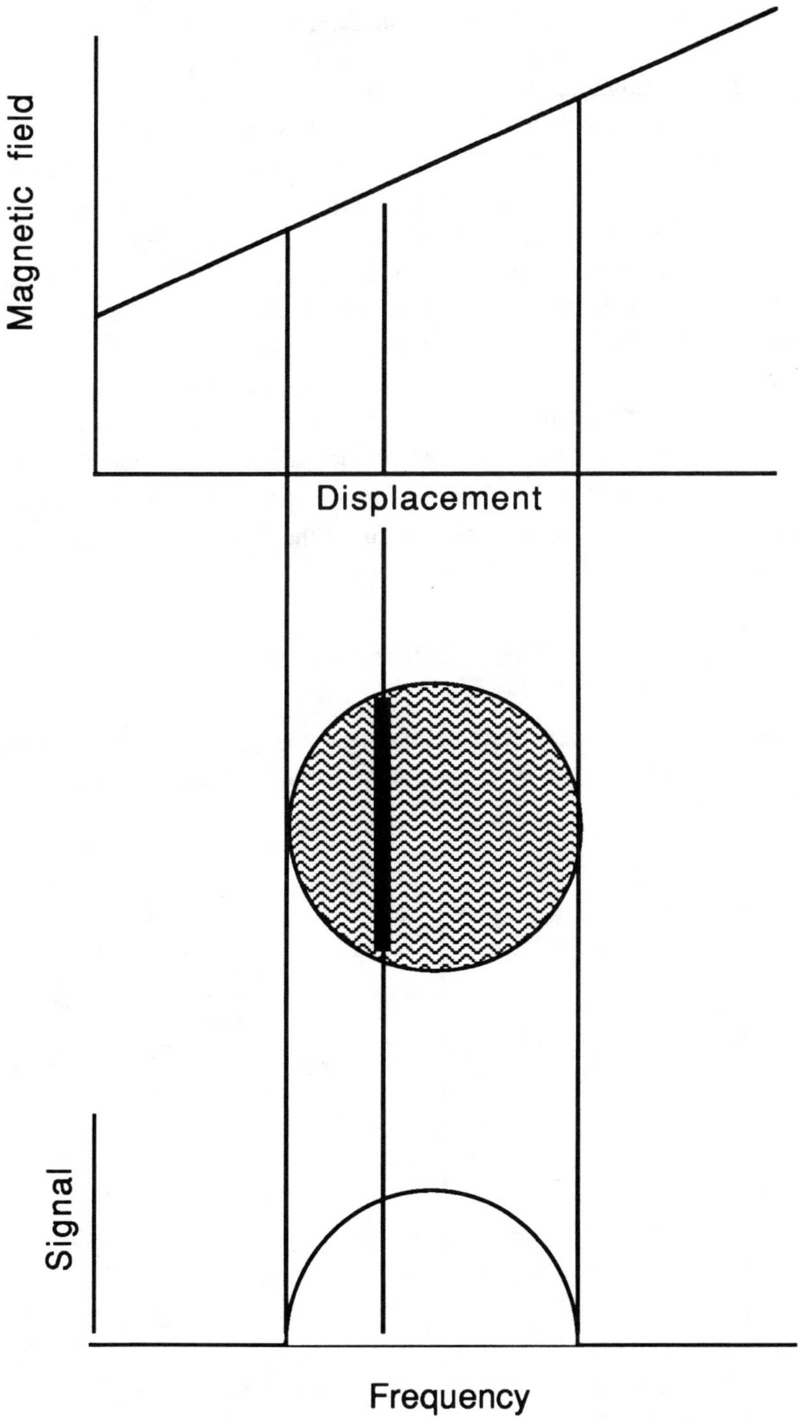

Fig. 1. Application of a linear field gradient to a test-tube sample.

In the case of simple objects with uniform spin density, these one-dimensional projections can reveal much about the object. For 'real' subjects, an image can be reconstructed by combining a number of such projections, recorded in gradients of different strength and orientation. The mathematical foundation was laid in connection with gravitational theory by Radon in 1917[9], although the first image reconstruction techniques were not developed until 1956, when Bracewell[10] used them to map the distribution of microwave emissions from the solar surface. These techniques were independently redeveloped for the determination of molecular structure from transmission electronmicrographs, for optical applications, and, finally, for NMR by Paul Lauterbur[1]. This is a perfect example of the failure of knowledge to cross scientific disciplines.

For the reconstruction of NMR images, many different approaches are in fact possible, and these account for the diversity of the early MRI literature[11]. They are best compared in reciprocal space[12]. If the spin density distribution of an object is $\rho(\mathbf{r})$, an equivalent description can be given in terms of the Fourier amplitudes $P(\mathbf{k})$ of the spatial frequencies $\mathbf{k}$ where:

$$P(\mathbf{k}) = \int p(\mathbf{r}) \exp(i\mathbf{k}.\mathbf{r})\, d\mathbf{r} \, .$$

In order to completely describe an object, it is necessary to map out the whole of **k**-space. Different imaging schemes can be distinguished by the manner in which they achieve this (Fig. 2.)

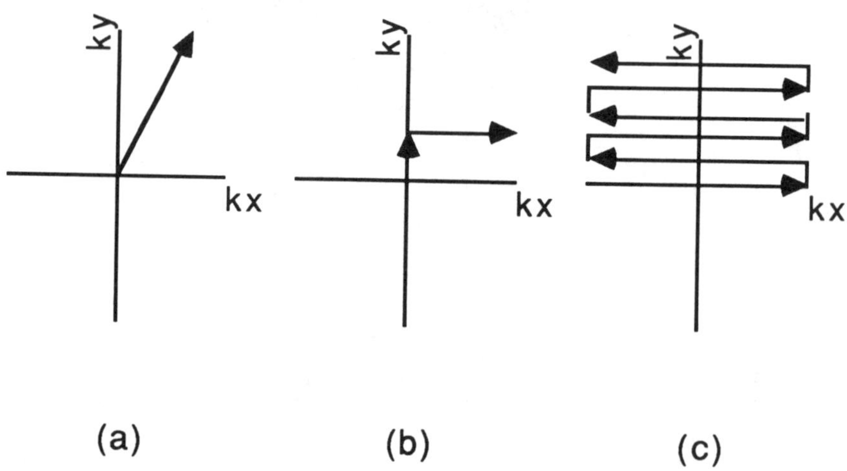

Fig. 2. **k**-Space description of projection reconstruction (a), Fourier imaging (b) and echo-planar imaging schemes.

Commonly used MRI techniques acquire a single line of **k**-space during each sampling period. For most purposes, the Fourier methods are preferable to projection reconstruction techniques as **k**-space is sampled uniformly on a rectangular grid, rather than radially. It is also possible to cover **k**-space in a single transient using the technique of echo-planar imaging (EPI)[13]. EPI is still the only genuine real-time MRI technique, generating two- or even three-dimensional images in times as short as a few tens of milliseconds.

2.2.2 Image contrast

The signal in (^1H) MRI is derived from mobile protons. We are composed of approximately 70% water and each water molecule has 2 protons. The images therefore primarily reflect the water distribution, with lesser contributions from the more mobile lipids, for example, the triglycerides in adipose tissue. There is about a 15% variation in water content amongst the different soft tissues, providing a limited degree of spin density contrast. However, other, potentially more important sources of image contrast can be exploited in MRI experiments.

When the nuclear spins are excited in an NMR experiment, the Boltzman distribution between the two spin states is disturbed and takes some time to recover. This process involves an interchange of energy with 'the lattice' and the time constant is known as the spin lattice relaxation time T_1. The variation in T_1 between different soft tissues is much greater than the variation in water content and, at low operating frequencies, can be as much as a factor of 3-4. The NMR signal is detected by the EMF induced in a radiofrequency receiver coil oriented orthogonal to the direction of the magnetic field. This signal decays by non energy-dependent processes with a time constant T_2 which shows a similar tissue variation to T_1. The sequences developed by high resolution spectroscopists for measurement of T_1 and T_2 have been taken over by the imaging community to generate images which show relaxation time, rather than spin density contrast. The contrast in T_1 images is the reverse of that observed in T_2 images: a long T_2 means that the NMR signal decays more slowly, whereas a long T_1 means that it takes longer to recover. Inevitably, both T_1 and T_2 effects will be present to some degree in all images. Since the effects tend to cancel, it is important to ensure that one relaxation time or the other should dominate and one speaks of T_1- or T_2-weighted images.

2.3 MRI instrumentation

The rapid development in NMR instrumentation, which has taken place over the last two decades, reflects, in part, an increasing demand from the scientific community. Equally significant has been the availability of minicomputers, which are now an essential feature of all NMR spectrometers, and which permit the full multiplex advantages of a Fourier transform approach to be realised. The importance of cheap computing power should not be underestimated: certainly it would not have been possible to have developed MRI techniques without it.

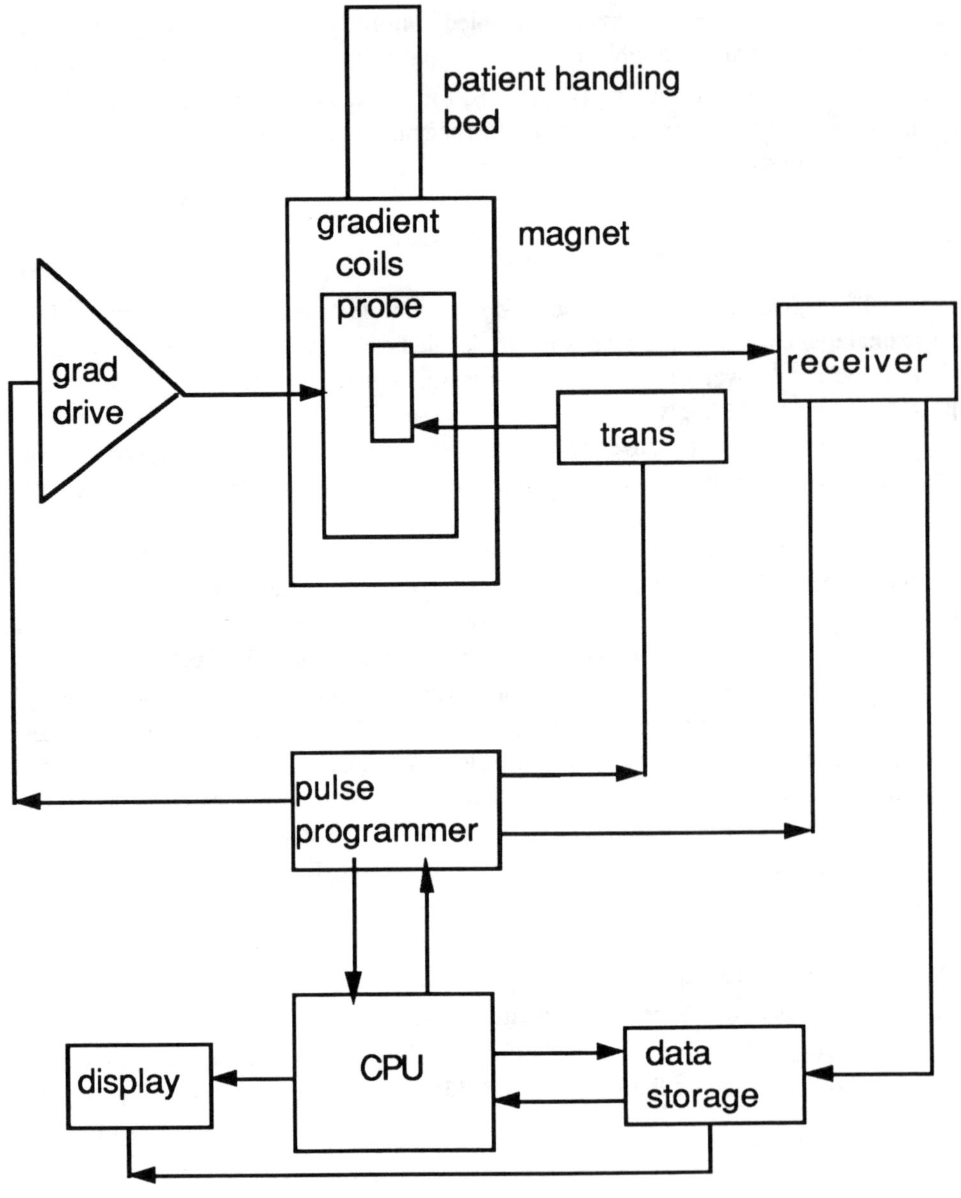

Fig. 3. Schematic diagram of apparatus for MRI.

The instrumentation requirements for MRI are broadly similar to those for conventional NMR spectrometers with the following additions: x, y and z gradient coils and drivers, image display and a radiofrequency transmitter capable of generating shaped pulses. The apparatus is illustrated schematically in Fig. 3.

2.3.1 Magnet system

Most of the early development MRI systems and many of the first commercial systems employed resistive magnets. These are unstable and cannot easily generate magnetic fields in excess of about 0.25 T. (Most clinical systems now operate in the range 0.3 - 1.5T.) Since 1982 the world MRI market has been dominated by superconducting systems to the extent that, of the 3000 or so MRI installations to date, some 85% have superconducting magnets[14]. Superconducting magnets are inherently stable and high field whole-body systems (currently up to 4T) are readily available from a relatively small number of manufacturers of which the market leader is (and has been from the outset) Oxford Magnet Technology. The drawback with superconducting systems is their cost - both capital and running. Typically, the magnet for a clinical MRI system accounts for about 1/3 of the total system cost of upwards of £1M, and the annual cryogen (liquid nitrogen and helium) expenses, which arise from the need to maintain the magnet windings below the critical temperature (Tc) for superconductivity, amount to some tens of thousands of pounds. Siting can also be problematic and expensive, particularly for very high field (>1.5T) unshielded magnets.

The very high cost of superconducting magnets distinguishes MRI from other imaging modalities and restricts its application to relatively few of the more prestigous healthcare institutions. In order for the technology to become more accessible, and the evidence of its clinical efficacy strongly suggests that there is such a need, magnet costs must be substantially reduced. In the longer term, the development of high Tc superconducting materials could have a profound effect on the industry. In the shorter term, magnet manufacturers are looking for acceptable compromises in magnet bore, field strength and homogeneity - the problem being that the NMR is an inherently insensitive technique (when compared to other branches of spectroscopy) and image quality is easily impaired.

Permanent magnets are attracting increasing attention. Constructed from ferrite material, they can be made to generate a field of 0.2T with a homogeneity of about 40 ppm over a spherical volume of diameter 36 cm, though with a substantially increased weight (8.5 tonnes compared to the 3.5 tonnes of a typical 0.5T superconducting magnet). Such a system has recently been released in Japan by Siemens-Asahi. Over the period 1988-90, about 100 similar sytems were installed in Japan by Hitachi. We can certainly anticipate substantial world growth in this area.

Apart from cost, another problem limits the wider application of MRI, namely siting difficulties. High field magnets generate substantial peripheral fields which represent a potential hazard to patient and equipment (particularly devices employing cathode ray tubes) alike. Permanent magnets solve this problem by being self-shielding; manufacturers of other magnet systems seek to reduce the problem by employing passive[15] or, more recently, active shields.

Conventional MRI systems require the patient to be slid deep within the bore of a superconducting magnet. There is considerable risk of claustrophobia and a relatively high patient refusal rate results. This can be alleviated by sympathetic counselling in which the genuine non-invasiveness of the technique (compared, for example, to explorative surgery!) is emphasized. Oxford Magnet Technology have proposals for flatbed (one-sided) magnets which would greatly improve accessibility and reduce patient anxiety. Unfortunately, a high price would be paid in terms of peripheral field.

2.3.2 Gradient coils

Nearly all the MRI and many of the MRS techniques (see Section 5) employ magnetic field gradients to achieve localization. The switching on and off of these gradients, which sometimes has to be effected rather quickly, induces eddy currents in nearby metal structures, most notably the stainless steel bore of superconducting magnets. These eddy currents decay rather slowly, with time constants of some tens of milliseconds which are comparable to typical tissue T_2 relaxation times, and which lead to image distortions. The magnetic fields generated by the eddy currents have, in general, a different spatial distribution to those generated by the gradient coils themselves and so cannot be fully compensated by pre-emphasis pulses applied to these coils - the method normally adopted by manufacturers of MRI equipment. Because of the cylindrical geometry of the magnet apertures, it is usual to wind the gradient coils on a cylindrical surface, coaxial with the magnet bore. It is possible to surround this gradient coil set with a second set of current-carrying conductors distributed over the surface of a cylinder of larger radius, such that the field outside the outer surface is cancelled, thereby eliminating eddy current effects[16]. The best way in which to design these actively compensated gradient systems, is by using the 'target field' approach in which the positions of the conductors for both inner and outer coils, are calculated from the field distribution required[17,18]. Using a similar approach it is also possible to design gradient sets which have low inductance, permitting faster gradient risetimes to be achieved[19].

The principal advantage conferred by these advanced design techniques is that of speed. This is useful for slice selection (see below) in samples which have short T_2 relaxation times and, particularly, in the application of high speed MRI techniques such as EPI. The improved design methods also permit gradient linearity to be maintained over a larger volume than conventional designs.

2.3.3 Selective pulses

Magnetic field gradients are the key to spatial discrimination, making different regions resonant at different frequencies. Often it is necessary to image a limited region of the sample, for example, a single slice, or set of slices. This can be accomplished with a narrowband radiofrequency pulse which interacts only with those spins resonating over a narrow range of frequencies, and hence, which lie within a narrow spatial region. The early design of such radiofrequency pulses followed a straightforward approach based on Fourier theory. Unfortunately the NMR spin system is not a linear one and so this method leads to imperfect selectivity, particularly for large flip angles[20].

Given the difficulty of obtaining analytic solutions to the selective excitation problem, many workers in the field have turned to computer optimization methods, with techniques such as optimal control theory and simulated annealing[21,22] being particularly popular. More rational approaches based on linearization of the Bloch equations, which govern the motion of the spin system, have also been extremely successful[23].

As well as leading to improvements in the quality of slice definition in conventional MR images, new possibilities have been opened up. For example, pulses which are spatially selective in two-dimensions have been generated[22-24]. These permit high resolution images to be obtained from small regions within a larger object. Such a region-of-interest method is particularly valuable where the total number of image points is restricted either by computer memory, or, as in the case of EPI, by more fundamental hardware limitations. Three-dimensional selectivity is also theoretically possible and the ultimate development of this methodology would be the ability to selectively excite an arbitrarily shaped volume. Such pulses would be used to record NMR spectra from specific regions, identified as 'of interest' by MRI immediately prior to the spectroscopic investigation.

3. NEW MRI TECHNIQUES

In conventional MRI sequences, the signal from fast moving blood becomes 'scrambled' and lost so that major vessels are seen as structures with apparently hollow lumens, sometimes referred to as 'flow voids'. However, it is possible to preserve the signal from the moving blood, and difference images (angiograms) showing the vascular network can be constructed[25,26]. These are finding important clinical applications in the head and neck, and also, more recently, in the peripheral vasculature. Using similar techniques, employing pulsed magnetic field gradients, it is also possible to quantify flow[27,28]. Measurements of flow velocities through defects in the cardiac wall have proved clinically useful but the "Holy Grail" of the measurement of flow through the coronary arteries has proved elusive.

As well as macroscopic flow, there is growing interest in the measurement of tissue perfusion and diffusion[29]. A clean separation of these processes is difficult to achieve, but it is certainly possible to generate images reflecting these properties. Of great interest is the anisotropic diffusion readily observed in the brain in regions where the myelinated nerve bundles run unidirectionally.

The diffusion measurements are greatly assisted by fast imaging methods such as echo-planar imaging[13] which continue to evolve. At the present state-of-the-art, it is possible to generate 256×256 images using this technique. Single shot volume imaging (echo-volume imaging) is also possible through a straightforward extension of the two-dimensional EPI experiment[30]. Echo-volume imaging is capable of generating images at an astonishing rate - potentially a set of 64 every few tens of milliseconds. This presents a storage problem, and, more importantly, a major problem of assessment. It is simply not practicable for a radiologist to search through several thousand conventional images for a single patient. Interactive displays, data reduction methods and eventually auto-analytical procedures will become essential in the relatively near future.

4. NMR MICROSCOPY

From its inception the main thrust of the developments in MRI has been directed towards clinical applications. However, from the outset, it was appreciated that it was equally possible to scale the apparatus down to produce an NMR microscope. Indeed, the seminal paper on MRI by Mansfield and Grannell[2] had in mind 'NMR diffraction' from atomic planes in crystal structures, rather as X-rays undergo Bragg reflection. The sensitivity of the NMR technique is such that it will never be possible to directly image molecular structure (although three-dimensional structures can be inferred from COSY and NOESY-type high resolution NMR experiments - see the companion paper by Dr Feeney). Rather, the predicted isotropic resolution for ^{1}H NMR microscopy, for the highest magnetic field strengths currently available (14 T) is on the order of a few μm[31], somewhat better if slices are selected whose thickness exceeds the in-plane resolution. The resolution will thus approach, though is unlikely to exceed that of the optical microscope: cellular resolution is possible, but detailed studies of subcellular organelles appear to be ruled out. Given the inherent limitation of the technique it is reasonable to ask what advantage, if any, the technique offers.

The principal merit of NMR microscopy (and of MRI as a whole) is that no specimen preparation is required: intact samples can be observed in their natural state without being frozen, dried, sectioned, fixed or stained. The selected image plane can lie deep within the sample and observation is not limited to surfaces as it is for optical microscopy (although the recent development of the conformal microscope has improved the situation somewhat for relatively thin transparent specimens). The orientation of the plane can also be controlled with ease, or indeed, a complete volume image can be obtained. Resolution is in principle infinitely variable, though, in practice, is limited on sensitivity grounds. It should also be recognised that the parameters being imaged (proton density, relaxation times, etc) are different from those of other imaging techniques and that it is possible to alter the relative dependence of the image on these parameters, i.e. the image contrast.

The image that first alerted the scientific community at large to the possibilities of NMR microscopy was that of a single cell (albeit a large one - an ovum of the African clawed toad *Xenopus laevis*), imaged at a resolution of $10 \times 13 \times 250$ μm[32]. Since then, rather higher in-plane resolution (4 μm) has been achieved in plant tissue[33]. Microinjection of an NMR contrast agent (gadolinium-DTPA-dextran which greatly reduces the ^{1}H T_1) has enabled demonstrative studies of the mapping of cell lineage to be performed[34].

To improve the spatial resolution will require the use of still higher field magnets, yet to become available, but under active development, driven by the demands of the high resolution macromolecular NMR community to extend the molecular weight range amenable to NMR study. Oxford Instruments, the company which pioneered superconducting magnets for NMR (and other) applications, remains at the forefront of this development. It is unlikely that any dramatic improvements will be achieved as diffusion effects become limiting as the resolution is increased[35]. However, these diffusion effects can also be made the subject of imaging experiments. Such 'q-space' images[36] can be used to study morphology in porous systems (oil-bearing rocks for example) or water translocation in plants[37].

5. MAGNETIC RESONANCE SPECTROSCOPY

The clinical application of localized NMR spectroscopy stems from small scale studies conducted at roughly the same time as the demonstrative imaging experiments. The successful demonstration that high resolution ^{31}P spectra could be recorded from isolated but intact skeletal muscle[38] led to the widespread use of this technique for monitoring tissue bioenergetics. Clinical studies were pioneered in Oxford by Professor Radda and were initially targetted towards the diagnosis of in-born errors of muscle metabolism for which purpose they proved extremely successful.[39]. Other early successes include the monitoring of neonates suspected to have been exposed to an ischaemic episode (birth asphyxia). Measurement by ^{31}P NMR of the ratio of the concentrations of phosphocreatine to inorganic phosphate in these children proved to be an accurate indicator of survival prospects and of possible brain damage[40]. The ^{31}P NMR spectrum is an excellent indicator of ischaemia/hypoxia: it is therefore extremely attractive to use ^{31}P NMR to perform regional studies in adult brain and heart. Such studies require the use of rather more sophisticated localization techniques than were employed in the early studies on muscle and infantile brain[41].

^{13}C NMR is well suited to the elucidation of metabolic pathways and the measurement of fluxes through them. Recently, human studies involving the incorporation of ^{13}C label from 1-^{13}C glucose into the C1 of muscle glycogen have shown that muscle is responsible for the uptake of virtually all the excess glucose during hyperglycaemia and that, in patients with non-insulin-dependent diabetes mellitus, it is a failure of glucose transport, rather than a defect in glycogen synthase, that is responsible for the reduced rate of glycogen synthesis compared to control subjects[42]. Attractive though ^{13}C studies are, they are both expensive and relatively insensitive. Much attention is therefore now also being focused on ^{1}H MRS[43].

Drug metabolism can be monitored *in situ* by NMR provided the drug accumulates to sufficient concentration and has an appropriate 'NMR label'. For example, the metabolism of 5-fluorouracil, a drug widely used in the treatment of disseminated tumours of the colon, breast and ovary, can be monitored in animal models[44] and also in patients undergoing therapy. There is a good correlation between the presence of active metabolite, detected by ^{19}F NMR, and therapeutic response[45].

6. SOLID STATE APPLICATIONS OF MRI

Although the human body may seem a solid enough entity, from the NMR viewpoint it has decidedly liquid-like characteristics. Real solids, in which the spins under study are not undergoing rapid isotropic motion, give rise to very broad lines due to strong unaveraged dipolar interactions. NMR studies of materials falls into two categories. At one extreme one can study a mobile phase present within the solid material, for example ingress of water, or oil in an oil-bearing rock. Here conventional MRI procedures for static imaging, flow imaging or 'q-space' imaging can be used with great effect[46,47]. Alternatively, direct imaging of a solid material, to identify internal defects in a polymer sample, for example, requires the use of line-narrowing procedures which remain at a relatively early stage of development[48].

7. REFERENCES

1. P. C. Lauterbur, "Image formation by induced local interactions: examples employing nuclear magnetic resonance," Nature, vol. 242, pp. 190-191, 1973.
2. P. Mansfield and P.K. Grannell, "NMR 'diffraction' in solids?," J. Phys. C, vol. 6, pp. L422-L426, 1973.
3. G.N. Hounsfield, "Computerized transverse axial scanning (tomography): Part I. Description of system," Brit. J. Radiol., vol. 46, pp. 1016-1022, 1973.
4. P. Mansfield and A. A. Maudsley, "Medical imaging by NMR," Brit. J. Radiol., vol. 50, pp. 188-194, 1977.
5. W.S. Hinshaw, P. A. Bottomley and G. N. Holland, "Radiographic thin-section image of the human wrist by nuclear magnetic resonance", Nature, vol. 270, pp. 722-723, 1977.
6. R. Damadian, M. Goldsmith and L. Minkoff, NMR in cancer: XVI. FONAR image of the live human body," Physiol. Chem. Phys., vol. 9, pp. 97-100, 1977.
7. P. Mansfield, I. L. Pykett, P. G. Morris and R. E. Coupland, "Human whole body line-scan imaging by NMR," Brit. J. Radiol., vol. 51, pp. 921-922, 1978.
8. I. R. Young, A. S. Hall, C. A. Pallis, N. J. Legg, G. M. Bydder and R. E. Steiner, "Nuclear magnetic resonance imaging of the brain in multiple sclerosis," Lancet, vol. ii, pp. 1063-1066.
9. J. Radon, "On the determination of functions from their integrals along certain manifolds," Ber. Saechs. AKad. Wiss. Leipzig, Math. Phys. K1, vol. 69, pp. 262-277, 1917.
10. R. N. Bracewell, "Strip integration in radioastronomy," Austral. J. Phys., vol. 9, pp. 198-217, 1956.

11. P. G. Morris, Nuclear Magnetic Resonance Imaging in Medicine and Biology, chapter 4, Oxford University Press, 1986.

12. S. Ljunggren, "A simple graphical representation of Fourier-based imaging methods," J. Magn. Reson., vol. 54, pp. 338-343, 1983.

13. P. Mansfield, "Multi-planar image formation using NMR spin echoes," J. Phys. C, vol. 10, pp. L55-L58,1977.

14. D. Hawksworth, "New magnet designs for MR," Magn. Reson. Med., vol. 17, pp. 27-32, 1991.

15. E. R. Andrew, "Passive magnetic screening," Magn. Reson. Med., vol. 17, pp. 22-26, 1991.

16. P. Mansfield and B. Chapman, "Active magnetic screening of gradient coils in NMR imaging," J. Magn. Reson., vol. 66, p. 573, 1986.

17. P. Mansfield and B. Chapman, "Multi-shield active magnetic screening of coil structures in NMR," J. Magn. Reson., vol. 72, pp. 211-233, 1987.

18. R. Turner, "A target field approach to optimal coil design," J. Phys. D, vol. 19, pp. L147-L151, 1986.

19. R. Turner, "Minimum inductance coils," J. Phys. E, vol. 21, pp. 948-952, 1988.

20. P. G. Morris, D. J. O. McIntyre, D. E. Rourke and J. T. Ngo, "Rational approaches to the design of NMR selective pulses," NMR in Biomed., vol. 2, pp. 257-266.

21. S. Kirkpatrick, C. D. Gelatt, Jr. and M. P. Vecchi, "Optimization by simulated annealing," Science, vol. 220, pp. 671-680, 1983.

22. C. J. Hardy, P.A. Bottomley, M. O'Donnell and P. Roemer, "Optimization of two-dimensional spatially selective NMR pulses by simulated annealing," J. Magn. Reson., vol. 77, pp. 233-250, 1988.

23. J. T. Ngo and P. G. Morris, "General solution to the NMR excitation problem for non-interacting spins," Magn. Reson. Med., vol. 5, pp. 217-237, 1987.

24. J. Pauly, D. Nishimura and A. Macovski, "A k-space analysis of small tip-angle excitation," J. Magn. Reson., vol. 81, pp. 43-56, 1989.

25. V. Wedeen, R. Meuli, R. Edelman, S. Geller, L. Frank, T. Brady and B. Rosen, "Projection imaging of pulsatile flow with magnetic resonance," Science, vol. 230, pp. 946-948 , 1985.

26. C. Dumoulin, S. Souza, M. Walker and W. Wagle, "Three-dimensional phase contrast angiography," Magn. Reson. Med., vol. 9, pp. 139-149, 1989.

27. P. van Dijk, "Direct cardiac NMR imaging of heart wall and blood flow velocity," J. Comput. Assist. Tomogr., vol. 8, pp. 429-436, 1984.

28. G. L. Nayler, D. N. Firmin and D. B. Longmore, "Blood flow imaging by cine magnetic resonance," J. Comput. Assist. Tomogr., vol. 10, pp. 715-722.

29. D. Le Bihan, "Intravoxel incoherent motion using steady-state free precession," Magn. Reson. Med., vol. 7, pp. 346-351, 1988.

30. P. Mansfield, A. M. Howseman, and R. J. Ordidge, "Volumar imaging using NMR spin echoes: echo-volumar imaging (EVI) at 0.1T," J. Phys. vol. 22, pp. 324-330.

31. P. G. Morris, A. Jasinski and D. J. O. McIntyre, "NMR microscopy of plants," Modern microscopies: techniques and applications, P. J. Duke and A. G. Michette, pp. 171-183, Plenum Press, New York, 1990.

32. J. B. Aguayo, S. J. Blackband, J. Schoeniger, M. A. Mattingly and M. Hintermann, "Nuclear magnetic resonance imaging of a single cell," Nature, vol. 322, pp. 190-191, 1986.

33. R. W. Bowtell, G. D. Brown, P. M. Glover, M. McJury and P. Mansfield, "Resolution of cellular structures by NMR microscopy at 11.7T," Phil. Trans. R. Soc. Lond. A, vol. 333, pp. 457-467, 1990.

34. C. B. Ahn and Z. H. Cho, "A generalized formulation of diffusion effects in mm resolution nuclear magnetic resonance imaging," Med. Phys., vol. 16, pp. 22-28, 1989.

35. Z. H. Cho, C. B. Ahn, S. C. Juh, J.M. Jo, R. M. Friendenberg, S. E. Fraser and R. E. Jacobs, "Recent progress in NMR microscopy towards cellular imaging," Phil. Trans. R. Soc. Lond. A, vol. 333, pp. 469-475, 1990.

36. P. T. Callaghan, C. D. Eccles and Y. Xia, "NMR microscopy of dynamic displacements: **k**-space and **q**-space imaging," J. Phys. E, vol. 21, pp. 820-822, 1988.

37. C. F. Jenner, Y. Xia, C. D. Eccles and P. T. Callaghan, "Circulation of water within wheat grain revealed by NMR micro-imaging," Nature, vol. 336, pp. 399-402, 1988.

38. D. I. Hoult, S. J. W. Busby, D. G. Gadian, G. K. Radda, R. E. Richards and P. J. Seeley, "Observation of tissue metabolites using ^{31}P nuclear magnetic resonance," Nature, vol. 252, pp. 285-287, 1974.

39. G. K. Radda, "The use of NMR spectroscopy for the understanding of disease," Science, vol.233, pp. 640-645, 1986.

40. P. L. Hope, A. M. Costello and E. B. Cady, "Cerebral energy metabolism studied with phosphorus NMR spectroscopy in normal and birth-asphyxiated infants," Lancet, vol. ii, pp. 366-370, 1984.

41. P. A. Bottomley and C. J. Hardy, "Strategies and protocols for clinical ^{31}P research in the heart and brain," Phil. Trans. R. Soc. Lond. A, vol. 333, pp. 531-544, 1990.

42. G. I. Shulman, D. L. Rothman and R. G. Shulman, "^{13}C NMR studies of glucose disposal in normal and non-insulin-dependent diabetic humans," Phil. Trans. R. Soc. Lond. A, vol. 333, pp. 525-529, 1990.

43. D. G. Gadian, "Proton studies of brain metabolism," Phil. Trans. R. Soc. Lond. A, vol. 333, pp. 561-570, 1990.

44. A. N. Stevens, P. G. Morris, R. A. Iles, P.W. Sheldon and J. R. Griffiths, "5-Fluorouracil metabolism monitored in vivo by ^{19}F NMR," vol. 50, pp. 113-117, 1984.

45. W. Semmler, P. Bachert-Baumann, F. Guckel, F. Ermark, P. Schlag, W. J. Lorenz and G. van Kaick, "Real-time follow-up of 5-fluorouracil metabolism in the liver of tumor patients by means of F-19 MR spectroscopy," Radiology, vol. 174, pp. 141-145, 1990.

46. P. A. Osment, K. J. Packer, M. J. Taylor, J. J. Attard, T. A. Carpenter, L. D. Hall, N. J. Herrod and S. J. Doran, "NMR imaging of fluids in porous solids," Phil. Trans. R. Soc. Lond. A, vol. 333, pp. 441-452, 1990.

47. J. J. Attard, T. A. Carpenter, A. de Crespigny, S. L. Duce, L. D. Hall, R. C. Hawkes, N. J. Herrod and R. J. Hodgson, "A survey of some applications of NMR chemical microscopy," Phil. Trans. R. Soc. Lond. A, vol. 333, pp. 477-485, 1990.

48. J. B. Miller, D. G. Cory and A. N. Garroway, "Line-narrowing approaches to solid state NMR imaging: pulsed gradients and second averaging," Phil. Trans. R. Soc. Lond. A, vol. 333, pp. 413-426, 1990.

Part III
National Concerns

Introduction to Part III

The papers in this section describe and reflect current policy debates on instrumentation. Stine traces the discussion in the United States around programs to support instruments in chemistry. Kruytbosch outlines the more general development of instrumentation as a policy issue, and compares it with the evolution of policy on research facilities. Georghiou and Rocher present data from recent assessments of the state of the instrument base in academic institutions in the United Kingdom and France.

The contrast in perspective could not be greater between the case studies presented in the last section and the policy issues outlined here. The few basic themes of instrumentation policy overlap little if at all with the features of instrument use that emerge from an examination of scientific practice. One overlap concerns the theme of scale. But in the policy discussions, a stripped down version of the concept of scale appears. Scale is of interest only because it involves higher costs, not because it either derives from industrial modes of production (a la Mendelsohn and Schaffer) or leads to changes in social organization (a la Krige). The other major themes of the policy discussion also involve money, for example, aggregate resources devoted to instruments within a nation; the pattern of distribution of instruments; the ratios of government spending to laboratory size as measured in people. When seen through the lens of government spending, then, all users and uses of instruments appear "standardized" (a la Blume). The viewpoint is uncompromisingly consumption oriented.

The gap between the case studies of previous sections and the policy issues of this section is the measure of what science studies has to contribute to science policy in this area. Georghiou assigns the sociology of science the technical role of determining "need"—that is, how little can the government spend without sacrificing scientific progress? The historians and sociologists who contributed to earlier sections did not attempt this task, but instead did much more: They changed the perspective on scientific instruments radically by bringing the innovative process into view and by tracing their connection to the larger economy. A science studies perspective, then, takes instrumentation policy out of the esoterica of "policy for science" and places it firmly within more general economic policy.

What specific features of instrument use should policy be taking into account, according to the case studies in the earlier sections? The main one is this: that an important function of research instrumentation is to provide the site for development of technological know-how and skill. To serve this function, however, researchers and students must tinker with their instruments; black boxes do not serve the purpose. The know-how function of instruments provides a common base for researchers and technologists in other areas of practice, including industrial practice, and is a major factor stimulating interaction and exchange of ideas and techniques across institutional boundaries. This is why the strategy of "following the instruments" is so effective in tracing research-industry connections.

This whole set of observations suggests that the consumption-oriented instrument policy which is standard in the United States and Europe is addressing the expensive part of the instrument-use pattern—the conspicuous consumption described by Blume in the last section—while missing the less expensive and more influential tinkering component. The full import of Blume's hypothesis on standardization becomes clearer in this light. If he is right, too-lavish government funding for instruments may be feeding the (insatiable) demand for conspicuous consumption while failing to encourage or support the tinkering style of experiment which contributes directly to the national industrial base. Surely, this is a strategy which few policy makers would consciously choose.

-SEC

Chapter 12

Scientific instrumentation as an element of U.S. science policy: National Science Foundation support of chemistry instrumentation

Jeffrey K. Stine
National Museum of American History

1. BACKGROUND

Two fundamental trends have characterized scientific research in the United States since World War II: the first is the steady and dramatic rise in its costs, and the second is the federal government's willingness to underwrite ever larger percentages of those costs. Many factors have helped multiply the expenses of research, and few loom larger than the growing costs of science's capital base, including instrumentation. Despite the critical role of such instrumentation, however, academic research administrators have not always given it high priority, sometimes preferring--when federal R&D funding has temporarily declined--to defer capital investments in favor of unbroken support for personnel.

This investment pattern has led to fluctuations in the degree of access to advanced instrumentation and has challenged science policy makers to devise the best means to distribute instrumentation among the nation's research institutions. It has also led to inconsistent characterizations of its importance to the health of science overall. Although a perennial element in U.S. science policy, instrumentation periodically draws cries of great concern, as exemplified by the spate of reports, studies, and articles that appeared during the late 1970s and early 1980s chronicling the "crisis" state of instrumentation in U.S. colleges and universities and warning of the dire consequences if the situation was not soon corrected.[1] From the moment of the National Science Foundation's establishment in 1950, calls for greater funding of equipment were heard. Examinations of the problem, however, tended to be anecdotal, or based on unsystematic surveys. Not until the mid-1980s--following the outburst of warnings and complaints of the previous ten years--were extensive, systematic data collection on and evaluations of instrumentation needs undertaken.[2]

The quality of instrumentation--that is, the research equipment other than routine apparatus--is undeniably important to investigators who want to stay at the cutting edge of advanced research. Unfortunately for these scientists, while the pace of technological improvements has resulted in dramatic leaps in the precision, sensitivity, capacity, and speed of instruments, the costs of acquisition, operation, and maintenance of the equipment have soared. Moreover, this process has also hastened the onset (or perception) of obsolescence as the effective life spans of state-of-the-art equipment have shortened.[3]

To provide an example of how the U.S. government dealt with such issues and with academic science's instrumentation needs from 1950-1990, this essay focuses on the National Science Foundation's support of chemistry instrumentation.

2. NATIONAL SCIENCE FOUNDATION AND CHEMISTRY INSTRUMENTATION

Following World War II, scientists in many fields became increasingly dependent on sophisticated instruments, although these needs varied widely among disciplines. High-energy physics, for example, the classic "Big Science" discipline of the postwar period, required such enormous capital investment that government funding was essential both to build and operate the facilities that incorporated linear accelerators, cyclotrons, synchrotrons, and the like. Moreover, the expense of such operations meant that few could be funded, and physicists found themselves sharing instrumentation as a matter of course. Oceanography and astronomy, too, had high capital costs, and researchers in those fields routinely shared equipment.[4]

In the 1940s and early 1950s, research in chemistry proceeded with relatively few high-cost instruments, but by the mid-1950s, chemists were increasingly relying on the progression of advanced instrumentation.[5] Linus Pauling explained that chemists were simply "attacking harder problems than they were in the early days, and these harder problems require more complex apparatus and facilities," adding that "there are few chemists now who work just with test tubes and beakers."[6] Writing in 1986, John K. Taylor of the National Bureau of Standards' Center for Analytical Chemistry observed that advances in modern chemical instrumentation involved "measurement principles that were little known and even unknown to chemists half a century ago."[7] He listed the revolutionary effects of such analytical tools as mass spectrometry, chromatography, nuclear magnetic resonance, and electron spin resonance. The acquisition of instrumentation, Taylor added, "has come to be a major item in a laboratory's budget and the rate of its obsolescence and advances in the state-of-the-art are major concerns to laboratory management." The need for sophisticated laboratory facilities and physical plants capable of accommodating state-of-the-art equipment grew along with advances in instrumentation. Many of the new instruments needed more precise temperature and humidity control, as well as increased need for clean rooms, special handling facilities for radiochemicals and hazardous wastes, and the like.[8]

Despite the growing importance of instrumentation for nearly all fields of science, the National Science Foundation (NSF) did not isolate instrumentation as a separate policy issue until the mid-1950s, when senior government science administrators began responding to complaints about the sorry state of academic research facilities and equipment. The Foundation broke out research facilities and equipment from its overall budget for the first time in fiscal year 1956. NSF Director Alan T. Waterman explained to the Senate Appropriations Committee that "in many areas of science there have been no major research facilities constructed since World War II, even though new techniques have been discovered and even though many of the older facilities have become obsolescent."[9] The Foundation's annual report elaborated on the program's rationale, claiming that "as the frontiers of knowledge are pushed forward, basic research becomes more and more complicated. Machinery and facilities for performing research become increasingly complex and costly. Meanwhile, resources of the universities, where basic research is an essential part of the educational process, have not kept pace with increasing costs."[10]

The NSF annual report added that "as time goes on, it is anticipated that the future progress of research will depend in some fields of science upon an increasing ratio of research tools to scientific manpower."[11] The goals of this program were to increase funding for research facilities and equipment that were scarce and costly; to promote geographic distribution of such facilities and equipment; and to foster both research and training of scientific personnel through the provision of advanced facilities and equipment.[12]

The NSF fiscal year 1957 budget proposal contained $7 million for a new facilities program, which included construction of a radio astronomy observatory, preliminary studies for the construction of an optical astronomical observatory, and partial support for the acquisition of small nuclear research reactors and digital computers for use at universities.[13] The Foundation expanded its research facilities program the following year by adding funds for oceanographic research vessels, a Southern Hemisphere astrograph, and biological research facilities.[14]

Upon learning of the proposed new research facilities program, the Bureau of the Budget (BoB) requested NSF to prepare "a report on the current status and future needs for research facilities [and major equipment], against which annual requests for Federal support funds could be evaluated."[15] The Foundation obliged, submitting its report to the BoB in 1957. The report explained that when it came to fostering science, "Government obligation and participation tends to differ and become more complicated when large-scale physical facilities and major equipment are involved." One of the primary complications for science policy makers was the question of "program balance as between the support of investigators and the support of facilities."[16] The NSF asserted that for many fields of science the greatest limitation to the advancement of research (and, by connection, subsequent advances in defense, medicine, agriculture, and industry) came from "the sheer expense of such instrumentation." Because "the accelerating pace of scientific discovery and innovation is tending to shorten the useful life of much of the scientific equipment and instrumentation now in place," the NSF stated, the national interest calls for increased federal support of the capital base of science.[17]

The NSF report highlighted the strain placed on the agency's funding mechanisms by the escalating costs of instrumentation. During its first years of operation, the Foundation funded the purchase of instrumentation through its regular grants and contracts for scientific research. Investigators applying for NSF support requested funds in their proposal budgets to purchase the instruments necessary for their research. Rarely were grants or contracts awarded solely for the purchase of instrumentation. The NSF reported that by the mid-1950s budgetary restrictions "would not permit the awarding of a sizeable volume of grants or contracts for the purchase of expensive research instruments without interfering seriously with the support of integrated research projects."[18]

In chemistry, the types of instruments that created such problems in 1957 included nuclear magnetic resonance spectrometers (falling--in 1957 dollars--in the cost range of between $30,000 and $50,000 each), mass spectrometers ($15,000 to $50,000 each), X-ray diffraction equipment ($15,000 to $20,000), high-pressure high-temperature equipment ($25,000 to $50,000), ultrasonic equipment ($15,000), and nuclear resonance magnetometers ($8,000 to $10,000).[19] Nevertheless, despite the coverage of 15 areas of instrumentation needs in 10 disciplines, there was no section of the NSF report devoted specifically to chemistry.

The Foundation was organized into several divisions, among them the Mathematical, Physical, and Engineering Sciences (MPE) division, which included chemistry within its jurisdiction.[20] It was customary for the NSF's various divisions to maintain advisory committees consisting of distinguished, practicing scientists in the fields represented by the divisions; the MPE, for example, established its own external advisory committee in 1952. Sometimes a division would have more than one advisory committee, as was true of the MPE which maintained a separate advisory panel for its Chemistry Section.[21]

From the beginning, chemists on the MPE advisory committee repeatedly emphasized the crucial role of instrumentation to chemical research. By the late 1950s, however, the advisors worried that rising instrumentation costs were not being met by NSF grants to independent investigators in chemistry. Believing that instrumentation should be separated from other costs, such as salaries, travel, and supplies, the advisors recommended that dedicated instrumentation grants be established.[22] Ralph A. Beebe, chairman of the NSF's Advisory Panel for Chemistry, stated that "in comparison with certain other disciplines it seems that chemistry has no need for very expensive facilities such as radioastronomy observatories, reactors, computers, or ships for oceanographic research. However, there is a very definite need for research instruments in the cost range from $5,000 to $50,000." To help meet this need, the NSF introduced an experimental program in October 1956; it allocated $300,000 (half from the NSF Chemistry Program and half from NSF's MPE Division) for support of chemistry instrumentation for use on a department-wide basis. Of the 77 qualifying applications, 39 grants were approved for funding, most for instrument purchases at university departments where several investigators would share the equiment.[23]

Despite the strong recommendations of the NSF Advisory Panel for Chemistry, the first two awards were not granted until 1959. In 1960, the Foundation funded three other chemistry instrumentation proposals. The MPE's Chemistry Section responded by earmarking five percent of its fiscal year 1961 and 1962 research budgets for instrumentation.[24] Although this increase brought the number of chemistry instrumentation grants up to 18 in 1961 and 21 the following year, NSF's Walter R. Kirner, head of the Foundation's Chemistry Program, claimed that such expenditures "were nowhere near sufficient to meet the demand." He believed that the chemistry instrumentation program should be treated as an equivalent program to the large facilities programs sponsored by the NSF's other disciplinary programs.[25] In 1964, Kirner informed his colleagues of the National Academy of Sciences' newly initiated study of U.S. chemistry, chaired by Frank Westheimer of Harvard University. The report was being done, Kirner said, "because of the complaints of many chemists, both established and young, as to the inadequacy of the support now being provided, especially by NSF. These complaints are being voiced by the chairmen of many of the Chemistry departments throughout the country to their Senators and Representatives, to the Director of the National Science Foundation, to the President's Science Advisor, and to the Chairman of the Board of the American Chemical Society."[26]

In August 1962, William von E. Doering, Whitehead Professor of Organic Chemistry at Yale University and chairman of the NSF's Advisory Panel for Chemistry, told NSF Assistant Director Geoffrey Keller that the chemistry advisory panel had "responded to the urgent need for modern scientific equipment in chemistry departments throughout the country by allocating the entire anticipated increase in available funds expected in FY 1963 to proposals for the purchase of equipment." Doering relayed the panel's strong endorsement that the NSF provide "immediate large-scale relief" to address the problem. He outlined the background of the "present distress" in chemistry, stating that investigators needing specialized equipment for research in biochemical and nuclear chemical areas were generally able to fulfill their instrumentation needs through grants provided by the National Institutes of Health and the Atomic Energy Commission, but that investigators in other areas who required sophisticated, expensive instrumentation of the "most widely applicable sort"--instrumentation "without which no department could pretend to be supporting modern chemical research"--were left wanting. Doering contended that it was the NSF's responsibility to fill this gap. Although the Foundation had awarded some instrumentation grants to individual researchers, he charged that it had failed to meet the broader instrumentation needs of the nation's academic chemistry departments. The deficit in chemistry instrumentation was "made up of a great many relatively small items, any one of which, taken by itself, does not compare in size with some single items needed in physics, mathematics or geology." Large instruments attract attention, he pointed out; they "evoke individual response and can be handled as separate line items in the budget. In chemistry, the need is spread among many small items and is not easily appreciated in its entirety without the help of a survey."[27]

Having identified the necessity for a survey of chemistry departmental needs for instrumentation, Doering proceeded to conduct his own, not as the head of the NSF's Advisory Panel for Chemistry, but as an independent chemist at Yale. In February 1962, he mailed a questionnaire to the chairmen of all 121 chemistry departments listed in the American Chemical Society's *Directory of Graduate Research*, and received an astounding 111 responses.[28] He asked the chairs for data relating to their departments' need for instruments costing $2,000 or more and which would be shared by two or more researchers.[29] Doering summarized the results of his survey in a letter sent back to all respondents. "The extent of the responses as well as the nature of many individual comments," he concluded, "makes it clear that the overwhelming majority of departments, whether large or small, or in privately or state supported universities, are in urgent need of scientific apparatus of the larger, more generally used type." The total value of the equipment needs, he estimated, was $14,429,000, or "$130,000 per department on the average."[30] Keller, in response to Doering's letter and survey results, said simply: "At least it can be said that you are now in a strong position to hold your own in the face of the urgent requests from other disciplines!"[31]

The instrumentation needs of chemists received a more sympathetic hearing from Alan Waterman's successor, Leland J. Haworth, who assumed the NSF directorship in 1963. Following the recommendation of Glenn Seaborg, Haworth asked John D. Roberts for material relating to the remarks Roberts had made before the Chemistry Section of the National Academy of Sciences.[32] Roberts, who chaired the Division of Chemistry and Chemical Engineering at the California Institute of Technology and who sat as a member of the NSF's own Mathematical, Physical, and Engineering Sciences Advisory Committee, eagerly responded. "Speaking completely frankly," he began, "I do not believe that chemistry is coming off so well at the Foundation. Probably a major reason for this is an increase in allocation of research funds for large facilities such as oceanographic vessels, atmospheric research centers, Mohole, radio-telescopes, etc. Small facilities for chemistry are not being taken care of as well." He granted that the amount of money allocated for chemistry within the MPE had risen, but he pointed out that "the percentage for chemistry of the total of the MPE research including facilities has declined from some 47.2% in 1952 to 8.6% in 1963." The problem with this trend, he said, was that it was occurring precisely at the point when the NSF was receiving "a great increase in meritorious chemistry proposals," meaning that "matters are now at the point where the Chemistry Program has the highest or close to the highest declination rate of very good proposals in the MPE Division."[33]

One of the most problematic aspects of this downturn, Roberts asserted, was the toll being taken on the support of chemistry facilities and instrumentation. Chemistry was the only discipline, he said, which had not received "substantial support" through the NSF's National Research Programs, National Research Centers, or University Facilities grant program. "Chemical research is rather less of a teamwork affair than research in many other sciences and therefore does not often require large spectacular facilities," he explained. "Nonetheless, the need for small intra-departmental facilities in universities and colleges, particularly of chemical research instruments is reaching urgent proportions." Roberts's hope, along with that of his colleagues on the MPE Advisory Committee, was to establish a line item for chemistry instrumentation in the NSF budget. Compared to NSF's support of physics research, he said, such a program for chemistry was "almost ironically small." He added that "the growing tendency towards support of large fixed installations may well tend to lock out programs such as chemistry which involve flexible support through many small grants." He thought that "this tendency should be checked because new fundamental discoveries are most often made independently of the big fixed one-purpose installations." Because chemistry touches so many other branches of science, he said, progress made in chemical research because of sophisticated analytical instruments will expedite advances in other disciplines.[34]

Dissatisfied with the federal support given to the chemical sciences, the 13 members of the NSF's Advisory Panel in Chemistry submitted an impassioned report--entitled "Federal Support of Basic Chemistry"--to the Foundation's Director in October 1963. The panel was responding "to the pleading of large numbers of research chemists for a review of the adequacy of the support which chemistry receives in the United States," a pleading fueled in large part by the perceived high rate of rejection for quality research proposals submitted to the NSF by chemists. Not only were good proposals denied, the Panel asserted, but "the great majority of those supported were funded at a level quite inadequate for the proper execution of work."[35]

To illustrate this "appalling state of present affairs," the Panel pointed to the "callous insensitivity" toward "the innumerable pieces of equipment and apparatus demanded by modern scientific research." The panel members felt compelled to stress the central role of chemistry in American society. "The contribution of applied chemistry to the state of health, to the economic and material welfare, and to the security of the people of this country would not bear the briefest outline," they wrote, "were it not for the blinding excitement aroused by the appalling successes of applied physics in developing new techniques of destruction and in pioneering the exploration of near-space."[36] Chemists in the early 1960s often felt themselves in the shadow of physicists and space scientists, not only in influence within the higher levels of government, but also with regard to amounts of federal funding received for instrumentation and research in general. Chemistry was not, after all, "Big Science" in the traditional sense.

Nowhere else, the panel asserted, was the breakdown in federal support of chemistry better illustrated than in the area of the unmet need for advanced research equipment. During the early 1950s, NSF chemistry advisors "believed it was more important for the development of basic chemistry to use the inadequate and limited funds [of the NSF] to support the research activities of graduate students and postdoctoral assistants," hoping that chemistry department chairpersons would "find ways of raising enough money from sources other than N.S.F. to satisfy the need for capital research equipment." This belief dissolved in 1956, the panel contended, when they learned that "a serious discrepancy had developed between the amount of equipment needed to carry on basic research and the amount departments had been able to provide." NSF officials responsible for chemistry responded by reallocating funds set aside for research proposals to a new granting program aimed at purchasing instrumentation.[37]

The panel speculated that chemists had not put in for more equipment grants because they reasonably expected that their chances of winning support were extremely small. As the panel phrased it: "intelligent persons do not ask for what they need if the balance between the effort involved in asking and the chance of being given becomes ludicrously and frustratingly unfavorable." Moreover, they added: "It is one thing for the operators of 'Big Science' to expend great individual effort to raise large sums for the creation of entire research establishments; quite another for the typically 'Small' scientist of chemistry to devote a comparable amount of effort for a sum of money that averages $70,000 a year."[38]

Following up on the concerns of the chemistry research community regarding declining levels of funding for chemistry, Presidential Science Advisor Donald F. Hornig mailed a questionnaire to chemistry department chairs in May 1964, the results of which he passed on to NSF Director Haworth.[39] Hornig's action prompted several department heads to write directly to Haworth. For example, Carnegie Institute of Technology's chemistry department chairman, Robert B. Carlin, complained that "financial support for sound basic research in chemistry has declined alarmingly throughout the Country during the past two years." Echoing the concerns to many of his academic colleagues, Carlin claimed that there was a "state of malnutrition that now threatens fundamental research in chemistry."[40]

William White, chairman of the chemistry department at the University of Vermont, told Haworth of his concern over the seven percent decline in NSF allocations for basic research in chemistry over the past two fiscal years. "The fact that this cut was made," he said, "indicates that we must be continuously vigilant to avoid future actions of this type which may weaken the vitality of one of our nation's most fundamental and vigorous sciences." White speculated that the downward trend in chemistry funding is what led to the National Academy of Sciences' establishment of the Westheimer committee, which was undertaking an evaluation of chemistry in the United States.[41] J. F. Bunnett, chairman of Brown University's chemistry department, told Haworth that the hardship placed upon chemistry departments in the United States could be weathered in the short run, because chemical research could "be maintained at a modest level even with reduced federal support, by drawing on funds from private sources and by using equipment and materials purchased in the past," but that "such makeshift support can not long suffice."[42]

Haworth heeded the pleadings of the chemistry community and his own staff advisors, and became convinced of the need to expand the Foundation's support of specialized research facilities programs in the physical sciences at the nation's colleges and universities. "Continued progress in research is increasingly dependent upon the availability of modern instruments and facilities," he told Congress in 1964. Haworth explained that NSF support of such facilities had heretofore been restricted to physics, the atmospheric sciences, and oceanography. Beginning in fiscal year 1965, he said, the "Foundation plans to extend such assistance to the fields of chemistry, astronomy, and engineering."[43]

The 1965 Westheimer Report had proven extremely influential in drawing attention to the instrumentation needs of chemistry, and the report's emphasis on the neglected state of chemical instrumentation in the nation's universities was picked up by the NSF.[44] In 1967, Arthur F. Findeis, director of the NSF's Chemistry Research Instruments Program, submitted a staff report on the state of chemical instrumentation, which addressed the needs of academic chemistry departments by updating the equipment inventory presented in the Westheimer report. The NSF report claimed that "a graduate department of average size needs over one million dollars of major equipment to conduct quality graduate education which includes research on problems of current chemical interest." Findeis reported that the NSF Chemistry Research Instruments Program had funded about 14 percent of instrumentation costs in U.S. universities between 1964-1967, at a cost of $64 million. Given the projected growth of university chemistry faculties--which were expected to double in the next decade--and the fact that sophisticated instrumentation had an average useful lifetime of ten years, maintenance of NSF's 14 percent level required an annual budget of $7 million for chemistry instrumentation, plus an addition $20 million to correct the immediate needs of the current backlog.[45]

The question of improving the government's management of instrumentation procurement in the federal laboratories led the House Government Operations Committee's Subcommittee on Research and Technical Programs to hold hearings in August 1967. Two months later it issued a report in which the need for upgrading research instrumentation was emphasized, along with the warning that "too much money for equipment may also be damaging for research. Scientists, like other Americans, can become obsessed with gadgetry for its own sake, in having it or using it to pile up masses of data without adequate forethought to research design."[46] The Committee even quoted Vannevar Bush, who had testified before Congress four years earlier that "the costs of research go up very rapidly when one gets into hardware. When money comes easily there is a tendency to rush into use of complex equipment too fast and too far. . . . We need to use care that our American love of gadgetry does not lead us astray in our emphasis [on instrumentation]."[47]

As universities swelled in size during the late 1960s and as their research needs expanded correspondingly, the rising federal science budget leveled off, the consequence of other priorities within the Johnson and Nixon administrations. The NSF responded to its reduced budget allocations by cutting back on its institutional support programs, as well as in its support of facilities and large-scale instrumentation.[48] As Haworth told Senator John O. Pastore, chairman of the Appropriations Subcommittee on Independent Offices, in June 1969, "budgetary stringencies in recent years have made it necessary for the Foundation to defer the procurement of new and advanced equipment in favor of maintaining ongoing research programs, the interruption or serious curtailment of which would represent a serious loss in terms of both the acquisition of scientific information and the continuity of the work of individual scientists." He warned Congress that "continued short-term budget reductions in the facilities and instrumentation areas will surely impose very substantial long-term hazards to the excellence of American science."[49] When rising energy costs and severe inflation were added to the reductions in the level of federal aid to research, many colleges and universities began neglecting their capital investments in science, a management decision that sowed the seeds for later problems.

In 1970, the National Science Board (NSF's governing body) submitted a report to the Congress on the state of the physical sciences. Reflecting the recent slow-down in the growth of federal funding of science, the Board made several recommendations, among them that instrumentation be given high priority, as it represented "the pacing item for research in much of the physical sciences." The Board expressed an interest in shared facilities and instrumentation. "Basic responsibility for the creation and operation of such facilities can be vested in single institutions or in groups of institutions," it reported, and "the pattern of users groups, such as those operating in high-energy physics, should be encouraged to spread throughout the physical sciences."[50]

Under pressure from the research community to increase federal expenditures for instrumentation, the NSF contracted the National Academy of Sciences in 1971 to survey the nation's university laboratories and assess the state and trends for 10 science and engineering disciplines. The 42-member Academy panel confirmed that there was indeed an urgency to the equipment needs of the nation in all fields, including chemistry.[51]

Following up on the Academy's findings, the NSF contracted with the University of Utah's Department of Chemistry to survey "the distribution of major instruments and the various forms and management practices involved in instrumental support of chemical research" in the United States and three European countries. The Utah panel reached several conclusions, among them that "judicious government funding over the past two decades" had produced "an impressive array of sophisticated instrumentation in U.S. laboratories." The NSF had been most responsible for this support of chemistry instrumentation, and the panel found that within the agency's funding limitations, it had "met the major bonafide needs of the research community in an efficient manner." "As a result," it continued, "the latest developments in scientific instrumentation have been widely distributed, and most laboratories with vigorous research programs have been provided with access to sophisticated instrumentation."[52]

The panel said that future instrumentation needs would be high, and it found existing instrumentation centers to be an efficient way to distribute costly and expensive equipment among smaller institutions, and also an effective way to enhance sharing of such instruments even among researchers at larger institutions. The panel was concerned about utilization patterns for high-cost equipment, believing it prudent to ensure that such instruments were not left idle during normal working hours, nor during evenings, weekends, and holidays. The panel urged the NSF to follow the lead of the NIH's Research Resources Branch (which had supported instrumentation centers for biomedical applications) and establish--on an experimental basis--a small number of instrumentation research centers to promote the advancement of state-of-the-art instrumental analysis. Even so, the panel strongly recommended "the continuation of the present instrument grants program in the Foundation in its present format," which primarily funded the purchase and maintenance of chemistry instrumentation at the single investigator and departmental levels.[53]

In 1975, the National Science Board began prodding the NSF to expand its efforts to meet the instrumentation needs of the nation's colleges and universities. The Foundation responded by shifting more of its resources to instrumentation grants and by exploring alternative means to increase access to advanced research equipment.[54]

When Richard C. Atkinson became the new NSF Director in 1977, he appointed George Pimentel as deputy director. Pimentel, who was a prominent chemist at the University of California at Berkeley, took a special interest in the question of federal support of research instrumentation. Speaking before the House of Representatives, he identified the need to bring "a brand new arsenal of instrumental techniques" on line. Instrumentation "is not just an obsolescence problem accumulated over the past decade," he testified; it is much more sophisticated--"If we want to stay at the front of science, we must take advantage of computer techniques, laser techniques." He raised the possibility of establishing more regional instrumentation facilities, which would encourage "shared usage": "We are trying to set up nuclear magnetic resonance laboratories with wide regional access specifically to address this need to provide sophisticated instrumentation, expensive instrumentation, but to share as much as possible."[55]

The House Committee on Science and Technology found attractive the idea of establishing regional centers for instrumentation, but it was also concerned about avoiding needless duplication. In commending the NSF's regional instrumentation facilities plan in 1978, the Committee called on the Foundation to "conduct inventories of, and analyses of the needs for, scientific instrumentation." Moreover, the Committee recommended "that the NSF embark upon an inventory of available instrumentation nationwide, and thorough analysis of cost recovery policies involved in the operation of instrumentation service facilities."[56] In 1979, the Committee continued to press the NSF to gather data on "the age, type, and geographic distribution of R&D plant and equipment," in order to provide indicators that "would serve on a repetitive, periodic, or continuing basis [and] to assist in assessing the status and trends of instrumentation resources."[57]

One important aspect of the regional instrumentation facilities program--as least as far as chemistry was concerned--was that it was a separate item within the NSF's overall budget, something that had not existed for Foundation support of chemical instrumentation heretofore.

The NSF received pressure from many different sources to do many different things. With limited resources, the Foundation therefore had to determine how best to allocate its resources--to establish priorities, make trade-offs, etc. Instrumentation, which competed against other programs and expenditures, was not immune, and even within the broad category of NSF support for instrumentation, the scientific community--itself a divergent group--was fragmented on how best these funds should be allocated. The regional instrumentation facilities program represented a new approach; not surprisingly, it had both supporters and detractors.

3. REGIONAL INSTRUMENTATION FACILITIES PROGRAM

Regional instrumentation facilities represented one of various mechanisms that NSF tried as a way of helping to meet growing instrumentation needs. The idea began to gain political support during the 1970s. In the House Appropriations hearings for the NSF in 1977, for example, Michigan Democrat Bob Traxler asked, with regard to the NSF's support of chemistry instrumentation, whether it was "possible to centralize the chemistry research facility so as to accommodate more projects on the equipment that now exists?" Foundation representatives answered in the affirmative, adding that chemistry departments often receive grants for instrumentation to be shared by researchers within the department. Although the idea of regional instrumentation facilities was not mentioned, NSF spokesman E. Creutz claimed that "chemists rely so heavily on instrumentation, that they were one of the first groups of scientists to recognize the need for facilities which could be shared in this way."[58]

The regional instrumentation concept for chemistry held great appeal for the Foundation's leadership. NSF Director Richard C. Atkinson and his senior staff first presented a proposal for a Regional Instrumentation Facilities (RIF) program to the National Science Board in January 1978.[59] The RIF program, Atkinson summarized, was intended to meet "the growing need for wide user access to modern instrumentation." The RIFs would be limited to those instruments, such as nuclear magnetic resonance and mass spectrometry, that were "too sophisticated for most users to maintain" and "too expensive to be numerous." These instrumental techniques, he emphasized, "have become essential to forefront research in cross-disciplinary areas of science extending from the health sciences through biology, chemistry, materials research and physics." The RIF program was thus designed to meet a variety of science policy criteria and to appeal both to the scientific community and to the Congress. Cost-effectiveness would be achieved not only by encouraging individual investigators from various institutions to share instrumentation, but also by lowering start-up costs. Rather than starting from scratch, existing research centers would be built up, thereby exploiting the auxiliary equipment, support facilities, and technical personnel already in place."[60]

Aside from addressing the financial problems associated with providing up-to-date instrumentation to U.S. researchers, the RIF included other policy issues as well. Atkinson described them to the NSB: "First, it provides a rational way to begin addressing the issue of wider geographical distribution of basic research in the United States. Second, it affords the opportunity of promoting more interaction between industry and university scientists. And third, it addresses the critical need of young investigators and those at smaller institutions for access to very sophisticated instrumentation."[61] The issue of geographic distribution was a central element for the political success of the program in Congress. Foundation officials therefore went out of their way to assure that the regional instrumentation facilities program addressed this question.

Atkinson provided members of the National Science Board with a detailed background paper on the proposed RIF, wherein the rationale for the program was fully revealed. With regard to providing the most expensive and sophisticated instrumentation to U.S. researchers, the RIF was argued to be "the most equitable and cost-effective approach." Five reasons buttressed this assertion: "certain instruments (measurements) are essential to advances in several scientific disciplines; recent instrumental advances have been so revolutionary that completely new frontiers have been exposed; the dilemma created by these advances is that now the instrumentation is very costly to purchase, operate and maintain; the impact of these trends is greatest for young investigators and those at smaller institutions (including minority institutions); and when instruments are too costly for the individual scientist, then regional facilities provide an economical and equitable way for providing access to all qualified scientists." The report continued: "In addition to meeting a documented scientific need, regional facilities represent a rational approach to the question of proper geographical distribution. Moreover, the proposed model is economical, since the idea is to build on existing expertise in most cases."[62]

The National Science Board was assured that NSF-supported regional would be selected by the scientific community on the basis of merit, using the normal peer review, competitive process. Nevertheless, the report listed the types of instrumentation most likely to be funded--by and large, "those instruments and measurements that are of widespread interest scientifically, yet are now very expensive and require highly trained personnel to operate and maintain." These criteria were met, according to the report, by such instruments--or collections of instruments--as "nuclear magnetic resonance spectroscopy (NMR), mass spectrometry (MS), X-ray crystallography, lasers, surface science, and separation science."[63] In some areas, individual instruments were highly expensive, while in other areas the expense was not so much in the high costs of a single instrument but the requirement of a whole set of complementary instruments applied jointly on a research problem. The proposed surface science facility fitted the latter category, as the analysis of surfaces and small particles required an array of techniques, such as X-ray photoelectron spectroscopy, Auger spectroscopy, and secondary ion mass spectrometry.

The RIF was not without precedent, the Foundation staff stressed. Other government-sponsored centers for sharing instrumentation had fared well. The National Laboratories, for example, were administered by the Department of Energy and also received support from NSF. Beginning in the 1950s, the NSF had also sponsored a limited number of "national research centers," which catered to the extremely costly instrumentation and facilities needs of optical and radio astronomy (of which the Kitt Peak National Observatory and the National Radio Astronomy Observatory are examples) and astrospheric sciences. And the National Institutes of Health (NIH) had operated regional NMR and mass spectrometry facilities through its Biomedical Research Support Grants (two of the facilities were jointly sponsored by the NSF). Access to these facilities, however, were far more limited than that proposed for the RIFs. For one thing, the number of these existing facilities was small, and for another the range of their instrumental techniques was limited, and access to the facilities was restricted to a few disciplines. The NIH facilities, for example, were open only to biological research supported by the NIH. The RIFs would provide greater access, a broader range of instrumentation, and more encouragement to multidisciplinary participation than the NIH facilities.[64]

The NSF sought to heighten the RIFs' political viability in Congress by making geographic distribution an explicit criterion. Geographic distribution had been a major concern of the Congress since the establishment of the NSF; every member wanted to assure that his or her constituents received their fair share of government expenditures, and as research facilities became linked to regional economic development (through such well-known, high-tech corridors as California's Silicon Valley, Massachusetts's Route 128, and North Carolina's Research Triangle) and as the amount of money spent on research facilities elevated them into the category of significant "public works," the incentives to allocate federal research expenditures broadly became even greater. The NSF also recognized the equity question, intending ultimately to provide up to 25 RIFs in "broad enough geographic regions that all U.S. scientists with need for advanced measurements will have reasonable access, including those at smaller institutions."[65] Discussion of geographic distribution typically aroused concerns within the scientific community that the peer review system would somehow be weakened, as "political" criteria would take precedence over the criterion of "scientific" merit. In an attempt to shortcircuit that worry, the NSF stressed that the primary basis for selection would remain the "competitive review of proposals" based upon peer review.[66] This is all to say that the program was designed to be attractive to both Congress and the scientific community.

Within the NSF, the Chemical Synthesis and Analysis Section of the Foundation's Chemistry/MPE Division administered all the RIF awards, which eventually led to observations both inside and outside the NSF that the RIF program was a turf and resources grab by the agency's chemists, including Pimentel. Perhaps anticipating such criticism, the RIF project announcement emphasized how the shared instrumentation would benefit a wide range of disciplinary approaches, not just chemistry. As the announcement read: "The Foundation recognizes that the research tools of science are rapidly assimilated into many disciplines; and that research and instrumentation go hand in hand in the creation of new knowledge."[67]

The RIFs were to be funded by four-year continuing grants, with grant renewals based on standard NSF policies, plus positive reports on wide user access--that is, upon the extent and quality of service provided. The NSF's 1978 "Regional Instrumentation Facilities Announcement" solicited proposals from all interested parties. The response was outstanding: over 160 proposals requesting more than $200 million.[68] While the agency circulated its request for proposals, it contracted the American Council on Education to survey the shared use of scientific equipment within the nation's institutions of higher education, the results of which were published in 1979.[69]

The regional instrumentation center idea was new for most chemists, and its reception was expectedly mixed, given the strong tradition within chemistry of the primacy of the individual investigator. The chemists welcomed increased federal expenditures for instrumentation, of course, but the preference for many researchers was to have the instrumentation furnished on a department by department basis, if not on an investigator by investigator basis. In its 1980 report to the NSF on *The Scientific Instrumentation Needs of Research Universities*, the Association of American Universities (AAU) addressed the NSF's regional instrumentation centers program. While granting that some extremely expensive instrumentation regional centers were necessary, the AAU expressed concern about other categories of instrumentation that were less expensive. Many of the researchers interviewed by the AAU perceived the centers as fostering "conservative science" because "instruments committed to a broad range of user needs cannot also be modified to meet highly specialized research requirements" and the considerable investment of time and money involved sometimes discourages researchers from undertaking "high-risk experiments" and encourages them to propose "'safer,' less innovative research." The researchers interviewed by AAU also pointed out that the costs of traveling to distantly-located centers can hamper the training of graduate students.[70]

A 1981 evaluation of the status of research university instrumentation commented on the NSF's regional instrumentation facilities program. "At some level of cost," the authors wrote, "it is clear that regional or national facilities are necessary--that the instruments required to support the research are simply too expensive to remain under the purview of the individual researcher, a single department, or even an institution. The high energy physics community has faced this problem for years." In their opinion, however, several factors worked against the effectiveness of RIFs, including "the time and expense of travel, inadequacy of the centers for training, and delays in obtaining experimental results." Moreover, they claimed, "the facilities are also perceived by many to foster conservative science because the time and expense involved in their use creates a reluctance to undertake high-risk experiments."[71]

The NSF organized a site visit to the Northeast Regional Center for Laser Spectroscopy and Dynamics at MIT in September 1981. All four members of the visiting committee gave high rating to the research performed at the center, although they differed in their assessment of the value of such centers as a mechanism to further research overall. Veronica Vaida, then an assistant professor of chemistry at Harvard University, expressed her "concern that funding of centers does not compete with individual funding. It is the flexibility of individual researchers that made science in the U.S. so much more productive than anywhere else." She added that "if funding of centers will compete with funding of individuals in the future, the number of deserving projects that cannot be supported will increase. One must remember also that a center provides only a part of the necessary equipment. To ensure originality of research, individual funding must continue."[72] Bryan E. Kohler, chairman of Wesleyan University Department of Chemistry, concurred. "I believe that we really saw very little that wouldn't have come to pass under the aegis of normal research grants," he said. While lasers "are by no means cheap," he said, "they are not so exotic or expensive as to require the strict rationing that seems to me to be part of the rationale for the Regional Center concept."[73] The other two reviewers saw no such problems and were highly supportive of the NSF continuing its support of RIFs.[74]

The National Research Council held a workshop on scientific instrumentation in 1982, which also addressed the pros and cons of centralized research facilities. The Council's report parroted several of the AAU's findings, including the concern with the possible negative effect of regional centers on the training of graduate students. Despite its stated concerns, the Council found more to commend the regional centers than had the AAU. "Central facilities may provide the researcher with greater flexibility in that he is free to use the facility that best meets his needs," it reported. What concerned the participants most was the ability to provide state-of-the-art instrumentation; not to rely on "standardized commercial instruments," and it was here that they worried that centers might not allow the opportunity for modifying the instruments to gain new advantages--something more commonly done to instruments in laboratories of individual investigators.[75]

Harry Gray of the California Institute of Technology spoke at the Council's workshop on "the instrumentation problem in chemistry." His comments reflect some telling aspects of the culture and tradition of chemists, and help to explain their reactions to regional centers. "The chemistry community is coming from a slightly different place than the people in physics, who have been used to big instrumentation for a long time," he said. "Chemists are a tougher bunch of characters who tend to fight and to squabble a lot, don't trust each other the way physicists trust each other, and are basically small-time operators."[76]

Robert Rosenzweig of the Association of American Universities commented on the RIF program in 1982. By that time, 14 centers had been established at a cost of roughly $20 million. In addition to regional centers devoted to the anticipated instrumentation of nuclear magnetic resonance, mass spectrometry, surface analysis, lasers, and high-resolution electron microscopy, a center offering isotope-dating and trace-element analysis instrumentation was also established. Rosenzweig generally applauded the idea of shared instrumentation, but he raised several concerns about the possible negative effects of such an approach. "Even for interdepartmental sharing within the same university," he wrote, "investigators may be uninformed about instrumentation newly acquired by other departments and may object to the need to leave their own laboratories to work in another building. When facilities are located off campus, investigators complain about the inconvenience and cost of traveling to other institutions or to regional instrumentation centers." He went on to list many of the same objections raised in the 1980 NSF/AAU report, but he claimed that "the central policy issue in the question of shared facilities and instrumentation is how best to deliver technology at the cutting edge of science to those researchers who can make the fullest possible contribution to scientific knowledge," and toward that goal, recommended that the NSF weigh all the problems associated with sharing against the benefits of the RIFs.[77]

By 1985, only seven Regional Instrumentation Facilities remained under the jurisdiction of the NSF's Chemistry Division; the remainder either no longer funded or transferred to other NSF Divisions. To help it assess how well this administrative experiment fared, the NSF had Albert B. Harvey, head of the Naval Research Laboratory's Chemical Diagnostics Branch, detailed to the Foundation for six months to conduct an extensive evaluation of the RIFs. Harvey visited all seven facilities, read all the NSF site visit review-panel reports, and interviewed dozens of RIF staff and administrators, as well as host and visiting researchers. He found the scientific productivity--measured in quality and quantity of science produced--at the RIFs to be "extremely successful," which led him to conclude that the NSF "should not only continue to support these useful facilities" but should also "encourage this type of shared resource even more." He recognized the strain, however, created by chemists who preferred operating their own labs, to which he argued that "the user-facility concept is now well-recognized and established throughout the scientific community and while chemists continue to operate and may continue to elect operating largely in the 'cottage industry' mode, the need for selected types of shared facilities is a *MUST*, if all chemists with good ideas are to be afforded the opportunity to make use of increasingly expensive analytical equipment and the expertise needed to utilize them to their fullest potential." This "cottage industry" mode, he said, was promoted by chemists (many of them prominent) who held strongly to the "one-person-one-instrument approach." Harvey attributed this preference to "the fact that chemist's analytical tools are not beyond the purchasing power of at least *some* principal investigators or departments." One of the key strengths of the RIFs, he concluded, was the ability to push the development of instrumental technology, promoting steady increases in state-of-the-art instrumental development.[78]

Despite the favorable evaluations of the RIF program given by Harvey and others, the split within the chemistry community over the desirability of providing instrumentation on a shared basis, and the departure of the RIF's organizational champion, NSF Deputy Director George Pimentel, led the Foundation's leadership to discontinue the formal Regional Instrumentation Facilities program, and to disperse the remaining active grants among its established divisions.

4. CONCLUSION

"Big Science" disciplines such as physics, astronomy, and oceanography long ago embraced the idea of shared instrumentation and research facilities, in large part because they had few options--the capital costs have been so high in those fields that few, if any, research institutions have been able to bear the complete financial burden. Chemistry, on the other hand, has yet to reach the scale of Big Science, with the result that many investigators remain skeptical of providing chemistry instrumentation through shared facilities. Most advanced instruments in chemistry remain within the financial bounds of at least some individual research institutions; nevertheless, the costs of acquisition, operation, and maintenance of state-of-the-art chemistry equipment continues to out pace the rate of inflation.

The history of National Science Foundation support of chemistry instrumentation provides numerous and varied examples of attempts to meet the capital needs of the nation's academic chemists. The Regional Instrumentation Facilities program, an experiment to stretch the NSF's limited resources, may prove a more attractive approach in the future if the U.S. continues to face times of restricted federal budgets. Under conditions of financial austerity, chemists may be forced to compromise their desire for independence in order to obtain broad access to the instrumentation they require.

5. ACKNOWLEDGMENTS

The author would like to thank Maria Grotz and Lyn M. Lawrence for their help as researchers, and J. Merton England, John D. Holmfeld, Marcel C. LaFollette, George T. Mazuzan, and Nathan Reingold for their comments and criticism.

6. REFERENCES

1. For a survey of instrumentation as an element of U.S. science policy, see Jeffrey K. Stine and Gregory A. Good, "Government Funding of Scientific Instrumentation: A Review of U.S. Policy Debates since World War II," *Science, Technology, & Human Values*, vol. 11, pp. 34-46, Summer 1986. Examples of instrumentation studies from the late 1970s and early 1980s include: Bruce L. R. Smith and Joseph J. Karlesky, *The State of Academic Science: The Universities and the Nation's Research Effort*, Change Magazine Press, New York, 1977; Irene L. Gomberg and Frank Atelsek, *Expenditures for Scientific Research Equipment at Ph.D.-Granting Institutions, FY 1978*, Higher Education Panel Report Number 47, American Council on Education, Washington, 1980; Association of American Universities, *The Scientific Instrumentation Needs of Research Universities: A Report to the National Science Foundation*, n.p., Washington, 1980; and Association of American Universities, *The Nation's Deteriorating University Research Facilities: A Survey of Recent Expenditures and Projected Needs in Fifteen Universities*, n.p., Washington, 1981. See also, Robert M. Rosenzweig, *The Research Universities and Their Patrons*, pp. 84-108, University of California Press, Berkeley, 1982.

2. Legislation passed in 1979 required the NSF to develop research equipment indicators, although these studies did not appear for several years. See U.S. General Accounting Office, *Studies of U.S. Universities' Research Equipment Needs Inconclusive*, GAO, Washington, 1984; Susan E. Cozzens, "Instrumentation," *Science & Technology Studies*, vol. 5, pp. 51-52, Summer 1987; and Linda S. Wilson, "The Capital Facilities Dilemma in the American Graduate School," in Bruce L. R. Smith (ed.), *The State of Graduate Education*, pp. 121-149, The Brookings Institution, Washington, 1985.

3. For a discussion of these general issues, see L. Berlowitz et al., "Instrumentation Needs of Research Universities," *Science*, vol. 211, pp. 1013-1018, March 6, 1981. For the case of nuclear magnetic resonance, see George C. Levy (ed.), *NMR Spectroscopy: New Methods and Applications*, American Chemical Society, Washington, 1982.

4. For an overview of post-World War II science policy, see Jeffrey K. Stine, *A History of Science Policy in the United States, 1940-1985*, Science Policy Study Background Report No. 1, House Committee on Science and Technology (99th Cong., 2nd sess., 1986); and Bruce L. R. Smith, *American Science Policy since World War II*, The Brookings Institution, Washington, 1990.

5. For developments in chemistry instrumentation, see Ernest Child, *The Tools of the Chemist: Their Ancestry and American Evolution*, Reinhold Publishing Corporation, New York, 1940; Fred W. McLafferty, "Trends in Analytical Instrumentation," *Science*, vol. 226, pp. 251-253, October 19, 1984; "The History of Chemical Instrumentation: An Introduction," Information

Pamphlet No. 8, Beckman Center for the History of Chemistry, Philadelphia, 1987; Herbert A. Laitinen and Galen W. Ewing (eds.), *A History of Analytical Chemistry*, American Chemical Society, Washington, 1977; and John T. Stock and Mary Virginia Orna (eds.), *The History and Preservation of Chemical Instrumentation*, D. Reidel Publishing Company, Dordrecht, Holland, 1986.

6. Remarks by Linus C. Pauling in *The Crisis Facing American Science*, edited proceedings of a town meeting, p. 6, New York Academy of Sciences, New York, 1968. See also, Philip H. Abelson, "Chemistry Without Test Tubes," *Science*, vol. 224, p. 1297, June 22, 1984.

7. John K. Taylor, "The Impact of Instrumentation on Analytical Chemistry," in Stock and Orna, *The History and Preservation of Chemical Instrumentation*, p. 3. See also, McLafferty, "Trends in Analytical Instrumentation," pp. 251-253; and Dean Stanley Tarbell and Ann Tracy Tarbell, *Essays on the History of Organic Chemistry in the United States, 1875-1955*, pp. 335-352, Folio Publishers, Nashville, TN, 1986.

8. Taylor, "The Impact of Instrumentation on Analytical Chemistry," p. 5.

9. Waterman's statement in U.S. Senate, Committee on Appropriations, *Independent Offices Appropriations, 1957*, p. 380 (84th Cong., 2nd sess., 1956). See also, Waterman's statement in U.S. House of Representatives, Committee on Appropriations, Subcommittee on Independent Offices, *Independent Offices Appropriations for 1958*, Part 2, pp. 1337-39 (85th Cong., 1st sess., 1957).

10. National Science Foundation, *Sixth Annual Report for the Fiscal Year Ended June 30, 1956*, p. 54, NSF, Washington, 1956.

11. NSF, *Sixth Annual Report*, p. 55.

12. See NSF, *Sixth Annual Report*, pp. 55-59.

13. Senate Committee on Appropriations, *Independent Offices Appropriations, 1957*, pp. 380-381.

14. See U.S. House of Representatives, Committee on Appropriations, Subcommittee on Independent Offices, *Independent Offices Appropriations for 1958*, Part 2, p. 1337 (85th Cong., 1st sess., 1957).

15. National Science Foundation, *Federal Financial Support of Physical Facilities and Major Equipment for the Conduct of Scientific Research*, A Report to the Bureau of the Budget, p. 2, GPO, Washington, June 1957.

16. NSF, *Federal Financial Support of Physical Facilities and Major Equipment*, p. 2.

17. NSF, *Federal Financial Support of Physical Facilities and Major Equipment*, pp. 42, 47, 50.

18. NSF, *Federal Financial Support of Physical Facilities and Major Equipment*, p. 43.

19. NSF, *Federal Financial Support of Physical Facilities and Major Equipment*, pp. 43-46.

20. The number and names of these divisions changed over time. For example, the MPE became the Mathematical and Physical Sciences division in the mid-1960s when engineering broke off to form its own division. For the early history of the NSF, see J. Merton England, *A Patron for Pure Science: The National Science Foundation's Formative Years, 1945-57*, NSF, Washington, 1982.

21. For an example of the issues addressed by this panel, see Agenda, Fourth Meeting of Advisory Panel for Chemistry, 7 October 1954 (Office of the Director Subject Files/1951-56 folder, box 8, Record Group [RG] 307, National Archives [NA], Washington, D.C.).

22. For an example of the MPE's concern, see Minutes of the 12th Meeting of the Divisional Committee for Mathematical, Physical, and Engineering Sciences, 7-8 March 1957 (Office of the Director Subject Files/1960-61 folder, box 44, RG 307, NA).

23. Ralph A. Beebe, "Report on the Chemistry Program," in Minutes of the 13th Meeting of the Divisional Committee for Mathematical, Physical, and Engineering Sciences, 24-25 October 1957 (MPE/MPS Divisional Committee Minutes/1952-1965 folder, NSF Historian's file, National Science Foundation, Washington, D.C.).

24. See justification for the $15,000,000 chemical research instruments line item request in the fiscal year 1965 NSF budget prepared by MPE's Chemistry Section; copy attached to John D. Roberts to Leland J. Haworth, 13 August 1963 (MPS-Chemistry folder, box 3, NSF Office of the Director subject files, 307-75-051, Washington National Records Center [WNRC], Suitland, Maryland).

25. Minutes of the 21st Meeting of the Divisional Committee for Mathematical, Physical, and Engineering Sciences, 30-31 October 1961 (MPE/MPS Divisional Committee Minutes/1952-1965 folder, NSF Historian's file, NSF).

26. Walter R. Kirner, "Federal Support of Basic Research in Chemistry," March 27, 1964. (This unpublished report was an appendix to the Minutes of the Divisional Committee for Mathematical, Physical, and Engineering Sciences meeting, 27-28 March 1964; MPE/MPS Divisional Committee Minutes/1952-1965 folder, NSF Historian's file, NSF).

27. William von E. Doering to Geoffrey Keller, 1 August 1962 (Office of the Director Subject Files/1963 folder, box 76, RG 307, NA).

28. Doering discusses this in his 1 August 1962 letter to Geoffrey Keller.

29. A copy of Doering's 9 February 1962 survey letter is contained in Office of the Director Subject Files/1963 folder, box 76, RG 307, NA.

30. A copy of Doering's 16 May 1962 standard response letter is contained in Office of the Director Subject Files/1963, box 76, RG 307, NA.

31. Geoffrey Keller to William von E. Doering, 31 August 1962 (Office of the Director Subject Files/1963 folder, box 76, RG 307, NA). See also, William von E. Doering to Alan T. Waterman, 17 October 1962 (Office of the Director Subject Files/1963 folder, box 76, RG 307, NA).

32. Leland J. Haworth to John D. Roberts, 30 July 1963 (MPS-Chemistry folder, box 3, NSF Office of the Director subject files, 307-75-051, WNRC).

33. John D. Roberts to Leland J. Haworth, 13 August 1963 (MPS-Chemistry folder, box 3, NSF Office of the Director subject files, 307-75-051, WNRC).

34. Roberts to Haworth, 13 August 1963.

35. "Federal Support of Basic Chemistry," An Independent Report Prepared by Members of the Advisory Panel in Chemistry of the National Science Foundation, October 1963, p. 2. (The 16-page typescript report is in the MPS-Chemistry/1964 folder, Box 9, NSF Office of the Director Subject Files, 70A-3621, WNRC.)

36. "Federal Support of Basic Chemistry," p. 3. For a detailed account of these positions, see George B. Kauffman and H. Harry Szmant (eds.), *The Central Science: Essays on the Uses of Chemistry*, Texas Christian University Press, Fort Worth, 1984.

37. "Federal Support of Basic Chemistry," pp. 11-12.

38. "Federal Support of Basic Chemistry," p. 13.

39. See, for example, Earl Frieden to Donald F. Hornig, 22 July 1964; and Hornig to Frieden, 28 August 1964 (MPS-Chemistry/1964 folder, box 9, NSF Office of the Director subject files, 70A-3621, WNRC).

40. Robert B. Carlin to Leland J. Haworth, 9 July 1964 (MPS-Chemistry/1964 folder, box 9, NSF Office of the Director subject files, 70A-3621, WNRC).

41. William N. White to Leland J. Haworth, 2 June 1964 (MPS-Chemistry-/1964 folder, box 9, NSF Office of the Director subject files, 70A-3621, WNRC).

42. J. F. Bunnett to Leland J. Haworth, 16 March 1964 (MPS-Chemistry/1964 folder, box 9, NSF Office of the Director subject files, 70A-3621, WNRC).

43. Leland J. Haworth's statement in U.S. House of Representatives, Committee on Appropriations, Subcommittee on Independent Offices, *Independent Offices Appropriations for 1965*, Part 2, p. 757 (88th Cong., 2nd sess., 1964).

44. The Westheimer Report is the informal name for: Committee for the Survey of Chemistry, National Academy of Sciences/National Research Council, *Chemistry: Opportunities and Needs*, National Academy of Sciences, Washington, 1965.

45. Draft memorandum, Arthur F. Findeis to NSF Chemistry Section Head, 10 October 1967 (A-C folder, box 1, NSF Office of the Director subject files, 307-75-051, WNRC). See also, memorandum, M. Kent Wilson to NSF Director, 23 July 1968, (MPS-Chemistry folder, box 3, NSF Office of the Director subject files, 307-75-051, WNRC).

46. U.S. House of Representatives, Committee on Government Operations, *Better Management of Research Equipment Procurement and Utilization in Federal Laboratories*, Report No. 867, p. 4 (90th Cong., 1st sess., 1967). See also, "Report of the CASE Subcommittee on Facilities," attached to letter of Nathaniel H. Karol to Leland Haworth, 20 July 1967 (CASE Subcommittee on Facilities/1967-69 folder, Box 5, NSF Office of the Director subject files, 307-75-052, WNRC).

47. Vannevar Bush before the U.S. House of Representatives, Select Committee on Government Research (November 21, 1963), quoted in House Committee on Government Operations, *Better Management of Research Equipment*, p. 4.

48. See Smith, *American Science Policy*, pp. 82-83.

49. Leland J. Haworth to Senator John O. Pastore, 30 June 1969, reprinted in U.S. House of Representatives, Committee on Appropriations, Subcommittee on Independent Offices and Department of Housing and Urban Development, *Independent Offices and Department of Housing and Urban Development Appropriations for 1970*, Part 2, p. 1021 (91st Cong., 1st sess., 1969). For examples of the internal debate at the NSF over the possible effects of reduced budgets on the Foundation's support of instrumentation, see memorandum, Leland J. Haworth to Members of the National Science Board, 27 January 1969 (Advisory Committee for Mathematical and Physical Sciences folder, box 2, NSF Office of the Director subject files, 307-75-052, WNRC).

50. National Science Board, *The Physical Sciences*, pp. xi-xii, GPO, Washington, 1970.

51. National Academy of Sciences, "An Assessment of the Needs for Equipment, Instrumentation, and Facilities for University Research in Science and Engineering," a report submitted to the National Science Foundation, September 1971 (a copy of this unpublished report is contained in the Instrumentation Survey folder, box 4, NSF Office of the Director subject files, 307-77-080, WNRC).

52. Cheves Walling, David M. Grant, Jean H. Futrell, and Ronald J. Pugmire, *A Study to Improve the Management of Costly Instrumental Centers*, a report to the National Science Foundation, pp. ii and 1, University of Utah, Department of Chemistry, Salt Lake City, n.d. (probably 1975).

53. Walling et al., *A Study to Improve the Management of Costly Instrumental Centers*, p. 5.

54. See U.S. House of Representatives, Committee on Science and Technology, Subcommittee on Science, Research and Technology, *1976 National Science Foundation Authorization* (94th Cong., 1st sess., 1975); and U.S. House of Representatives, Committee on Appropriations, Subcommittee on HUD-Independent Agencies, *Department of Housing and Urban Development-Independent Agencies Appropriations for 1976*, Part 1 (94th Cong., 1st sess., 1975).

55. Pimentel's statement in U.S. House of Representatives, Committee on Appropriations, Subcommittee on HUD-Independent Agencies, *Department of Housing and Urban Development-Independent Agencies Appropriations for 1979*, Part 5, pp. 87-88 (95th Cong., 2nd sess., 1978).

56. U.S. House of Representatives, Committee on Science and Technology, *Authorizing Appropriations to the National Science Foundation*, Report No. 95-993, p. 24 (95th Cong., 2nd sess., 1978).

57. U.S. House of Representatives, Committee on Science and Technology, *Authorizing Appropriations to the National Science Foundation*, Report 96-61, p. 30 (96th Cong., 1st sess., 1979). See also, U.S. General Accounting Office, *Minimum Requirements Are Needed for Colleges and Universities to Justify Research Equipment Purchases*, GAO, Washington, 1978. In 1981, NSF Director John B. Slaughter told the Committee: "It is difficult to quantify the instrumentation problem in very precise terms. We have at this Committee's request started to develop better indicators of the state of scientific instrumentation. . . . But while precise quantification may not be feasible, there is a great deal of evidence that the problem is both real, and large. Various estimates of the magnitude of the need have been made, but the *minimum* amount necessary to catch up is probably several hundred million dollars." U.S. House of Representatives, Committee on Science and Technology, Subcommittee on Science, Research and Technology, *1982 National Science Foundation Authorization*, p. 710 (97th Cong., 1st sess., 1981).

58. Traxler's and Creutz's statements both appear in U.S. House of Representatives, Committee on Appropriations, Subcommittee on HUD-Independent Agencies, *Department of Housing and Urban Development-Independent Agencies Appropriations for 1978*, Part 1, p. 377 (95th Cong., 1st sess., 1977).

59. See National Science Board, 195th Meeting Book, 19-20 January 1978 (NSB Records, National Science Foundation, Washington, D.C. [hereafter cited as "NSB, 195th Meeting Book"]).

60. Memorandum, Richard C. Atkinson to Members of the National Science Board, 4 January 1978 (NSB, 195th Meeting Book).

61. Atkinson to Members of the National Science Board, 4 January 1978 (NSB, 195th Meeting Book).

62. "Regional Facilities," p. 1, contained in NSB, 195th Meeting Book.

63. "Regional Facilities," p. 2, in NSB, 195th Meeting Book.

64. See "Regional Facilities," p. 2, in NSB, 195th Meeting Book; draft "Project Announcement: Regional Instrumentation Facilities," in NSB, 195th Meeting Book; and Rosenzweig, *The Research Universities and Their Patrons*, p. 106.

65. "Regional Facilities," p. 2, in NSB, 195th Meeting Book.

66. "Regional Facilities," p. 3, in NSB, 195th Meeting Book.

67. Draft "Project Announcement: Regional Instrumentation Facilities," in NSB, 195th Meeting Book.

68. See National Science Board, 209th Meeting Book, 20-21 September 1979, p. 79-310-3.

69. See Frank J. Atelsek and Irene L. Gomberg, *Shared Use of Scientific Equipment at Colleges and Universities, Fall 1978*, Higher Education Panel Report No. 44, American Council on Education, Washington, November 1979.

70. Association of American Universities, *The Scientific Instrumentation Needs of Research Universities*, p. 8.

71. Berlowitz et al., "Instrumentation Needs of Research Universities," p. 1018.

72. Veronica Vaida to A. F. Findeis, 28 September 1981 (MIT/Michael S. Feld/CHE-8217705 folder, box 17, accession 307-89-269, WNRC).

73. Bryan E. Kohler to Arthur F. Findeis, 28 September 1981 (MIT/Michael S. Feld/CHE-8217705 folder, box 17, accession 307-89-269, WNRC).

74. See W. Happer to Arthur F. Findeis, 21 September 1981; and William M. Jackson to Arthur Findeis, 22 September 1981 (MIT/Michael S. Feld/CHE-8217705 folder, box 17, accession 307-89-269, WNRC). For other positive assessments of the RIFs providing access to lasers, see "Labscan: The MIT Regional Laser Center," *Laser Focus*, pp. 31-34, April 1982; and Lewis Holmes, "Regional Centers Support Laser-Based Research," *Laser Focus*, pp. 78-84, November 1985.

75. National Research Council, *Revitalizing Laboratory Instrumentation*, p. 4, National Academy Press, Washington, 1982).

76. Harry Gray's comments in National Research Council, *Revitalizing Laboratory Instrumentation*, p. 40.

77. Rosenzweig, *The Research Universities and Their Patrons*, pp. 107-108.

78. Albert B. Harvey, "Special Report on Regional Instrumentation Facilities," October 1985 (my thanks to Albert Harvey [currently with the NSF's Division of Electrical & Communications Systems] for providing me a copy of this unpublished report to the NSF).

Chapter 13

The NSF surveys of academic research instrumentation and academic research facilities: a study in data collection and analysis and policy formulation

Carlos Kruytbosch
Science Indicators Ltd. & National Science Foundation

Abstract

This paper tracks the evolution of public support for scientific instrumentation from an undifferentiated part of general growth of publicly supported R&D, to a separate and publicly visible sector of R&D policy and support. Comparisons are made to parallel policy developments in the area of academic research facilities.

The emergence of a complex data system about instrumentation is described. An Appendix displays examples from the NSF 1989/90 survey of the kinds of policy-relevant information now regularly generated about the supply, utilization and condition of U.S. academic research instrumentation.

Introduction

A "rational" model of the policy formulation and decision process can be depicted as: problem sensing --> data gathering --> data analysis --> formulation of policy options/solutions --> cost/benefit analyses --> decision. An examination of the sequence of events during the 1970's and 1980's concerning public policy about provision of support for academic research instrumentation suggests a rather different process. Further, a comparative look at the recent evolution of Federal Government policy in the area of academic research facilities yields yet another pattern of data gathering and decision making.

Academic Research Instrumentation as a Policy Issue

This discussion deals with developments in instrumentation policy during the 1970's and 1980's on the grounds that it was only during this period that the issue became truly a part of a public policy debate. Prior to this time there had also been important developments in instrumentation policy and programs, but it can be argued that these matters were confined almost exclusively to the scientific communities involved. The role of instrumentation in the grants programs at NSF during the first two decades of the Foundation's existence is well documented in the paper by Jeffrey K. Stine elsewhere in this book.(pp. 238) Additional material on chemistry instrumentation is in W.Henry Lambright, "The National Science Foundation and Decisions on Chemistry Instrumentation", unpublished report to the National Science Foundation, June 1983.

The major landmarks in the emergence of academic research instrumentation as a public policy issue are depicted in Chart 1.

The nearly two-decade long period of continuous expansion of Federal research funding came to an abrupt and unexpected halt in 1968. By the 1973 recession, taking inflation into account, growth in Federal (as well as industrial) R&D support actually became negative. These trends occasioned much worry in academic and science policy circles. They were comprehensively documented in an NSF supported study carried out by the American Association of Universities (Bruce Smith and Joseph Karlesky, The State of Academic Science, New Rochelle, NY: Change Magazine Press, 1977). The study was cited in President Carter's State of the Union Message in 1978 as part of his program to continue restoration of adequate growth in funding for basic research -- a process already begun under President Ford in 1975.

CHART 1

INSTRUMENTATION AS A PUBLIC POLICY ISSUE: SOME KEY DATES

1950 - 1968 Continuous growth in Federal R&D funding

1968 - 1974 Real decrease in Federal R&D support -- esp. for academic R&D facilities and trainee/ fellowships

1971 National Academy of Sciences study of equipment, instrumentation and facilities needs in ten disciplines.

1976 - 1978Internal NSF studies on instrumentation support show decline to well below 10% of total grant funds.

1977 Smith and Karlesky, The State of Academic Science, documents funding trends, and focuses attention on instrumentation and facilities. Cited in President Carter's State-of-the-Union Address.

1977 - 1980 Congress supplies $50-100M "add-on" budget annually for instrumentation. NSF initiates new programs - "Regional", "Departmental" "Biological" Instrumentation Programs.

1980 Congress directs NSF to conduct a definitive survey of academic instrumentation needs.
NSF conducts interim NSF/AAU instrumentation study

1981 - 1983 Academic research instrumentation programs launched by ONR and DOE

1984 First systematic NSF data available.

1986/87 Subsequent NSF instrumentation 1989/90 Surveys completed

The Smith and Karlesky study devoted considerable attention to the issue of the decline of supporting resources for research, especially "emerging instrumentation needs" (see pp.166-172). They cited internal NSF studies and the National Science Board's Science Indicators - 1974 showing significant declines in the proportion of both NSF and NIH (National Institutes of Health) grant funds allocated for purchase of permanent equipment -- e.g., NSF 11.2% and NIH 11.7% in 1966 to NSF 5.4% and NIH 5.7% in 1974 (p.169). Smith and Karlesky hypothesized that, "The drop in support for equipment may be viewed as a natural outcome of recent trends. When funds are tight, agency officials are tempted to allocate money to essential personnel costs and to postpone acquisition of new equipment. Investigators, too, have been reluctantly willing to postpone replacement or to neglect adequate maintenance.... [thus]...with a research life of three to six years, middle-range instruments are wearing out and becoming obsolete". (p.169)

Similar conclusions had been reached earlier in a 1971 study conducted by a panel of the National Academy of Science, "Survey of Research Equipment Needs in Ten Academic Disciplines", Evaluation Staff, OBPA/NSF, February 1972.According to Stine (p.24) the study was initiated by the NSF in response to pressures from the research community for increased Federal support for instrumentation. The study was never published and apparently did not receive much public notice. It was not even cited in the Smith and Karlesky work. As I recall in discussions in the NSF Director's Planning and Policy Office (in which I was a staff member) during the late 1970's about how to design a study of instrumentation needs, the NAS study was seen as lacking in credibility because the materials on each of the 10 disciplines was assembled and presented by a committee of persons from that discipline. Thus, there was no check on the vested interest of that discipline in maximizing its needs.

The cumulative effect of rising "noise" levels from the research community combined with several partial studies showing reduced funding levels for instrumentation in recent years, persuaded the Congressional Committees overseeing NSF to include substantial budget add-ons of $50 to $100 million for instrumentation from 1978 to 1980. By the latter year, however, the House Committee on Science, Research and Technology began to experience difficulties in defending the instrumentation increases before their colleagues during the full House floor debates. For example, one theme of the critics was that the increases merely reflected academic "gold-plating". The Committee returned to NSF with demands for data and explicit instructions to undertake a full-fledged survey of the supply, utilization and needs for scientific research instrumentation.

A complete survey would require considerable lead time. The NSF managers sensing the urgency in the Congressional demands launched a quick survey in collaboration with the American Association of Universities -- see American Association of Universities, The Scientific Instrumentation Needs of Research Universities, Washington, D.C., June 1980. Multidisciplinary teams of interviewers visited over 100 leading university laboratories, and for comparison several commercial, government and national laboratories were also visited. Comparisons of the age of equipment in the industrial and academic laboratories showed that the equipment being used in the top-ranked universities had a median age twice that of the instrumentation available in leading industrial laboratories. Further, the report contained evidence of the rapidly increasing costs of equipment attributable to the "sophistication factor". The costs of equipping a laboratory for organic chemistry capable of conducting forefront research increased by a factor of 4 (or two in constant dollars) between 1970 and 1980.

The budgetary add-ons for instrumentation in the late 1970's made it possible for NSF to initiate several new instrumentation programs -- augmentation of the Chemistry Departmental Instrumentation Program, a new Regional Instrumentation Program which was heavily weighted toward chemistry departments, and a new Program for Biological Instrumentation.

During the period from roughly 1980 to 1983 the rising tide of "noise" from the research community and pieces of evidence from special studies were also accompanied by a new drive on the part of military research agencies to rebuild their ties with the academic research community, which had languished since the so-called "Mansfield Amendment" of 1968. Amongst other academically oriented initiatives, the Office of Naval Research mounted a widely advertised competitive program for academic research instrumentation. A similar program was launched in the Department of Energy, Division of Basic Energy Sciences.

Meanwhile, driven by the Congressional mandate of PL 96-44, the NSF's Division of Science Resources Studies was devising and implementating an innovative study design to capture data on the supply, utilization, maintenance and needs for academic research instrumentation. It was fortuitous for the study that a new phase had just been completed in the continuing saga of the relationships between the research universities and the Federal Office of Management and Budget concerning the regulations for reimbursement of indirect costs of Federally sponsored research (OMB Circular A-21). It so happened that a revision in this Circular required universities to maintain central inventories of equipment purchased with Federal funds. This made possible the collection of rich information on the characteristics of instruments in the inventories. (See also U.S. General Accounting Office, Studies of U.S. Universities' Research Needs Inconclusive. Washington, D.C.: U.S.G.A.O., April 3, 1984. GAO/RCED-84-105.)

The NSF instrumentation survey of 1983/84 initiated a triennial cycle of surveys, with follow-up studies being conducted in 1986/87 and 1989/90. The subsequent surveys showed significant gains being made in the total inventory of academic research equipment, as well as the average age and state of repair of existing stock. It seems fair to conclude that the period of crisis in academic research equipment had been successfully surmounted. The system had now entered a period of stable funding for this function at an appropriate level, and this level was being systematically monitored by a sophisticated survey mechanism.

It is interesting to note that Congress had already taken key measures to restore instrumentation funding levels before the availability of comprehensive data on the topic. The NSF surveys essentially served to validate and corroborate decisions already made.

A brief comparison with the policy history of the issue of academic research facilities shows quite a different pattern of development.

The Case of Academic Research Facilties

Chart 2 presents some key events in the recent (since 1980) history of the issue of academic research facilities, and compares them with parallel phases in the instrumentation issue.

Around 1980 there was a discernible increase in the level of "policy noise" from the academic community about research facilities. This coincided with a historic low in capital funds expenditures for academic science and engineering (See National Science Board, Science and Engineering Indicators - 1989, Washington, D.C.: U.S. Government Printing Office, 1990. Pp. 112, 310). Total expenditures in 1979 were only 28% of the level of expenditures in 1968, in constant dollars. The Federal share in 1979 was just over 20%, as compared to over 30% in 1968.

The 1980 AAU/NSF study of academic research instrumentation inadvertently turned up some spectacular instances of literally "crumbling" academic research facilities. These findings prompted the AAU to undertake an independent pilot study of facilities needs at 15 research universities. (See, American Association of Universities, The Nation's Deteriorating University Research Facilities, Washington, D.C.: AAU, July 1981).

CHART 2

COMPARISON OF THE EVOLUTION OF "INSTRUMENTATION" AND "FACILITIES" AS POLICY ISSUES AND AS OBJECTS OF DATA COLLECTIONS

INSTRUMENTATION	FACILITIES
PHASE I +/- 1975 "Noise" Testimony Small agency studies	+/-1981 "Noise" Small studies: AAU '81, "Crumbling Academe" '84, ISCARF '84, DOD '84. NAS Symposium '84
PHASE II 1977-1980	1983-1986
Congress responsive--appropriates $50-100M to NSF. Congress demands data 1979.	a)Federal agencies not responsive-- NSF caught between academia and OMB NSF/AAU study in 1980 b) OMB: facilities not a Federal role. c) Congress responds with "Floor Amendments" => $300 million a year.
PHASE III 1983	1987
Data base established, but problem is resolved Enter period of routine monitoring	Congress directs policy NSF/SRS to do facilities survey. NSF initiates nominal Facilities program Data base established

Policy problem remains.

The following three years saw an efflorescence of partial studies (University of California, The Federal Role in Higher Education Facilities Funding: A Case Study of the University of California. 1981. Ad Hoc Interagency Steering Committee on Academic Research Facilities, Adequacy of Academic Research Facilities: A Brief Report of a Survey of Recent Expenditures and Projected Needs in Twenty-Five Academic Institutions. Washington, D.C.: NSF, April 1984. National Science Board, Academic Science & Engineering Physical Infrastructure, Vol.I, Issues and Problems, and Vol.II, Background Materials. Washington, D.C., National Science Board, June 1984. NSB-84-160. H.Kaiser, Crumbling Academe, Washington, D.C.: Association of Governing Boards of Universities, 1984. Department of Defense, Selected University Laboratory Needs in Support of National Security. April 29, 1985. Armand Hammer Foundation/American Cancer Society, Survey of Cancer Research Facilities Needs. April 1985. U.S. House of Representatives, Committee on Science and Technology, Science Policy Task Force, Hearings on, "The Federal Government and the University Research Infrastructure", May 22, 1985.

This phase of the issue development culminated in a widely attended Conference on Academic Research Facilities in July 1985, sponsored by the National Academy of Sciences. All the studies concluded that there was indeed a pressing need for renewal of the academic physical plant for research, and most stated that the Federal government should play a leading role. A review of the partial studies in the Proceedings of the Academy Conference concluded that they all came up with a five-year need for expenditures in the region of $1.5 billion annually.

As in the case of instrumentation, the policy noise level became deafening. However, the issues diverged at this point. In instrumentation, the Federal policy establishment as a whole had been unanimous, even before the existence of comprehensive data, that there was a critical need for instrumentation and that it was appropriate for Federal R&D supporting agencies to play a major role. With facilities, all agreed upon the need. However, while the Congressional committees generally were sympathetic with the pleas of their academic constituents, the President's Office of Management and Budget rejected a significant Federal role in supporting construction of academic research facilities, except in certain limited cases of "national" facilities. The National Science Foundation thus truly found itself "between a rock and a hard place". There was no way it could please its three major reference groups -- the Congress, the OMB and the academic community. NSF saw a major danger in the possibility of Congress passing legislation requiring significant expenditures on facilities, but not appropriating the necessary funds, thereby necessitating NSF's regular research support programs to "eat" the difference.

Unlike the case with instrumentation, no major facilities program was launched. A variety of responses occurred. The NSF was permitted by OMB to initiate a nominal facilities program with budget of a few million dollars. The Congress responded to the pleas of its constituents by passing numerous "floor amendments"directly appropriating funds for particular academic R&D facilities within individual Congressional districts. When summed, these projects add into the hundreds of millions of dollars. (See, e.g., C. Kruytbosch, "The Role and Effectiveness of Peer Review", in Ciba Foundation, The Evaluation of Scientific Research. Chichester, U.K.: Wiley, 1989, esp. pp.74-78.)

All parties were agreed upon the need for a credible data base. Congress ordered such a study in its NSF Authorization Act of 1985 (PL 99-159, Section

108), and under the watchful eye of OMB, NSF's Division of Science Resources Studies produced its first report, Science and Engineering Research Facilities at Doctorate-Granting Institutions. Washington D.C.: National Science Foundation, September, 1986. The results of that preliminary study, as well as of NSF's subsequent facilities surveys in 1988 (NSF 88-320) and 1990 (NSF 90-318), are interpreted by the parties as validating their original positions. Those who reject a Federal role point to the large increases in facilities investments being made from state, private and institutional sources, and claim the problem is being solved without major Federal intervention. The pro-Federal role parties maintain that the problems are not being solved rapidly enough, and that the absence of duly funded Agency facilities programs has fostered the Congressional practise of bypassing Federal agency budget proposals before Committees and making appropriations by "floor amendment". In other words, the policy disagreements persist, and the development and availability of credible data about the phenomenon have not changed policy positions.

In conclusion, this brief review of the history of the two issues suggests that the positions of the parties on the issues did not change after the creation of credible data bases. In each case the new data was seen as validating earlier positions. However, despite this outcome in the short run, it may be that in the longer run, the ability to track future development of the phenomena could foster changed perceptions and definitions of the policy issues.

APPENDIX

Selected Tables from the 1989/90 NSF Academic Research Equipment Surveys

Source: National Science Foundation, Academic Research Equipment and Equipment Needs in Selected Science and Engineering Fields: 1989/90. (NSF 91-311), May 1991.

The survey data are based on multi-stage samples of 11,575 instruments, from 909 departments and research centers at 79 institutions -- 55 colleges and universities and 24 medical schools.

A. Trends in expenditures for equipment

1. Trends in annual expenditures for academic research equipment, 1982-83 to 1988-89.

2. Expenditures for purchase of academic research equipment as a percentage of total R&D expenditures, by field, 1982-83 to 1988-89.

B. Trends in sources of funds for equipment

8. Sources of funds for acquisition of academic research equipment, 1982-89.

C. Departmental priority equipment needs

10. Cost and intended use of departments' top-priority research equipment needs, by field, 1989-90.

D. Trends in numbers and prices of equipment systems

13. Trends in aggregate ammounts and average prices of academic research equipment, by type, 1985-86 to 1988-89.

Table 1. Trends in annual expenditures for academic research equipment, 1982-83 to 1988-89

Type of expenditure and field	Survey		
	1982-83	1985-86	1988-89
	(Dollars in millions)		
Purchase of nonexpendable research equipment costing $500+, total	400	678	831
Engineering	96	174	253
Agricultural sciences	28	34	46
Biological sciences	132	194	246
Chemistry	39	81	84
Computer science	20	49	45
Environmental sciences	33	55	55
Physics/astronomy	52	91	102
Maintenance/repair of existing research equipment, total	101	149	175
Service contracts and field service	*	72	87
Other (salaries, tools, etc.)	*	77	87
Operation of existing research equipment (supplies, technician salaries, etc.), total	*	*	403

*Not ascertained in that survey.

NOTE: Details may not sum to totals due to rounding.

SOURCE: National Science Foundation, SRS

Table 2. Expenditures for purchase of academic research equipment as a percentage of total R&D expenditures, by field, 1982-83 to 1988-89

Type of expenditure and field	Percent of total R&D for purchase of research equipment*		
	1982-83	1985-86	1988-89
Total	9	11	10
Engineering	9	12	12
Agricultural sciences	3	3	4
Biological sciences	10	11	10
Chemistry	13	19	15
Computer science	13	17	11
Environmental sciences	6	8	6
Physics/astronomy	12	14	12

*Total R&D expenditures for 1982, 1985, and 1988 were obtained from *Academic Science/Engineering: R&D Funds, Fiscal Year 1988*, National Science Foundation, NSF 89-326, p. 17.

SOURCE: National Science Foundation, SRS

Table 8. Sources of funds for acquisition of academic research equipment, 1982-89*

Source of funds	1982-83	1985-86	1988-89
	(Dollars in millions)		
Total (aggregate purchase price)	$1,311	$2,044	$3,177
	(Percent of total)		
Federal, total	53	49	48
NSF	18	17	15
NIH	14	14	14
Department of Defense	8	7	10
Department of Energy	5	5	4
Other (NASA, USDA, etc.)	7	5	4
Non-Federal, total	47	51	52
Institution funds	30	30	31
State government	5	6	9
Private/industry	11	12	11
Other	2	2	1

*Base = aggregate purchase price of all instrument systems costing $10,000-$999,999 that were in research use during the specified year.

NOTE: Details may not sum to totals due to rounding.

SOURCE: National Science Foundation, SRS.

Table 10. Cost and intended use of departments' top-priority research equipment needs*, by field, 1989-90

Field	Median price	Intended use			
		Total	Replace an existing instrument	Expand capacity (more copies of existing equipment)	Upgrade capabilities (to perform experiments you cannot do now)
		(Percent of aggregate price of all reported needs)			
Total	$80,000	100	17	25	58
Engineering	100,000	100	13	30	57
Agricultural sciences	35,000	100	15	32	54
Biological sciences	50,000	100	18	28	54
Chemistry	150,000	100	17	21	62
Computer science	200,000	100	18	26	56
Environmental sciences	100,000	100	27	17	56
Physics/astronomy	200,000	100	13	11	76

*From lists of three items per department/center of top-priority research instruments in $10,000-$1 million range. Needs lists include some items over $1 million and some that call for multiple copies (e.g., 10 workstations-$250,000).

NOTE: Percentages may not sum to 100 due to rounding.

SOURCE: National Science Foundation, SRS.

Table 13. Trends in aggregate amounts and average prices of academic research equipment, by type, 1985-86 to 1988-89*

Type of equipment	Number of systems			Aggregate purchase price		Mean price per system	
	1985-86	1988-89	Percent change	1985-86	1988-89	1985-86	1988-89
				(Dollars in millions)		(Dollars in thousands)	
Total	53,390	78,950	50%	$2,040	$3,180	$38	$40
Computers and data handling equipment	10,530	17,420	65	520	780	49	45
Graphics/CAD/Image analysis	960	2,540	160	40	100	38	40
Other systems ≥ $200,000	250	490	100	90	180	378	354
Other systems/components	9,320	14,390	50	390	510	42	35
Spectrometers and light measurement equipment	8,910	12,510	40	450	700	50	56
NMR/magnets	1,120	1,060	10	100	160	93	152
Mass/GC-MS	640	920	40	70	120	115	133
X-ray (diffractometers, etc.)	840	1,220	50	50	90	64	75
Electron/auger/ionscattering	400	960	140	50	100	123	100
Other spectroscopy equipment	5,910	8,350	40	170	230	28	28
Bioanalytical instruments	18,450	25,450	40	520	820	28	32
Centrifuges	4,780	6,770	40	110	200	24	29
Electron microscopes	1,250	1,220	0	130	150	105	120
Other/unspecified microscopy	2,950	4,230	40	70	110	22	25
Chromatographs (except GC-MS)	3,700	5,280	40	70	120	20	22
Protein synthesizers/sequencers	680	1,440	110	30	100	48	70
Scintillation/radiation counters	4,120	4,290	0	80	80	19	20
Growth/environmental chambers	800	1,880	130	20	50	26	26
Cell sorters, cytometers	180	330	80	10	20	61	72
Other	15,500	23,580	50	550	870	36	37
Lasers, motion/optical analyzers	2,200	4,270	90	100	200	44	47
Electronics (cameras, recorders, etc.)	3,210	4,910	50	90	130	27	26
Temperature and pressure control/measurement equipment	2,140	3,840	80	70	90	31	24
Major prototypes systems ≥ $100,000**	250	590	130	70	150	287	248
Other***	7,700	9,970	30	230	300	30	30

*Estimates refer to research equipment systems costing $10,000 to $999,999.

**Includes all sysems costing $100,000 to $999,999 not elsewhere classified.

***Includes all systems costing $10,000 to $99,999 not elsewhere classified.

NOTE: Details may not sum to totals due to rounding.

SOURCE: National Science Foundation, SRS.

Chapter 14

Equipment funding at the Centre National de la Recherche Scientifique

Yves-André Rocher
Ecole des Mines de Paris (France)

1. INTRODUCTION

Funding of research equipment is a central aspect of the scientific policy of countries, institutions, and laboratories. In this paper, we wish to bring new insights to the question, using information related to the main French research institution, the Centre National de la Recherche Scientifique (CNRS).

Because of needs for equipment expressed by the scientific community during the past decades, equipment funding exhibits a constant and rapid growth. Three main reasons explain this trend:

- equipment is used in most laboratories, in all domains of science;
- the concept of equipment includes many different things, from measurement instruments, through oceanographic vessels and computers.
- instruments become more and more efficient and powerful, with increases in costs that have not been controlled by either economies of scale or reduction in electronics and computer prices.

The last is often referred to as the sophistication factor. Two examples illustrate. In the seventies, the highest performance nuclear magnetic resonance (NMR) spectrometer, a basic instrument for the study of molecular structures and therefore routine equipment in biochemistry, was worth about $ 170 000; nowadays, the best NMR spectrometers are five times as expensive. An even more striking example is the SuperACO machine, a powerful source of synchrotron radiation, built in Orsay between 1982 and 1985 and using the existing linear accelerator as an injector. The original instrument cost 130 million francs (MF), including civil engineering and buildings (although not the linear accelerator). In contrast, the European Synchrotron Radiation Facility (ESRF), now under construction in Grenoble, will cost some 2400 MF in investment, including civil engineering and buildings.

In French terminology, the NMR is referred to as *"équipement mi-lourd"* (upper middle range cost equipment), whereas the ESRF is *"très grand équipement"*. People responsible for the science policy as well as research investigators agree that these large instruments are of strategic importance for the construction of scientific knowledge: depending on whether or not they have such instruments at their disposal, researchers will be at the frontiers of scientific progress, or restricted to research of marginal interest.

In domains such as astronomy or nuclear or high energy physics, these statements are common sense. They are less obvious and even controversial in other domains of science: in social sciences and humanities, of course, but also in physics, chemistry, engineering sciences, and life sciences. It is a matter of fact, indeed, that in many laboratories, daily research is performed with the help of much simpler and cheaper instruments, in the cost range of a few thousand francs up to tens of thousands of francs. Such instruments, *"équipements moyens"* as they are called in France, retain a major place in the practice of research. For personal convenience, many of the best scientists seem to stick to light experimental methods. Although such an attitude is often referred to as nostalgia, it may well be that the flexibility and rapid implementation such methods provide are also prominent factors in scientific progress.

There are many ways to carry out research. Each category of instrument is associated not only with a need for performance in relation to the nature of the object of study, but also with a conception, a philosophy of the art of research. One can approach solid mechanics using a complex deformation machine, or numerical simulation techniques, or a combination of the two. The nature and scope of the results will be different. Knowing the instruments that scientists, laboratories, or institutions buy or fund gives direct access to their problematics and primary objectives.

This chapter is to be understood as a contribution to the production of indicators allowing for an assessment of research strategies. It covers the 1976/1990 period, during which the use of instruments has become widespread among all disciplines of science. It focuses on the so-called light or medium range areas of science as opposed to big science, where heavy equipment penetrated long ago. We want to analyse how the pressure of equipment demand has been taken into account by a large research institution such as CNRS. First, we shall look at the position of the three different classes of equipment that we defined above (*moyen, mi-lourd, très grand*) in the CNRS budget. We also examine how other budget headings were severely reduced in order to meet the increasing demand, and how extra funding had to be found through partnerships with various public and private institutions. At last, we attempt to compare the situation of equipment in French and American academic laboratories, with emphasis on the middle range equipment that appears to be strategically important.

Most of the data presented will refer to inputs. Of course, the question arises at once: is there any connection between the situation of equipment in laboratories and their scientific results? The answer is not straightforward and requires detailed analysis of both input and output, output being measured, for instance, on the basis of bibliometric techniques. Such a study is beyond the scope of this paper.

Before turning to the data, let us look briefly at what CNRS is: its missions, its modes of intervention, and its place within the French research system.

2. BRIEF PRESENTATION OF CNRS

The leading French institution for basic research, supported by the Ministère de la Recherche et de la Technology, CNRS covers the whole scope of scientific domains, from nuclear and high energy physics to social sciences and humanities.

In accordance with its statutes, CNRS has a twofold task: it must both carry on research on its own, and fund research in other organizations. Under the first task, CNRS initiates, develops, and organizes themes or programs which deserve to be carried on within the framework of a national policy. This is done mainly through its own laboratories but may also result from joint action with a partner - university, public research organization, industry - in joint laboratories or joint programs. As a funding agency, after assessing projects and research groups, CNRS distributes funds and manpower to other research establishments, especially to universities. Through associate laboratories and various procedures, it supports the most dynamic groups in these institutions and helps establish new teams.

In order to facilitate the emergence and development of new themes, CNRS extends its policy towards laboratories with a number of interventions that provide additional support in priority domains (via a special incentive budget heading); these special programs also encourage collaboration among teams in coordinated and multidisciplinary programs. These interventions are often coupled with the distribution of specific instruments to leading laboratories. Examples of such interventions include submicroelectronics and inertial fusion in engineering sciences; genome studies, and microbiology or immunology in life sciences.

CNRS is organized in seven scientific departments coordinated by a general directorate: nuclear and high energy physics; mathematics and basic non nuclear physics; engineering and computer sciences; chemistry; astronomy, marine and earth sciences; life sciences; humanities and social sciences. It employs over 26000 persons, including 11000 scientists and 15000 technicians, engineers and administrative staff. It covers some 1500 laboratories, some its own and some in associate status, with an overall academic staff of about 25000 full time equivalent scientists. In 1990, its budget reached 10.9 billion francs, i.e., 22% of total civil research expenditures in France.

3. CATEGORIES OF EQUIPMENT, BUDGET STRUCTURE, AND RESEARCH POLICY

This chapter uses primarily CNRS budget data[1], and also data drawn from a study of overall funding for middle range equipment by CNRS and its partners during years 1984 and 1985. Several points have been checked using the CNRS inventory, a data base with more than 600 000 entries, all financed with funds coming from CNRS. This approach is justified because there is a strong connection between on one hand, the budget structure and the way decisions are reached, and on the other hand, the practice of research.

The structure of the CNRS budget reflects the three categories of equipment mentioned above:

- *low and medium cost equipment*, with values up to 360 thousand francs (KF) in 1990. This heading is included in the general support allocation granted to laboratories.

- *upper middle range cost equipment*, with costs ranging from 360 KF to 10 MF. This heading is listed as such in the budget, and each instrument is identified by name (although this is less and less the case as years go by).

- *very large equipment*, funded in a national or international framework, with costs above 10 MF. In this category, one also finds regional or national *large computer facilities*. For this heading and for this one only, operating costs are also budgeted as well as investment costs. Such equipment is included in multiyear planning.

These categories cover the highest level in the allocation of funds: laboratory directors distribute resources between operating and small or medium investment costs. Allocation of funds for the purchase of upper middle range equipment falls to the heads of the seven scientific departments of CNRS, whereas decisions regarding very large equipment go to the Directeur Général of CNRS, and for the supersystems to the Minister himself.

Equipment categories also relate to the practice of research as it goes on in laboratories, the state of knowledge in different disciplines, and also (as we said above) to each scientist's conception of the art of research.

Other sources of funds are also used to cover equipment expenditures, for instance, incentive support granted to laboratories for a particular investigation or within the framework of a regional or national program on a specific theme. In a few cases, such interventions have resulted in the implementation of coherent and complementary instrument investments in all laboratories active in a priority domain.

Studying budgets over the long range gives access to the evolution of resources allocated to the different categories of equipment, for CNRS as a whole and for broad domains, allowing for comparisons between countries or scientific fields. It also reveals the priorities and strategies chosen by the leadership of a prominent institution such as CNRS to orient and organize research: How do they prepare for the long term? By putting incentives on new themes or by directing investments? Is it better to help "normal" laboratories in their development or should one favor big national centers?

4. EQUIPMENT IN THE CNRS BUDGET

4.1. The main headings in the CNRS budget

For CNRS as a whole and for every scientific department, Table I gathers data concerning the division of funds between the main headings of the institution's budget, considering only funds directly allocated for the practice of research (i.e., without manpower and without administrative expenditures; from now to the end of the paper, the word *budget* will always imply that restricted definition).

These data have been collected for every budgetary year in the 1976-1990 period. They are completed by an analysis of the expenditure per scientist (Fig.1). During that period, the total number of scientists increased from 7 289 budget posts in 1976 to 11 255 in 1990. As a consequence of such a growth (which, of course, is felt in a positive way both by the institution and by scientists), the total expenditure per scientist exhibits, according to periods, either a more rapid decrease, or a slower increase than the direct budget: as the latter increases in constant volume from base 100 in 1976 up to 121 in 1990, the budget per scientist decreases from 100 down to 78.5.

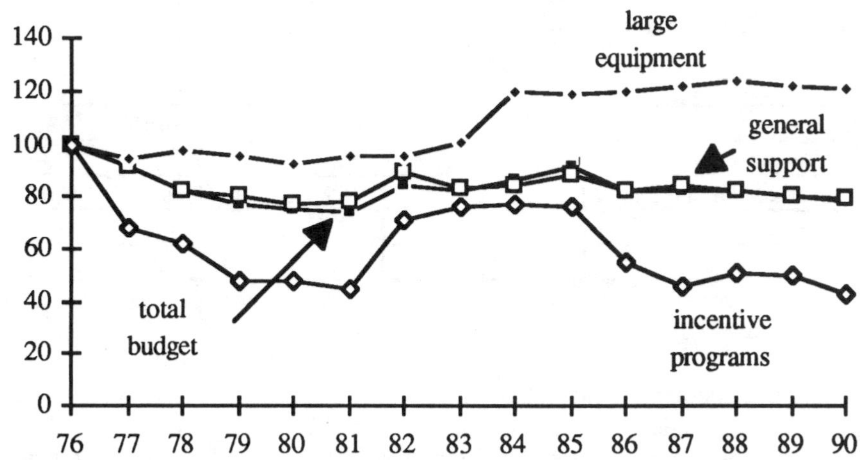

Fig 1. Evolution in volume of average resources per scientist at CNRS during the 1976-1990 period (base 100 in 1976). Total budget per scientist and general support decrease similarly and slowly. On the contrary, the budget for large equipment increases, whereas incentive funds exhibit a sharp fall, except for the "good" budget period, 1982-1985.

Considering the budget of CNRS as a whole, the **first observation** is the quasi constancy of the part of the budget allocated to the general and recurrent support of laboratories. The incompressibility of this heading appears here, resulting in a limitation of the institution's ability to invest in the long range. Successive budget reductions, associated with the increase of scientific manpower, resulted in a continuous erosion of laboratory daily resources, which could not be tolerated beyond a certain point: that is why the fraction of the budget allocated to general support had to be periodically increased. This occured whenever budgets were generous (for instance, in 1982 and 1985), although laboratories never recovered the level of resources they were granted in the golden sixties. Contrasting with this global evolution, the different scientific departments built specific policies, according to their proper needs and to their capacity, as well as that of laboratories, to find additional resources outside the institution. Life sciences have increased the part of their budget granted to general support (89% of the budget in 1988) in order to face growing operational costs and demand for small and medium cost equipment. Conversely, engineering sciences decreased that same fraction (56% of their budget in 1988) to make room for incentive programs and upper middle range cost equipment.

Knowing the expenditures for small and medium cost equipment in domains such as the life or social sciences is a key point in understanding their evolution. As we wrote before, general support for laboratories provides resources for this category of equipment. Until 1983, that fraction of general support was handled as a specific heading of the CNRS budget, amounting to some 35% of general support. An analysis of the CNRS inventory for recent years confirms that ratio. However, due to the multiplicity of possible external resources for this type of equipment, determining the level of actual expenses is not an easy task.

The **second observation** is the continuous and considerable increase of funds for large equipment (upper middle range and very large equipment):

%, total budget	1976	1982	1984	1987	1989	1990
computer facilities*	3,7	5,0	6,5	9,3	9,1	8,8
upper middle range eqpt.		3,8	5,2	5,4	6,1	8,0
very large equipment*	16,7**	14,5	15,0	16,7	17,4	16,4
total	20,4	23,3	26,7	31,4	32,6	33,2

*for computer facilities and very large equipment, budgets include operational costs;
** in 1976, this 16,7% also includes funding of upper middle range cost equipment.

If one compares annual budgets for equipment in 1982 and 1990, it appears that the budget for upper middle range cost equipment has increased in volume by a factor 2.3, the budget for computer facilities by a factor 2, and the budget for very large equipment by a factor 1.25. During the same period, the general support for laboratories has slightly decreased, by 1.4%.

An analysis of the budget per scientist shows a first period - 1976/1983 - during which the budget for large equipment (upper middle range cost plus very large equipment) was roughly held constant near its 1976 level, and then, a second period starting in 1984, characterized by a sharp increase of the budget by a factor of 1.2, to a level where it has remained.

The **third observation** is that large equipment, as a whole, has always held a priority position, whether budgetary years were good or poor. In good years, as we saw, general suport, which has been the first victim of budgetary shortages, together with incentive programs, was restored. 1982 was such a good year: general support was increased by 34.5% with respect to 1981.

Considering specifically each department (Fig.2, Tables II and III), in almost every case, we find both high levels of investment and high growth: for instance, the engineering sciences, which granted only 1.8% of their budget to large equipment in 1976, grant it 20% today. On the contrary, in nuclear and high energy physics, the part of the department budget devoted to large equipment (essentially very large equipment) shows a sharp decrease in the last two years. In the social sciences and humanities, budgets for middle range equipment stayed low: in this department, most equipment acquisitions are microcomputers, funded through the general support allocation.

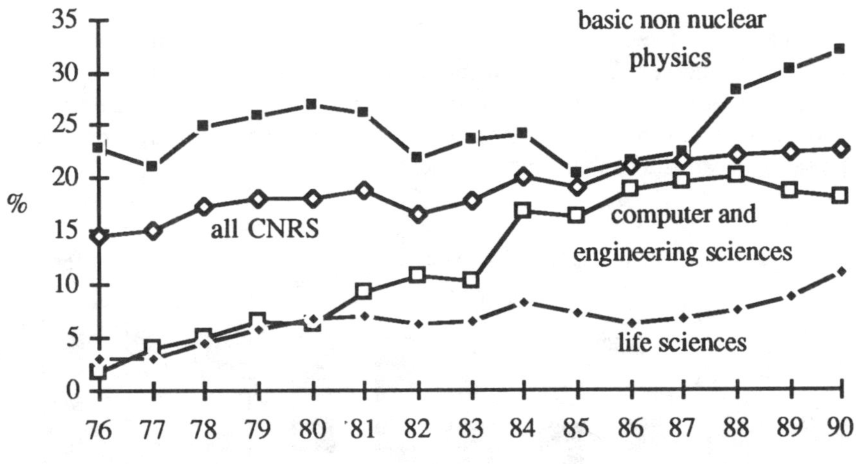

Figure 2. Evolution of resources for investment in large equipment, presented as parts of the total budget for all CNRS and for three scientific departments: basic physics, life sciences, computer and engineering sciences.

Correlatively, the **fourth observation** is that, except for the short period 1982/1985, when better budgets allowed for a renewal of incentive programs, the part of the budget dedicated by CNRS as a whole to incentive programs showed a sharp decrease (Fig. 3). In 1990, the average budget per scientist fell, indeed, to a little less than 40 % of its 1976 level.

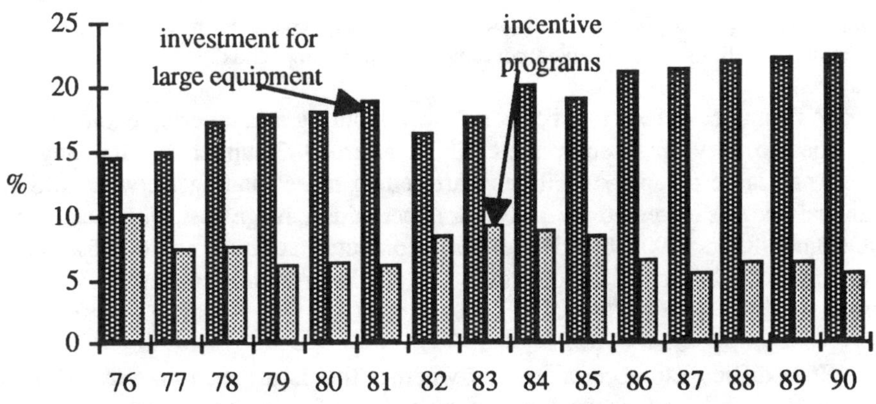

Fig. 3 Evolution of parts of the budget (in %) used for investment in large equiment on the one hand, and for incentive programs on the other. There is an obvious correlation between the increase of the former, and the fall of the latter.

The shift of budget proportion in favor of large equipment and at the expense of incentive programs is the clearest mark of the pressure exerted by the demand for large equipment. Except for social sciences and engineering sciences, the operating procedures at CNRS have thus widely changed.

4.2. Evolutions and trends in the funding of equipment

The **fifth observation** relates to the distribution of funding between upper middle range cost and very large equipment, including large computer facilities (Fig. 4). It appears that the latter dominates the former in scientific departments where those are present to some extent: nuclear and high energy physics; basic physics; chemistry; life sciences; earth, ocean, atmosphere and space sciences (Table III).

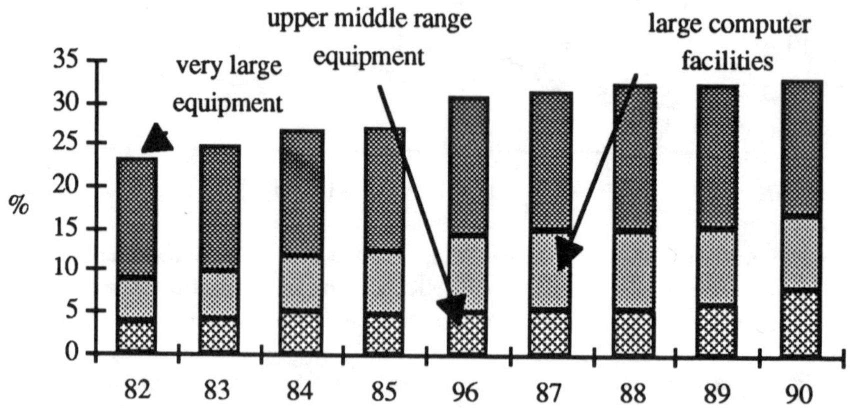

Figure 4. Evolution of resources granted by CNRS for large equipment, as % of the budget. For upper middle range cost equipment, only investment is taken into account. For very large equipment and computer facilities, operational costs are also taken into account. Compared to very large equipment, which stands at a high but rather constant level, upper middle range cost equipment and large computer facilities have received continuously increasing funds.

As a matter of fact, two periods show up: during a first period ending in 1982, priority goes to very large equipment. Then, after 1983, upper middle range cost equipment and large computer facilities start getting more funding: between 1982 and 1990, annual budgets increased by 132% for upper middle range cost equipment (60% in volume from 1985 to 1990), by 95% for large computer facilities, and by 25% for very large equipment. This observation holds for all departments concerned with large equipment.

It is difficult to ascertain today whether this is a long range trend. Indeed, budgets for 1989 and 1990 as well as first drafts for 1991, indicate new progress of the upper middle range cost equipment budget compared to budgets for large computer facilities and very large equipment. Although the latter seems to have reached a stable share of the budget since 1985, upper middle range equipment faces a rough competition: besides the fact that its operational costs are steadily growing, big science is the special ground for international cooperation to which the scientific community is deeply attached. In principle, large European or intercontinental accomplishments are allowed a permanent priority, warranted by treaties and agreements between countries and by multiyear planning, and therefore are protected from the ups and downs of budgets such as those French research encountered in 1983, 1984, or 1986. An eternally programmed machine seems to work to privilege the most basic, but also the most expensive, areas of science. Today, the great adventure is in Grenoble with ESRF, a gigantic machine with fantastic performances, which interest physicists, chemists, and biologists as well. Hence we see an increase of the budget for very large equipment in the corresponding departments.

The problem lies here. Large machines follow one another, more and more powerful, more and more expensive, with comparable lifetimes. In the family of synchrotron radiation machines, new projects take place at ten year intervals: ACO was brought into service in 1965 and lasted 23 years, then DCI was started in 1975 with a planned lifetime of 22 years, and SuperACO followed in 1987 planned for another 20 year period; finally, ESRF should get started in 1995. These are not mere sumptuous presents rewarding meritorious scientific communities. Among many other accomplishments, ESRF will enable studying structural evolutions taking place in matter during ultra fast transitory phenomena in real time, be they associated with industrial processes of transforming matter or with the modification of biological substances. How could one do without such a tool?

When costs reach impossible heights, international cooperation must take over from strictly national funding, but expenditures per country still grow. Since France has decided to stand at the top level of the most advanced science, very large equipment will keep dominating its science budgets.

In 1985, expenses for CNRS as a whole were as follows for the different categories of equipment:

	amount (MF, tax incl.	% of budget	% of equipment expenses
small and medium eqpt	390 MF	20%	42.4%
upper middle range eqpt	92 MF	4.7	10
very large equipment and computer facilities (operational cost incl.)	438 MF	22, 6	47.6
total	920 MF	47.3%	100%

These data can be compared with the results of a survey [2] carried out in 1983 and 1984 in American universities on behalf of the National Science Foundation (NSF): excluding "supersystems", i. e., very large equipment, which were excluded from the survey, it appears that expenditures for acquisition of instruments in the cost range of $ 75 000 to 1 000 000 (just about the cost range of *équipement mi-lourd* in France), comprised 40% of the total expenditures for instruments in these universities: this proportion is sligthly more than double what appears in the CNRS budget, where upper middle range cost equipment represents 19% of the total of that category plus small and medium class equipment.

Comparing French and U.S. policies for very large equipment is more difficult, due to lack of homogeneous data. What we can compare is annual investment in France and stock in the United States. In 1985, the amount of large equipment investment by CNRS rose to 210 MF, i.e., 30% of the total amount of investment for equipment during the same year, including upper middle range cost equipment but excluding large computer facilities. In the United States, the NSF survey estimates the "supersystems" stock at $ 250 million, or less than 15% of the total stock, including $ 1600 million for all other categories of equipment. If the comparison holds, it would seem that funding for very large equipment is at a much higher level at CNRS than in U.S. academic organizations.

One observes indeed a very sensible difference in the proportions allocated to different categories of equipment in the CNRS budget on the one hand, and in expenditures of U.S. academic institutions on the other hand. Of course CNRS is not the only source of funding for French laboratories in the academic system; universities, industrial firms, miscellaneous agencies, local authorities, etc, also contribute to acquisitions of equipment. It is therefore not very easy to know total resources allocated for equipment acquisitions. We believe, however, the picture drawn from that analysis of CNRS budgets is very close to reality.

As will be shown below, a correct figure of the total resources has been obtained for upper middle range cost equipment, and also for small and medium cost equipment, but for only a few domains.

4.3. Stability and transfers in the CNRS budget

Our analyses converge to show a relative stability of general support within the total budget. In this way, the institution helps laboratories face the erosion of their resources.

Before 1983, very large equipment received the main benefit of the priority granted to large equipment in the CNRS budget; after 1984, upper middle range cost equipment and large computer facilities became the main beneficiaries of the acceleration in this heading. This trend has been steadily confirmed since then, especially in the 1991 budget. Priority to equipment thus appears as a continuous aspect of CNRS policy, and in recent years investment in "normal" laboratories, where every scientist can use it, has received priority compared to investment for large national laboratories where more restricted communities can use it. The policy towards large computer facilities goes in the same direction: these are concentrated means, but everyone can use them through computational networks. Very large national and international instruments still weigh heavily in the budget, but they appear as a stable share since 1985. For how long?

In contrast, resources for incentive programs have been on a continuous downward slope, except for the 1982/1985 four year period when budgets improved. The investment policy for equipment has been possible only at their expense. At the same time, the CNRS board of directors has increased its direct influence in the choice and allocation of resources, as the practice of putting peer committees in charge of incentive programs has declined. The long term is receiving priority in research planning, a goal which is now being achieved by investment in equipment rather than by well-publicized themes.

Direct sponsorship of laboratories is thus favored compared with coordination of the scientific community through multidisciplinary themes. Of course, this is standard behavior, from which the different scientific departments may more or less diverge: in this respect, engineering and life sciences seem to stand in opposite positions.

5. JOINT FUNDING OF EQUIPMENT

After the golden sixties, research as a whole and CNRS in particular, experienced a period of budgetary stagnation for a decade, making it difficult and sometimes impossible to renovate the stock of older instruments and to invest in new, high performance equipment. Catching up with the gap created at that time has been one of the main elements of science policy since the beginning of the eighties. All institutions involved in the problem tried to organize and coordinate co-funding of upper middle range cost equipment on a multiyear basis, in order to meet both demands and capacity of use of the scientific community with maximum efficiency, within the scope of a national science policy. This called for cooperation between scientific departments at CNRS, as well as cooperation between public and sometimes private actors: ministries of research and of education, public research institutions, research agencies, local authorities, industrial firms, etc. Naturally, such behavior has been variable among institutions and even among their constituting parts.

On a national scale, during the 1983/1986 period, the so-called (from its founder) Landesman Committee at the ministry of research played a major role in the cooperative effort, although this committee actually was never institutionalized.

CNRS was also a solid partner in that policy.

5.1. A champion of cooperation

The chemistry department at CNRS perfectly illustrates such cooperative behavior, as illustrated by data collected for the 1981/1987 period concerning its policy on co-funding upper middle range cost equipment.

Until 1983, the cooperation was weak, sporadic, and unorganized, or else only in preliminary stages; only a few projects were funded every year, 30 between 1981 and 1983. The funding level was rather high: 714 000 F (tax incl.) on average per project.

Cooperation actually started in 1984: the number of projects funded every year quadrupled (153 during the 1984/1987 four year period). As the mean annual budget for this category of equipment doubled, the mean funding by CNRS per decision cut in half, to 380 000 F per decision, on average. Grants went more often to associate laboratories (half of the CNRS total funding for this purpose, instead of one third during the preceding period, the rest going to CNRS's own laboratories). Also, grants often went to small laboratories; this was quite exceptional before.

42 projects[3] involving upper middle range cost equipment were supervised by the department of chemistry and funded during years 1984/1986, for a total amount of 69,3 MF. Besides the chemistry department which contributed 20.8 MF (30% of the total), partners in co-funding were: other scientific departments of CNRS, who brought 8,8 MF (12,7% of the total), and different institutions outside CNRS (ministries, research institutions, agencies, industrials, local and specially regional authorities), who

altogether brought 39.7 MF (57,3% of the total). Thus cooperation gave the chemistry department more than twice its own contribution, for the benefit of a policy of which it was the main instigator and leader.

Two questions come to mind:
- Has cooperation modified the global amount of investment?
- Did it change the funding schedule?

Answering the first question is difficult, especially because the setting of this cooperative policy coincided with a sharp increase in public research resources. On qualitative grounds though, one may point out two effects of cooperation: the first one is an improved efficiency of investment, inasmuch as the coordinated implementation of resources was done for the benefit of coordinated scientific objectives. The second is the relation between location of new equipment and the partnership between CNRS and local and regional authorities. As the latter are newcomers in the funding of research, cooperation should encourage rearrangements in the mapping of research across French regions. Fifty-five per cent of French public research is located in the Paris area. From this point of view, it is important that 25 projects amongst the 42 were implemented in the provinces, of which 19 were actually funded by Regional Councils. As a result, these Councils contributed 26.7% of the total funding of projects in the provinces, a figure which is a measure of the part they now play in the organization and funding of research.

There is one point, however, where the cooperation policy produced negative results. In most cases delays in the complete implementation of equipment projects grew considerably longer. In the chemistry department alone, new projects were implemented in a single year in most cases before 1983. After cooperation was set up, an important proportion (more than one third in 1985) were only implemented in two or three years To these delays due to the department alone, one must add delays coming from other partners: among the 42 projects, only two were implemented in one year, and 34 in two years; 6 required three years. These delays seem to be the the main drawback of the cooperation policy.

5.2. What about other scientific departments?

A majority of CNRS scientific departments followed the pattern of the chemistry department. For CNRS as a whole, and for the years 1984 and 1985 together, acquisitions of upper middle range cost equipment rose to 420 MF. Of these, CNRS brought in only 163 MF through its own budget (39.3% of the total), plus some 60 MF coming from extrabudgetary resources for special programs. Other partners, already listed above, brought in the rest, slightly less than 200 MF. Through co-funding like that in the chemistry department, CNRS as a whole has been able to raise external resources totalling more than one and a half times its own contribution, for the benefit of a science policy that it instigates and leads for the main part.

Finally, we would like to emphasize a last point which might modify our preceding analysis of the benefits of cooperation. The association in the cooperation of scientific and political partners has a political character, the consequence of which is to favor the choice of visible and standardized equipment. We are thinking, for instance, of high performance measurement instruments (NMR spectrometers, electron microscopes,

X-ray diffractometers, etc.), as opposed to less standard equipment used in original experimental projects. Sometimes indeed, partners agree almost at once on funding such instruments, just because they are sure to pick the political fruit of their decision rapidly. Of course, inasmuch as this behavior is known, scientists orient their demands towards this category of equipment. Therefore, there is a real risk of developing a more standardized type of science, at the expense of a more original research.

6. A COMPARISON WITH THE FUNDING OF UPPER MIDDLE RANGE COST EQUIPMENT IN THE U.S.

American science has long been a model for European science and in particular for French science and CNRS. Therefore it is interesting to attempt a comparison with the situation in the U.S. academic system, where NSF carried out the survey we mentioned above. Such a comparison is possible for upper middle range cost equipment, inasmuch as overall funding for this category of instruments is known in detail in both cases (apart from resources that come from outside CNRS and outside the cooperation set up by the Landesman Committee, and those can be neglected). CNRS data are available for 1984 and 1985, whereas the NSF survey covers 1982 and 1983. We feel this shift has no important effect for our comparison.

The U.S. study covered all scientific fields CNRS is involved in, except mathematics, nuclear and high energy physics (inasmuch as supersystems were left outside the scope of the NSF survey), social sciences and humanities,. However, we had to restructure data so that similar domains would match each other in both countries. We thus had to aggregate computer plus engineering sciences on the one hand, and basic non nuclear physics plus astronomy plus environmental sciences on the other.

After this preparatory work, the comparison is as follows regarding annual expenditures of laboratories for the upper middle range cost category of equipment:

	NSF inquiry				CNRS			
	instrument expenses		scientific staff		instrument expenses		scientific staff	
	M$	%	nb.	%	MF	%	nb.	%
physics+ astronomy + environmental sciences	44,2	31,7	9210	22,0	55	26,3	5710	28,0
computer + engineering sciences	45,8	32,9	11850	28,3	79,5	37,7	4690	23,0
chemistry	17,5	12,6	3110	7,4	34,5	16,4	4100	20,1
life sciences	31,8	22,8	17650	42,2	41,5	19,7	5860	28,8
total	139,3	100	41820	100	210,5	100	20360	100

In spite of the simplifications we had to put in our analysis, we come to the following observations:

- The sectorial structure of expenditures for upper middle range cost equipment is somewhat similar in both cases: one must note, however, that less money was spent at CNRS in the fields of basic physics, astronomy, environmental sciences, and life sciences. The structure of the scientific staff, however, is very different from one country to the other.

- Normalized with respect to scientific staffs, expenditures are much lower in France. For all disciplines on the average, with a dollar worth 6.10 francs, U.S. scientists are granted resources about twice as high as their French colleagues: 20320 F against 10340 F, with large differences by domain:

expenditures per full time equivalent scientist	CNRS (F)	U.S.A. (F)	CNRS/U.S.A. (%)
chemistry	8415	34300	24,5
basic physics+ astronomy +environmental sciences	9720	29300	33,2
computer+ engineering sciences	16930	23600	71,7
life sciences	7080	11000	64,5

It appears that expenditures for upper middle range cost equipment at CNRS are only one third or one quarter of what they are in U.S. academic institutions in the fields of chemistry, basic physics, astronomy and environmental sciences. In contrast, the ratio of CNRS versus U.S.A. expenditures rises to 70% in the field of computer plus engineering sciences and life sciences. Similar results appear in a broad survey [4] carried on by the ministry of research and technology in France with respect to the stock of upper middle range cost equipment in the country.

But within these large domains, situations may be quite different by field. Considering computer and engineering sciences, it appears that expenditures in the fields of mechanical, civil, and chemical engineering are higher at CNRS (26 000 F per scientist as opposed to 13 000 F per scientist in the United States), whereas resources in the fields of computer science, electronical, and electrical engineering are much lower at CNRS: 12 000 F per scientist instead of 44 000 F in U.S. institutions. Here too, sectorial structures of expenditures and of scientific staffs are very different from one country to the other:

	USA				CNRS			
	expenses		staff		expenses		staff	
	M$	%	nb.	%	MF	%	nb.	%
computer science, electronics, electrical engineering	24,7	53,9	3430	28,9	36,4	45,8	3020	64,4
chemical, mechanical, civil engineering	21,1	46,1	8420	71,1	43,0	54,2	1670	35,6
total	45,8	100	11850	100	79,4	100	4690	100

The difference in the structure of expenditures is moderate; but the structure of scientific staffs is completely different. One could say that in a group of domains, investments are manpower-rich in one country and equipment-rich in the other, with the reverse situation in the other group of domains.

Looking at these comparisons, one is tempted to explain the rather poor situation of upper middle range cost category of equipment in some sectors of CNRS by the considerable effort in favor of very large equipment. In 1984/1985, very large equipment received 75% of total investments by CNRS for large equipment in the group of domains including basic physics, chemistry, astronomy and environmental sciences (80% for basic physics, 55% for chemistry, 81% for astronomy plus environmental sciences), 25% going to upper middle range cost equipment. From the comparison we attempted above between annual expenditures in France and stock in the United States, it would seem that very large equipment have a smaller relative weight in the United States.

Life sciences also spend large amounts of money for very large equipment, although to a lesser extent than in the domains we just quoted. Their more comfortable situation most likely results from a smaller demand for upper middle range cost equipment: according to the NSF survey, this demand is only one third of what it is in chemistry.

In the engineering sciences, the ability to find resources outside CNRS through contracts with miscellaneous organizations explains their rather high level of expenditures for upper middle range cost equipment.

If we look now at annual expenditures per laboratory, the comparison turns out in a somewhat different way:

	NSF		CNRS	
	staff/lab	MF/lab	staff/lab	MF/lab
physics + environmental sciences + astronomy	21,0	0,62	21,3	0,21
engineering + computer sciences	17,4	0,41	32,3	0,55
chemistry	17,7	0,61	21,0	0,18
life sciences	14,7	0,16	19,3	0,14
total	16,2	0,33	21,9	0,23

As research appears to have a higher degree of structuration at CNRS (the staff per laboratory is larger), looking at expenditures per laboratory leads to a somewhat different comparison. In what corresponds to the department of engineering plus computer sciences at CNRS, average expenditures per laboratory are higher in France than in the United States, but in other domains the gap remains high. If equipment is used very much, the French situation remains poor; however, when equipment is used to a lesser extent, organization at CNRS may lead to saving resources.

What about small and medium equipment? For this category, using the CNRS inventory, we have been able to estimate the annual expenditures in CNRS own laboratories, and in the fields of physics and life sciences[5]. These expenditures turn out to be very much the same as they are in U.S. universities: on the average, near 32 KF per scientist per year, in both domains.

We also attempted a comparison with data collected in the United Kingdom[6], although the objectives of the British instrument inquiry are much further away from ours. However, this study also brings information on equipment expenditures in U.K. universities during year 1987. Within the limits of such a comparison, it appears that resources at disposal of U.K. universities for the acquisition of upper middle range cost equipment would be some 30 to 40% higher than in CNRS laboratories. It also appears that for the same category of equipment, the contribution of industry is far larger in the United Kingdom, three to four times what it is in France. Conversely, the contribution of local authorities, very high in France as we emphasized, is low in the United Kingdom.

7. CONCLUSION

In a continuous way during the whole period we have studied, 1976/1990, CNRS devoted increasing resources to the acquisition of equipment, especially to large equipment: in 1990, the resources for large equipment rose to 33,2 % of the total budget (without salaries and administration expenditures); in 1976, the same proportion was 20,4 %. As very large equipment - accelerators, oceanographic vessels, telescopes, etc. - have stood at a rather constant level, with 16 to 17 % of the budget, two other headings have been the main beneficiaries of the increase in resources: large computer facilities and upper middle range cost equipment, each of them receiving 8 to 9% of the budget in 1990.

Owing to the strategic importance of these upper middle range cost instruments for which demand increases very fast, CNRS had to cooperate with different partners in a co-funding joint policy which started in 1984. This enabled CNRS to double its budgetary plus extrabudgetary resources for this category of equipment. However, a comparison with expenditures in U.S. academic institutions for the same type of equipment shows that total resources per scientist available in France are only one half of what they are in the United States, with the largest gaps in physics, chemistry, environmental sciences, and astronomy and better situations for engineering and life sciences.

It is difficult to ascertain whether this situation resulted in a severe handicap for the equipment-poor domains of French research. With respect to this point, one can bring another observation: the NSF inquiry has revealed that a large number of U.S. laboratories considered themselves as unable to carry on research in some advanced areas because of lack of adequate instruments. The domains that appeared the most concerned are computer science, electrical engineering, material science, physics, and chemistry; and in all these domains, the main demand is for the category of upper middle range cost. These domains are also those for which the gap of resources was the highest when we compared French and U.S. institutions. It is therefore highly plausible that the competitiveness of French laboratories has been weakened in the same areas of science.

Is it possible to confirm that point by output measurements? We shall give a quick answer to the question. A bibliometric study of the French system of research[7] carried on in 1987 showed, for instance, that the position of French research was relatively weak in domains such as computer science and material science; on the contrary, the position of physics is fairly good. We only remark that the situation of upper middle range cost instruments is poor in all three domains; however, the situation of physics is very good when one thinks of very large equipment. Giving a thorough answer to the question of input/output correlation requires a far more detailed study of inputs and outputs. It is beyond the scope of this paper.

As a matter of fact, the structure of funding for equipment in the CNRS budget has long been determined by the absolute priority that the most basic research has been given in this institution. But things appear to be changing nowadays, with a spectacular turn towards the needs of the economic and industrial sectors. This explains the sharp increase of funding for upper middle range cost equipment that can be noticed since 1985.

8. REFERENCES AND NOTES

1. Annual budgetary documents of CNRS;
Annual documents "bilan et position" edited by the scientific departments of CNRS;
CNRS inventory.
2. Academic research equipment in selected science/engineering fields, 1982-1983, NSF, August 1985.
3. The discrepancy between the earlier numbers and these arises from the fact most equipment are funded over several budgetary years. Also, several funding decisions of the chemistry department are concerned with equipment operations supervised by other scientific departments.
4. B. Jouffrey and M. Nauciel, "Les équipements scientifiques mi-lourds", Ministère de la Recherche et de la Technologie, 1988.
5. Estimating global expenses for small and medium equipment in CNRS laboratories is somewhat difficult, because external resources may be used for this purpose. These resources are known, but not the proportion actually used for purchasing equipment. However, in the fields of physics or life sciences, and in CNRS's own laboratories, most resources are provided by CNRS and external resources are relatively moderate, which allows for a fair estimate of expenditures for the type of equipment we are interested in.
6. L. Georghiou, P. Halfpenny, S. Hinder, "Survey of academic research equipment in the United Kingdom", Advisory Board for the Research Councils, Nov. 1989.
7. M. Callon et L. Leydesdorff, "La recherche française est-elle en bonne santé?", *La Recherche*, vol.18, p. 412, 1987.

TABLE I
Divisions of the CNRS budget

(%)

year	general support	large equipment	incentive program investment	others	volume index
1976	65,6	14,6	10,1	9,7	100
1977	65,9	15,1	7,5	11,4	95,7
1978	65,8	17,3	7,6	9,2	89,8
1979	67,3	18,0	6,2	8,5	88,6
1980	67,6	18,1	6,4	7,9	88,8
1981	69,2	18,9	6,1	5,8	90,4
1982	69,7	16,5	8,5	5,3	108
1983	66,5	17,8	9,4	6,3	109,2
1984	63,6	20,2	9,0	7,1	117,7
1985	63,7	19,1	8,4	8,8	127,4
1986	66,0	21,2	6,7	6,1	118,2
1987	66,9	21,5	5,6	6,0	121,1
1988	66,1	22,1	6,3	5,6	120,5
1989	65,7	22,3	6,3	5,8	121,1
1990	66,4	22,5	5,5	5,6	120,7

All scientific departments are included, but salaries and administration expenses are excluded;

Large equipment: upper middle range cost and very large equipment, including large computer facilities;

Others: international cooperation, transfer, scientific information, buildings and real estate acquisitions;

Volume index: change in volume of the budget, base 100 in 1976.

TABLE II
Proportion of their budget allocated to investments for large equipment by the scientific departments of CNRS
(%)

year	nuclear, high energy physics	basic physics	computer engineer. sciences	chemistry	astron., environ. sciences	life sciences	social sciences humanities	all CNRS	volume index
1976	17,1	22,9	1,8	9,6	26,0	2,9	2,6	14,6	100
1977	20,8	21,0	3,9	7,2	24,8	2,9	2,4	15,1	95,7
1978	28,2	24,9	4,9	6,2	26,0	4,4	2,5	17,3	89,8
1979	30,6	25,8	6,5	9,5	29,7	5,8	5,1	18,0	88,6
1980	28,6	26,9	6,2	9,3	33,8	6,8	2,9	18,1	88,8
1981	22,9	26,1	9,4	10,7	38,1	7,1	2,9	18,9	90,4
1982	15,6	21,9	10,9	14,5	36,0	6,2	2,6	16,5	108,0
1983	17,3	23,7	10,2	15,2	39,4	6,4	1,8	17,8	109,2
1984	21,2	24,0	16,7	18,7	39,7	8,3	2,5	20,2	117,7
1985	24,3	20,4	16,4	16,1	39,9	7,3	2,3	19,1	127,4
1986	28,3	21,6	18,8	16,3	38,1	6,3	2,9	21,2	118,2
1987	27,7	22,4	19,5	16,6	38,5	6,7	3,0	21,5	121,1
1988	28,6	28,1	20,0	19,8	35,8	7,5	2,9	22,1	120,5
1989	20,1	30,2	18,5	20,7	36,4	8,8	3,2	22,3	121,1
1990	14,9	32,0	18,1	22,9	38,4	11,1	3,0	22,5	120,7

TABLE III
Funds allocated by the scientific departments of CNRS for investments in large equipment
(current MF)

scientific department	equipt	1982	1983	1984	1985	1986	1987	1988	1989	1990
nuclear, high energy phys	very lar.	38,5	45,1	49,6	64,2	79,4	85,4	87,0	49,9	19,9
	up. mid.	-	-	11,9	13,0	8,0	8,0	7,0	9,0	24,3
basic physics	very lar.	28,2	28,8	34,0	32,2	30,7	35,6	46,8	56,7	55,2
	up. mid.	3,6	10,2	8,2	8,1	12,2	10,9	14,9	17,2	25,2
computer, engineering	very lar.	-	0,9	7,5	9,5	4,7	0,6	1,0	0,8	1,5
	up. mid.	10,0	9,1	12,4	13,3	24,0	30,0	30,0	30,0	30,0
chemistry	very lar.	14,4	15,9	18,9	18,6	18,3	20,8	28,6	32,5	42,2
	up. mid.	6,2	8,6	16,0	14,5	16,2	15,0	19,0	19,0	21,0
astronomy, environ. sc.	very lar.	47,7	57,7	67,4	72,0	69,1	71,5	68,8	66,9	66,8
	up. mid.	14,7	16,1	15,8	17,6	22,9	28,8	21,5	29,0	36,0
life sciences	very lar.	6,2	7,3	12,8	13,3	13,8	15,2	14,1	15,2	14,1
	up. mid.	9,1	10,5	15,3	15,8	9,9	11,0	15,8	21,8	34,8
social sc. humanities	up. mid.	2,5	1,9	2,6	2,8	3,6	3,6	3,6	4,5	4,5
incentive programs	up. mid.	4,9	5,4	6,9	7,0	-	-	-	-	-
CNRS total	very lar.	134,9	155,7	190,1	209,7	216,0	229,0	246,3	221,9	199,6
	up. mid.	51,0	61,8	89,1	92,1	96,6	107,3	111,8	130,6	175,9

Chapter 15

Models of sophistication or instruments of decline? Quantifying the state of academic research equipment

Luke Georghiou
Peter Halfpenny
John Yates
University of Manchester (United Kingdom)

ABSTRACT

Increased demand and constrained funding for research equipment have led to a requirement for quantitative data. This paper examines two aspects of the state of equipment: the sophistication factor, that is whether the real cost of equipment is rising because of demands of scientific progress; and the state of scientific equipment in British academic institutions. Findings are based on a census of academic research equipment performed in 1987 and a sample study performed in 1990. A positive but declining sophistication factor is noted. The period of study was characterised by significant acquisitions of equipment but scientists continue to perceive deficiencies. Recommendations are made on policies to improve the situation within the constraint of existing resources.

1. INTRODUCTION

The growth of scientific knowledge has been closely associated with the development of research instruments which have offered researchers continuous improvements in the quantity, quality and scope of the data which they collect and analyse. Developments in microelectronics and other technologies have enormously increased the capabilities of instruments while, in turn, scientific instruments have frequently represented the leading edge of technologies and have found subsequent applications in industry or medicine. Two trends have combined to increase the demands made by scientists for instrumentation. Progress within particular fields of science has led to demands for even higher performance standards for instruments. At the same time, the range of disciplines demanding use of instruments has widened, further increasing the overall demand for resources. The combination of these trends with a levelling off in the growth of funding for science in the past decade has focussed arguments about instrumentation onto questions of priority and efficiency. Consequently questions of funding for research instrumentation have been high on the science policy agenda for some time. Nevertheless, it is only recently that the debate in this area has begun to be informed by quantitative data on the nature and state of academic research equipment. In this paper, the British experience of collecting this data is described and two features of the debate are examined more closely:

- the 'sophistication factor', that is whether the real cost of equipment is increasing because of the demands made by scientific progress; and
- the state of scientific equipment in British academic institutions, focussing upon arguments that the ability of scientists to perform research is in decline, and that lack of adequate equipment is a contributory factor to that decline.

Following examination of these issues, policy implications and future information requirements are discussed.

1.1 The Sophistication Factor

The pioneering work on this topic was carried out by the Department of Education and Science in 1967[1]. This examined whether the cost of scientific work per scientist had become progressively greater by examining the budgets of several University and government laboratories. Substantial real-price growth rates per scientist were identified for major equipment items and these were collectively termed 'sophistication factors'. Since then the existence of sophistication factors has entered the conventional wisdom in debates over science funding. Futher evidence has tended to confirm the original findings. For example, a study by the Council for Scientific Policy in 1971 found a mean annual growth rate of the cost per scientist of research equipment between 20 and 30%. Other indicators such as the rise in the cost of each grant awarded by research councils and R&D price indices also revealed increasing R&D costs, though less emphatically. Some caution is needed when using figures based on expenditure since thus is often externally constrained, for instance in the situation where a grant applicant is required to reduce the proposed expenditure on equipment.

The assumption that the price of equipment rises faster than inflation is explained by the hypothesis that the trajectory of scientific development is demanding increasingly detailed or accurate measurements of physical phenomena. At one extreme this is clearly illustrated by the demand for ever more powerful particle accelerators in nuclear physics, but the effect is also said to exist in smaller scale activities. It is important to recognise that this is a demand-led phenomen. On the supply-side, if anything, sophistication has resulted in the price of a given effect being reduced by technical change. The extreme case here is that of computers. However, even if the unit cost of an effect is reducing, required increases in the absolute level of that effect may still result in a positive 'sophistication effect'.

How the required level comes to be set is a question which might be more appropriately answered by sociologists of science but clearly it is at least part determined by a combination of the trajectory of a sub-field, competitive pressure among research groups in that sub-field, the resources available for the development and purchase of instrumentation of the required type and the technical feasibility of advancing instrumentation in the direction demanded by the researchers.

Measurement of sophistication is particularly difficult. One approach would be longitudinal, monitoring the purchases in sub-fields and adjusting these by an appropriate deflator (the choice of which is in itself a source of problems). The advantage of this approach is that it reflects the dynamics of the situation. The disadvantages are that the result may be affected by some of the factors referred to above such as the groups' funding, changes in their relative status and differences in their research directions. In the studies undertaken by the authors[3], a cross-sectional approach was used which took the instrument as the unit of analysis. The approach relied on the assumption that researchers are aware of the costs of the various equipment options open to them. In the context of completing a questionnaire about the characteristics of research instruments, respondents were asked to provide four cost-estimates for each instrument:

- Original purchase price or construction cost - that is the actual cost when acquired by the department or institution, or the identifiable cost of construction.
- Replacement cost - the estimated cost to purchase the equipment, or equipment of roughly equivalent function and capability, at current prices.
- Cost of an enhanced item - defined the cost of equipment with capabilities which stand in similar relation to the current state of the field as the original item did at the time time of its pruchase.
- Cost of state-of-the-art equipment - the estimated cost of the most sophisticated instrument in the field.

The rationale for this approach was that the increase (or decrease) represented by replacement cost in relation to original cost represents a cumulative measure of inflation. The closest available measure of what we have termed the sophistication factor is the ratio between 'enhanced' and 'replacement' costs as the former incorporates an adjustment for advances in the expectations of those working in the field. The state-of-the-art cost was included in part to siphon away from 'enhanced' cost data any tendency to put down 'Rolls Royce' equipment to emphasise, for instance, perceptions of underfunding by the respondent. Nevertheless, the figure has some validity on its own as it indicated the premium necessary to be on the leading edge of equipment in a given area. All of these estimates include the effects of technical change or other factor price changes but the use of ratios to some extent cancels these out.

The disadvantages of this approach are that by referencing current equipment it tends to exclude instruments which are wholly new, or new to a sub-field. The increasing application of expensive instrumentation to biological sciences and to earth science may well involve acquisition of equipment not previously used in the area. In other words, sophistication due to incremental change is more likely to be covered than sophistication due to radical change.

1.2 The State of Equipment

As funding for science has come under strain, particularly in the United Kingdom, debate has turned to the ability of British science to keep pace with international competition. The rise of 'generic' technologies - information technology, biotechnology and new materials - has highlighted the relationship between scientific research and international industrial competitiveness[4]. The educational dimension of equipment has also been raised, with concern over whether post-graduate students are being trained on sufficiently up-to-date instruments to fulfil the future recruitment needs of industry.

Both sides of the dual-support system are under pressure. Research councils, responsible for funding larger items required for specific projects, face severe difficulties. One consequence has been deliberate paring down of support offered under each grant awarded. Economies have fallen disproportionately on new instrumentation[5]. In addition to this, basic instruments, the core equipment of a laboratory such as centrifuges, oscilloscopes and spectrometers, are not generally supported by research councils who consider that these should be funded by the Universities Funding Council (UFC) responsible for establishing well-found laboratories. Other sponsors such as industry and charities also tend to avoid supporting these items, preferring to fund large 'high-tech' pieces of equipment. Since the real value of UFC support has declined more rapidly than that of research councils a situation of imbalance has arisen.

Over the years a number of anecdotal accounts have been offered by the scientific community in attempts to convince policymakers that an ageing stock of equipment is leading to deficiences in key areas. Responses tended to imply that these were isolated examples or that adjustment of inefficiencies elsewhere in the system could compensate for the shortfall. The equipment surveys described in the next section have been a deliberate attempt by the Advisory Board for the Research Councils, the body responsible for putting the case for the science budget to ministers, to ground the debate on a foundation of data which describe the magnitude condition and useage of the national research equipment stock and indicate likely future requirements.

2. DESIGN OF THE SURVEYS

2.1 The International Experience

When commissioned to examine the feasibility of a survey addressing the above issues, the first step was to examine experience elsewhere. The seminal work in this respect was the first of a series of surveys carried out for the US National Science Foundation. The objective was to:

"...develop statistically reliable indicators of current national status of need for major research instruments in higher educational institutions, and of recent trends in the stock, cost, condition, obsolesence and utilisation of such instrumentation"

The questionnaire used by the NSF survey formed the basis of a study carried out in Australia and these two were a starting point in the design of the UK baseline survey undertaken by PREST/CASR in 1987-88. The NSF and Australian questionnaires were substantially modified following a pilot study. Nevertheless, one aim was to retain scope for international comparability, although there remains a need for a full international comparison of the state of instrumentation. In many ways this would offer a more concrete indicator of relevant levels of science support since the more conventional indicators such as funding inputs and manpower levels are particularly prone to the effects of countries' different institutional arrangements and accounting procedures.

A study published after the 1989 report on the baseline survey provides a useful starting point for an equipment-based comparison[7] and makes use of a report on middle range equipment in France[8] and an ongoing inventory of larger items in German universities[9].

2.2 Implementing the Surveys

Some lessons were derived from the NSF study, not least of which was that a large scale survey of this kind was feasible. (The American survey used a sample of 10,471 instruments out of a total of 22,479 in the cost range). An important difference between the UK and USA situations was that in the latter most universities had computerised inventories of equipment which greatly facilitated sampling of instruments. A general lack of such inventories in the UK meant that a sampling approach was not possible so it was decided to include all equipment, within the replacement cost range of £10,000 to £1 million used for research in science or engineering departments, excluding free-standing computers and facilities.

Definitional problems about which departments and equipment, items to include arose throughout the study making it necessary to offer a support service to respondents, generally through local survey co-ordinators in their own institutions. As each UK university department falls within a UGC/UFC 'cost-centre', which approximates to disciplinary boundaries, these cost-centres were used as the main unit of analysis. The British Standard Institute Root Thesaurus was used to classify the instruments. The degree of detail incorporated in this classification scheme varies considerably: for example, there is only one code for all types of microscope (other than electron microscopes), whereas there are numerous codes for different types of spectroscope. To classify all equipment in the baseline survey about 400 codes were used. Since the survey the scheme has been modified and updated, partly in the light of its use in the survey.

All but one of the 64 institutions invited to take part (all universities and five polytechnics in the UK) agreed to do this. 17,166 Equipment Questionnaires were returned with a further 730 separate questionnaires being filled in by heads of department. Subsequent checks indicated a response rate of at least 85%.

In late 1990 PREST/CASR was invited to perform a 'Dipstick Survey', that is, a limited exercise to examine changes which had occured since the baseline survey and to pilot a full scale update of the baseline survey. This study was confined to four large cost centres (Biochemistry, Chemistry, Physics and Mechanical, Aeronautics and Production Engineering) in six institutions. Because equipment is concentrated on both institutional and disciplinary bases, this constituted a sample of about 9% of the original population of instruments. Respondents were asked to update previously supplied information on equipment and to complete new forms for items acquired since 1988. Again, a very high response rate was achieved, offering for the first time a dynamic, if partial, picture of the equipment situation in the UK. The survey was supplemented by interviews with 39 out of the 40 heads of departments or their designates in the selected cost centres. Topics addressed included:

- delays to current research due to equipment needs;
- significant changes to nature and/or use of instrumentation in the intervening period; and
- the general state of equipment and laboratories.

3. FINDINGS.

1.1 The Sophistication Factor

Table 1 Ratios of Equipment Cost Estimates

		Original/ Replacement	Enhanced/ Replacement	SOA/ Replacement
Baseline Survey	All Items	0.54	1.25	2.15
(1987 Stock)	1990 Sample	0.50	1.31	2.29
1990 Survey	Current Stock	0.51	1.12	2.07

The ratios shown in Table 1 were calculated as averages of all cases where replacement cost and the relevant other cost were given for a particular item. The first two rows of the Table indicate the extent to which the departments selected for the 1990 study had differed from the overall figure in 1987. The third row then shows the ratios for current stock. This differs from the second row in that recent acquisitions are added and items no longer used for research are deleted.
The Original/Replacement ratio shows a fairly consistent figure of a doubling in price for cumulative inflation, although without detailed analysis by year of acquisition this cannot be decomposed. Of more interest to the issues raised in this paper is the Enhanced/Replacement ratio. Here a more volatile situation is evident with a higher than average ratio of 1.31 in 1987 declining to 1.12 in 1990. This may be due to a reduction in the sophistication factor for these fields, it may arise from changed perceptions about costs in the minds of respondents, and some of it may arise from changes in the instrument population. The fact remains that researchers do consider the sophistication factor to exist and are able to quantify it for individual instruments. That the Enhanced/Replacement ratio is greater than 100 demonstrates that increases in funding above the level needed to keep pace with inflation are necessary to maintain parity with the development of these subjects. State-of-the-art replacement ratio reveals that the premium to acquire state-of-the-art equipment remains over 100% for all cases, though it appears slightly reduced in 1990 compared with 1987.

3.2 The State of Equipment

3.2.1 The National Stock

The baseline survey had indicated a national stock with a replacement value in 1987 of about £712 million, with the average value per equipment item being £44,000. Centrifuges, microscopes and liquid chromatographs were the most frequent types of instruments, though by value electron microscopes and NMR's headed the list. The 'Dipstick Survey' sampled 1,277 of the instruments on the 1987 database. In the three years from 1987 to 1990, instruments with a value of 20% of the 1987 stock in the sample were acquired. The total number acquired was 216 instruments.

After allowing for deletions from the stock, there was a net increase of 2% in numerical terms. Equipment passing out of the system usually did so because of lack of sophistication rather than because it ceased to function. In both surveys, as would be expected, the distribution of items is skewed towards the lower end of the cost range but in terms of aggregate value expenditure on new items is largest on items in the £100k - £200k range (this category contains 32% of equipment by original cost). Distributions differ between subject areas, however, with the average value of equipment for biological subjects being about half of that for physical sciences and engineering.

3.2.2 State and Condition of Equipment

Aggregated statistics disguise the fact that equipment varies in its quality and state of repair. The former characteristic reflects the degree of obsolescence, the latter the physical condition of the equipment

Table 2 Technical Capability of Equipment (%)

		State-of-Art	Very Good	Adequate	Not Adequate	Sold/Transferred
Baseline Survey	All items	14	33	39	14	n/a
(1987 Stock)	1990 sample	7	29	40	15	9
Acquisitions 1987 - 1990 New Items		44	45	11	0	-

The first row of Table 2 shows the distribution of technical capability in 1987 for all items in the 1987 stock. The distribution of technical capability in 1987 for the four cost centres featured in the 1990 survey was virtually identical to that shown on the first row. The pattern of technical capability in 1990 (on the second row) shows, as one would expect, that the distribution has shifted with an overall decline in technical capability. Again as expected, an overwhelming majority of newly acquired items (89%) are at least 'very good'. Overall, the table shows that age is an important contributory factor to technical capability though it would appear that at a certain age the rate of decline of average technical capability is arrested.

Table 3 Change in Technical Capability Since Baseline Survey (%)

Declined	Static	Improved
21	75	4

Table 3 shows that the technical capability of three quarters of the sample remained static in the period from 1987 to 1990 with 21% declining. This is probably accounted for by a tranche of basic instruments which are less prone to obsolesence. Equipment in engineering was more frequently static, then in the other three cost centres.

Table 4 Working Condition of Equipment (%)

		Very Good	Adequate	Poor	Inoperable	Other
Baseline Survey	All Items	34	41	12	5	8
(1987 Stock)	1990 Sample	24	47	15	5	10
Acquisitions 1987 - 1990 (New Items)		84	13	1	1	1

As Table 4 indicates, working condition shows a similar pattern to that for technical capability. Again about three quarters of the sample of instruments did not change in working condition between the the surveys, while most of the remainder declined. A worrying 20% of equipment was in poor condition or inoperable.

The surveys provide a detailed account of what is happening but do not necessarily lend themselves to a simple interpretation. For example the scientific community tended to perceive the results in Tables 2-4 in the spirit of "Over half the equipment stock is only adequate or worse", while some policymakers perceived the message as "Over 88% of the equipment stock is at least adequate". Nevertheless, the figures emphasize the continuous process of change in the stock, and therefore the need for continued funding outputs.

To amplify the quantitative data, both surveys included exercises designed to allow heads of department to give an account of the equipment situation. It was from this data that a picture of a scientific community hampered by lack of resources for equipment emerged most strongly. In both cases over 80% of respondents expressed concern that research was either being delayed or prevented by current equipment shortfalls. Particular emphasis was placed upon the need for items which underpinned a range of projects, for example a DNA synthesis machine for biochemists requiring quantities of DNA of consistent purity.

Heads of Departments' comments also indicated an increased use and reliance on equipment in the past three years, though the strongest trend was towards computerisation, an area outside the scope of the two surveys.

As implied in Section 1.2, the dual-support mechanisms has been the object of particular criticism. Not only are basic items of equipment failing to be funded but increases in research council equipment funding (themselves in part a response to the first survey) have brought new strains. New items require maintenance and the cost of service contracts is said to be rising. The alternative of in-house maintenance is hampered by problems of retaining adequate technical support staff, because of uncompetitive salaries and freezes on appointments.

The physical environment also causes difficulties in many laboratories. Traditional problems of lack of space in many university departments have been joined by the increasing problem of providing specialised conditions and services for some of the more sophisticated instruments. Examples include air conditioning, clean room environments, specialised water/gas supplies and provisions for adequate fuming arrangements.

An overall measure of the Heads of Departments perceptions was obtained by asking whether they considered that they had well found laboratories. While only 28% considered that they did not have these by British Standards, this rose to 64% when international standards were applied, with a further 28% having mixed feelings.

4. POLICY IMPLICATIONS AND FUTURE INFORMATION REQUIREMENTS

British science has tended to suffer the extremes of trends which are present in most industrialised countries. Insofar as the studies performed in other countries are comparable, it is evident that the equipment problem is an international one. Science policies have emphasised the labour aspects of the production of scientific knowledge and neglected the capital dimension. Yet, the evidence of the results on the sophistication effect indicate that there is a trend towards increasingly capital intensive science. In a period of constrained or declining funds there is an understandable temptation to safeguard posts for scientists at the expense of investment in equipment, resulting in a sub-optimal approach to scientific advance.

Short of increasing funding for science on a large scale, a desirable but unlikely scenario in the short term in most countries, the way forward appears to lie in making better use of existing resources. A number of options suggest themselves:

- **Relax institutional rigidities in funding** to ensure that an optimum mix of equipment, technical support and facilities is available.

- **Explore non-traditional sources of equipment** by making more use of industrial facilities.

- **Promote greater use of existing equipment** through organised sharing.

- **Encourage innovation in equipment** particuarly where this involves cost reduction.

All of these have been tried to some extent but generally not on a sufficient scale to affect the overall picture. Sharing in particular offers considerable scope but many managerial and psychological barriers exist[10]. It would be easier to formulate policies for equipment if there was a better understanding, on a case-study basis, of its precise role in the generation of scientific knowledge. From the policy analyst's perspective one must hope that sociological studies of science will assist in this.

In the meantime, there is a continuing need to collect and collate quantitative data on the stock and state of academic research equipment. This provides the dual benefits of placing arguments about resources for funding on a sound footing and of providing a basis for the improved management of a stock of instruments in which the supply side is experiencing decline while demand continues for ever more sophisticated models.

5. ACKNOWLEDGEMENTS

The authors wish to thank the Advisory Board for Research Councils for supporting this work and all the scientists who contributed their time to the surveys. However, all views expressed here are those of the authors and do not necessarily represent those of the sponsors.

6. REFERENCES

1. Cohen A.V. and Ivins L.N. *Science Policy Studies 1: The Sophistication Factor in Science Expenditure*, HMSO, London, 1967

2. Bevan E.G. *An Analysis of Equipment Costs in University Science and Engineering Departments*, Council for Scientific Policy, London, 1971. This was based on questionnaire returns from 69 departments in 17 universities.

3(a). Georghiou L., Halfpenny P. and Hinder S., *Survey of Academic Research Equipment in the United Kingdom*, Advisory Board for the Research Councils, London, November 1989

(b) Yates J., Halfpenny P. and Georghiou L, *The Dipstick Survey,* Advisory Board for the Research Councils, (in press), London, 1991

4. Georghiou L. and Metcalfe J.S. 'Public science, intellectual property rights and research administration' in De La Mothe J. and Ducharme L.M. *Science, Technology and Free Trade,* Pinter, London 1990

5. Science and Engineering Research Council, Science Board, "Report of Science Board Instrumentation Panel", Swindon, 22 January 1986

6(a) Westat Inc., *Academic Research in the Physical and Computer Sciences and Engineering,* National Science Foundation, December, 1984

(b) Husman H.J. and Burgdorf K., *Academic Research Equipment and Needs in the Biological and Medical Science,* National Institutes of Health, Washington, April 1985

7. Atkinson H., Rogers P., and Bond R., Research in the United Kingdom, France and West Germany: A Comparison (Vol.1), Science and Engineering Research Council, Swindon 1990

8. Jouffrey B. and Nauciel M., Les equipements scientifiques mi-lourds, Ministere de la Recherche et de La Technologie, Paris, 1988

9. Wissenschaftsrat, Investitionen fur Grossgerate an den Hochschulen, Cologne, 1985

10. Hinder, S. and Halfpenny P. 'Research economies - Potential for sharing research equipment, *Science and Public Policy,* April 1990